FUNDAMENTALS OF QUANTUM MECHANICS

THIRD EDITION

FUNDAMENTALS OF QUANTUM MECHANICS

THIRD EDITION

J. E. HOUSE
Illinois Wesleyan University and Illinois State University

ACADEMIC PRESS
An imprint of Elsevier

Academic Press is an imprint of Elsevier
125 London Wall, London EC2Y 5AS, United Kingdom
525 B Street, Suite 1800, San Diego, CA 92101-4495, United States
50 Hampshire Street, 5th Floor, Cambridge, MA 02139, United States
The Boulevard, Langford Lane, Kidlington, Oxford OX5 1GB, United Kingdom

Notices
Knowledge and best practice in this field are constantly changing. As new research and
experience broaden our understanding, changes in research methods, professional
practices, or medical treatment may become necessary.

Practitioners and researchers must always rely on their own experience and knowledge in
evaluating and using any information, methods, compounds, or experiments described
herein. In using such information or methods they should be mindful of their own safety
and the safety of others, including parties for whom they have a professional responsibility.

To the fullest extent of the law, neither the Publisher nor the authors, contributors, or editors,
assume any liability for any injury and/or damage to persons or property as a matter of
products liability, negligence or otherwise, or from any use or operation of any methods,
products, instructions, or ideas contained in the material herein.

Library of Congress Cataloging-in-Publication Data
A catalog record for this book is available from the Library of Congress

British Library Cataloguing-in-Publication Data
A catalogue record for this book is available from the British Library

ISBN: 978-0-12-809242-2

For information on all Academic Press publications visit
our website at https://www.elsevier.com/books-and-journals

Working together
to grow libraries in
developing countries

www.elsevier.com • www.bookaid.org

Publisher: John Fedor
Acquisition Editor: John Fedor
Editorial Project Manager: Amy Clark
Production Project Manager: Maria Bernard
Cover Designer: Mark Rogers

Typeset by SPi Global, India

CONTENTS

PREFACE

The essential theme of this book continues that of the earlier editions. It is intended to be a presentation of several basic models in quantum mechanics and to show their application to chemical and physical systems. Throughout the development, emphasis is placed on showing sufficient mathematical details for the reader to follow the derivations. With the understanding that some readers may not have an extensive background in mathematics or perhaps may not have used some of the techniques recently, several mathematical procedures have been explained in detail. These include the solution of linear differential equations, the separation of variables, determinants, and series solution of differential equations. Such tutorials enhance the application of the book to self-study and review.

In this edition, there has been a major reordering of material so as to make what the author believes is a more logical flow of topics from the basic models to applications. Material from earlier chapters as well as several new sections have been collected in two new chapters on spectroscopy. The rationale for this change is that basic quantum mechanical models and the description of bonding in molecules are essential for interpreting various aspects of spectroscopy.

New sections appear in several chapters in order to illustrate the applications of quantum mechanics. In addition to new sections on topics in spectroscopy, others include band theory of metals, three-center bonds, and heat capacity of metals. Moreover, new problems have been included at the end of the chapters, while the practice of the second edition to include answers to some problems continues.

The appearance of the book has been improved by the inclusion of color illustrations. The majority of them have been altered or redrawn, and many additional illustrations have been included.

It is better to correct an oversight than to ignore it. Consequently, it is with deep gratitude that the author wishes to pay tribute to two professors whose efforts long ago made all editions of this book possible. The late Boris Musulin and John Eisele, through superb teaching, provided the interest in quantum mechanics that has lasted for so long. The background they gave the author provided the ability and desire to produce this book. Although both are long departed, their influence is still felt. In many ways, this book is a tribute to *their* success.

The author wishes to thank Academic Press/Elsevier and the editors for permission to reproduce and modify illustrations from his earlier books. Working with Amy Clark, John Fedor, and Katey Bircher has once again been a pleasurable experience, one with which the author hopes to be favored again. The loss of a computer at a most inopportune time made it necessary to retype all equations and remake almost all illustrations. Constant encouragement and understanding by these excellent editors helped to make that situation a hurdle rather than a roadblock.

Finally, the author wishes to acknowledge formally the enormous assistance of his wife, Kathleen A. House, in producing this edition. Her meticulous reading of the manuscript and numerous suggestions have been invaluable, and she has provided encouragement at all stages of the work.

CHAPTER 1

Origins of Quantum Theory

Quantum mechanics is a branch of science that deals with atomic and molecular properties and behavior on a microscopic scale. Although thermodynamics may be concerned with the heat capacity of a gaseous sample, quantum mechanics is concerned with the specific changes in rotational energy states of the molecules. Chemical kinetics may deal with the rate of change of one substance into another, but quantum mechanics is concerned with changes in vibrational states and structures of the reactant molecules as they are transformed. Quantum mechanics is also concerned with the spinning of atomic nuclei and the populations of atoms in an excited state. Spectroscopy is based on changes of quantized energy levels of several types. Quantum mechanics is thus seen to merge with many other areas of modern science.

An understanding of the main ideas and methods of quantum mechanics is important for developing an understanding of various branches of science, from nuclear physics to organic chemistry. This book attempts to develop that familiarity across the sciences.

The modern applications of quantum mechanics have their roots in the developments of physics around the end of the 19th and the early part of the 20th centuries. Some of the experiments, now a century or more old, still provide the physical basis for interpretations of quantum mechanics. The results of those experiments provide the foundations for "how we know what we know" and for modern theories. The names associated with much of this early work (Planck, Einstein, Bohr, de Broglie, et al.) are legendary in the realm of physics. Their elegant experiments and theories now seem almost commonplace to even beginning students, but these experiments were at the forefront of scientific development at the time. Therefore, it is appropriate for this book to begin with a brief review of a few of the more important early studies.

1.1 BLACKBODY RADIATION

When an object is heated to incandescence, it emits electromagnetic radiation. The nature of the object determines to some extent the type of

Fundamentals of Quantum Mechanics
http://dx.doi.org/10.1016/B978-0-12-809242-2.00001-2

radiation that is emitted, but in all cases a range or distribution of radiation is produced. It is known that the best absorber of radiation is also the best emitter of radiation. The best absorber is a so-called "blackbody," which absorbs all radiation and from which none is reflected. If this blackbody is heated to incandescence, it will emit a whole range of electromagnetic radiations whose energy distributions depend on the temperature to which the blackbody is heated. Early attempts to explain the distribution of radiation using the laws of classical physics were not successful. In these attempts, it was assumed that the radiation was emitted because of vibrations or oscillations within the blackbody. These attempts failed to explain the position of the maximum that occurs in the distribution of radiation and, in fact, they failed to predict the maximum at all.

Because radiation with a range of frequencies (ν) is emitted from the blackbody, theoreticians tried to obtain an expression that would predict the relative intensity (amount of radiation) of each frequency. One of the early attempts to explain blackbody radiation was made by Wilhelm Wien. The general form of the equation that Wien obtained is

$$f(\nu) = \nu^3 g(\nu/T) \tag{1.1}$$

where $f(\nu)$ is the amount of energy of frequency ν emitted per unit volume of the blackbody and $g(\nu/T)$ is some function of ν/T. This result is in fair agreement with the observed energy distribution at longer wavelengths, but it did not agree at all with the region of short wavelengths. In fact, this treatment predicted that the intensity would become infinite for the radiation of a shorter wavelength than visible light (the so-called "ultraviolet catastrophe"). Another relationship obtained by the use of classical mechanics is the expression derived by Lord Rayleigh,

$$f(v) = \frac{8\pi v^3}{c^3} kT \tag{1.2}$$

where c is the velocity of light (3.00×10^8 m/s) and k is Boltzmann's constant, 1.38×10^{-16} erg/molecule.

Lord Rayleigh and Sir James Jeans found another expression that predicts the shape of the energy distribution as a function of frequency, but only in the region of short wavelength. The expression is

$$f(v) = \frac{8\pi v^3}{c^3} \left(\frac{kT}{v} \right) = \frac{8\pi v^2 kT}{c^3} \tag{1.3}$$

Therefore, the Wien relationship predicted the intensity of radiation of high ν (short wavelength), and the Rayleigh-Jeans law predicted the intensity of low ν (long wavelength) radiation emitted from a blackbody. Neither of these relationships predicted a distribution of radiation that goes through a maximum at some frequency with smaller amounts emitted on either end of the spectrum.

In 1900, the problem with the explanation of the features of blackbody radiation was finally solved by Max Planck. Planck still assumed that the absorption and emission of radiation arises from some sort of oscillators. Planck made a fundamental assumption that only certain frequencies were possible for the oscillators instead of the whole range of frequencies that are predicted by classical mechanics. The permissible frequencies were presumed to be some multiple of a fundamental frequency of the oscillators ν_0. The allowed frequencies are then $\nu_0, 2\nu_0, 3\nu_0, \ldots$. Planck also assumed that energy is absorbed as the oscillator goes from one allowed frequency to the next higher one and that energy is emitted as the frequency drops by $n\nu_0$. Planck's explanation included the idea that the change in energy is proportional to the fundamental frequency ν_0. Introducing the constant of proportionality, h,

$$E = h\nu_0 \tag{1.4}$$

where h is Planck's constant, 6.63×10^{-27} erg s or 6.63×10^{-34} J s. The average energy per oscillator was found to be

$$\langle E \rangle = \frac{h\nu_0}{e^{h\nu_0/kT} - 1} \tag{1.5}$$

Planck showed that the emitted radiation has a distribution that is given by

$$f(\nu) = \frac{8\pi\nu_0^3}{c^3}\langle E \rangle = \frac{8\pi\nu_0^3}{c^3}\frac{h\nu_0}{e^{h\nu_0/kT} - 1} \tag{1.6}$$

This equation correctly predicted the observed relationship between the frequencies of radiation emitted and the intensity.

The successful interpretation of blackbody radiation by Planck provided the basis for energy being considered quantized, which is so fundamental to our understanding of atomic and molecular structure and our experimental methods for studying matter. Also, it established the familiar relationship between the frequency of radiation and its energy,

$$E = h\nu \tag{1.7}$$

These ideas will be seen many times as one studies quantum mechanics and its application to physical problems.

1.2 THE LINE SPECTRUM OF ATOMIC HYDROGEN

When gaseous hydrogen is enclosed in a glass tube in such a way that a high potential difference can be placed across the tube, the gas emits a brilliant reddish-purple light. If this light is viewed through a spectroscope or a prism (as shown in Fig. 1.1), the four major lines in the visible spectrum of hydrogen are seen. There are other lines that occur in other regions of the electromagnetic spectrum that are not visible to the eye.

In this visible part of the hydrogen emission spectrum the four lines have the wavelengths:

$$H_\alpha = 656.28 \text{nm} = 6562.8 \text{Å}$$
$$H_\beta = 486.13 \text{nm} = 4861.3 \text{Å}$$
$$H_\gamma = 434.05 \text{nm} = 4340.5 \text{Å}$$
$$H_\delta = 410.17 \text{nm} = 4101.7 \text{Å}$$

As shown in Fig. 1.2, electromagnetic radiation is alternating electric (E) and magnetic (H) fields that are perpendicular and in phase.

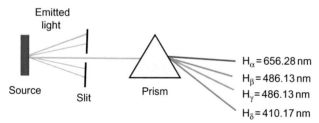

Fig. 1.1 Separating lines in the spectrum of atomic hydrogen.

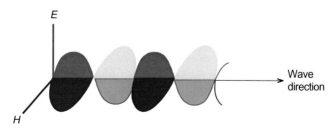

Fig. 1.2 Electromagnetic waves that are perpendicular and in phase. Areas that are shaded lighter represent displacement in the negative directions of the E and H axes.

Planck showed that the energy of electromagnetic radiation is proportional to the frequency ν so that

$$E = h\nu \tag{1.8}$$

where h is Planck's constant. Because electromagnetic radiation is a transverse wave, there is a relationship between the wavelength λ and the frequency ν. Frequency is expressed in terms of cycles per unit time, but a "cycle" is simply a count that carries no units. Therefore, the *units* of frequency are "cycles"/time or 1/time. The wavelength is a distance, so it has a dimension of length. The product of wavelength and frequency can be written dimensionally as

$$\lambda \times \nu = \text{distance} \times \frac{1}{\text{time}} = \frac{\text{distance}}{\text{time}} = \text{velocity} = \nu \tag{1.9}$$

In the case of electromagnetic radiation, the velocity of light is c, which is 3.00×10^{10} cm/s. Therefore, $\nu = c$ and

$$E = h\nu = \frac{hc}{\lambda} \tag{1.10}$$

In 1885, Balmer discovered an empirical formula that would predict the above wavelengths. Neither Balmer nor anyone else knew *why* this formula worked, but it did predict the wavelengths of the lines accurately. Balmer's formula is

$$\lambda \ (\text{cm}) = 3645.6 \times 10^{-8} \left(\frac{n^2}{n^2 - 2^2} \right) \tag{1.11}$$

The constant 3645.6×10^{-8} has units of cm, and n represents a whole number larger than 2. Using this formula, Balmer was able to predict the existence of a fifth line, which was discovered at the boundary between the visible and ultraviolet regions of the spectrum. The measured wavelength of this line agreed almost perfectly with Balmer's prediction.

Balmer's empirical formula also predicted the existence of other lines in the infrared and ultraviolet regions of the spectrum of hydrogen. These are as follows (shown with year of discovery and spectral region):

Lyman Series: $\quad n^2 / (n^2 - 1^2)$, where $n = 2, 3, \ldots$ (1906 – 14, UV)
Paschen Series: $\quad n^2 / (n^2 - 3^2)$, where $n = 4, 5, \ldots$ (1908, IR)
Brackett Series: $\quad n^2 / (n^2 - 4^2)$, where $n = 5, 6, \ldots$ (1922, IR)
Pfund Series: $\quad n^2 / (n^2 - 5^2)$, where $n = 6, 7, \ldots$ (1924, IR)

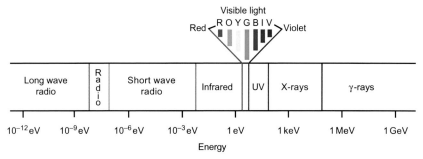

Fig. 1.3 The electromagnetic spectrum.

Balmer's formula can be written in terms of 1/wavelength and it is usually seen in this form. The equation becomes

$$\frac{1}{\lambda} = R\left(\frac{1}{2^2} - \frac{1}{n^2}\right) \tag{1.12}$$

where R is a constant known as the *Rydberg constant,* which is 109,677.76 cm^{-1}. The quantity $1/\lambda$ is called the *wave number,* and it is expressed in units of centimeters^{-1} (cm^{-1}). The empirical formulas can be combined into a general form

$$\bar{\nu} = \frac{1}{\lambda} = R\left(\frac{1}{n_1^2} - \frac{1}{n_2^2}\right) \tag{1.13}$$

When $n_1 = 1$ and $n_2 = 2, 3, 4, \ldots$, the Lyman Series is predicted. For $n_1 = 2$ and $n_2 = 3, 4, 5, \ldots$, the Balmer Series is predicted, etc. Other empirical formulas were found that correlated lines in the spectra of other atoms, but the same constant, R, occurred in these formulas. At the time, no one was able to relate these formulas to classical electromagnetic theory.

Although the discussion up to this point has been concerned with the line spectrum of atomic hydrogen, later sections of this book will deal with the interactions of other types of electromagnetic energy with matter. Spectroscopic techniques can make use of electromagnetic radiation of these types, which are illustrated in Fig. 1.3, so it is appropriate to show their relationship to energy.

1.3 ELECTRONS AND THE NUCLEUS

In 1911, Rutherford performed one of the revealing experiments in atomic physics that is now known as the gold foil experiment. Some radioactive

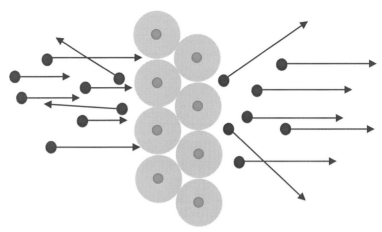

Fig. 1.4 A depiction of Rutherford's experiment.

heavy elements emit alpha particles (helium nuclei), and a beam of these particles was directed at thin gold foil, as depicted in Fig. 1.4.

Following Thompson's idea, it was believed at the time that atoms consisted of positive and negative charges that were distributed throughout the atom. Although most of the particles continued on their original paths, a small fraction of the particles were deflected through large angles or even reversed direction. However, when the results of the experiment were analyzed, it was concluded that some of the positive alpha particles must have encountered a region of the target atoms from which they were strongly repelled. That positive region occupied a small fraction of the volume of the atom, and it is now known that atomic nuclei have radii that are on the order of 10^{-13} cm. With radii of most atoms being in the range of 10^{-8} cm, it is seen that most of the volume of an atom is empty space, which explains why the majority of the alpha particles were unimpeded as they passed through the gold foil. The gold foil experiment provided our view of atomic structure and was a pivotal point in the development of our knowledge of atoms.

1.4 THE BOHR MODEL FOR THE HYDROGEN ATOM

It is not surprising that the spectrum of the hydrogen atom was the first to be explained because it is simplest atom. Rutherford had shown in 1911 that the model of the atom is one in which a small region of positive charge containing most of the mass is located in the center of the atom and the negative region surrounds it. Applying this model to the hydrogen atom, the single proton is positioned as the nucleus, whereas a single electron moves around it. Bohr incorporated these ideas into the first dynamic model of the hydrogen atom in 1913, assuming the electron is governed by the laws of classical or Newtonian physics. However, there were problems that could not be answered by the laws of classical mechanics. For example, it was shown that an accelerated electric charge radiates electromagnetic energy (as does an antenna for the emission of radio frequency waves). To account for the fact that an atom is a stable entity, it was observed that the electron must move around the nucleus in such a way that the centrifugal force exactly balances the electrostatic force of attraction between the proton and electron. As a result of the electron moving in some kind of circular orbit, it must constantly undergo acceleration and *should* radiate electromagnetic energy by the laws of classical physics.

Because the Balmer Series of lines in the spectrum of atomic hydrogen had been observed earlier, physicists attempted to use the laws of classical physics to explain a possible structure of hydrogen that would give rise to these lines. It was recognized from Rutherford's work that the nucleus of an atom is surrounded by electrons, which must always be in motion. In fact, no system of electric charges can be in equilibrium at rest.

Although the electron in the hydrogen atom must be moving, there is a major problem. If the electron circles the nucleus, it is undergoing a constant change in direction, as shown in Fig. 1.5. Velocity is a vector quantity that has both *magnitude* and *direction*. Changing direction constitutes a change in velocity, and the change in velocity with time is *acceleration*. The laws of classical electromagnetic theory predict that an accelerated electric charge should radiate electromagnetic energy. If the electron did emit electromagnetic energy, it would lose part of its energy. As it did so, it would spiral into the nucleus, and the atom would collapse. Also, electromagnetic energy of a *continuous* nature would be emitted, rather than just a few lines.

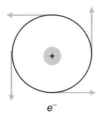

Fig. 1.5 The circular motion of an electron in a hydrogen atom.

Bohr had to assume that there were certain orbits (the "allowed orbits") in which the electron could move without radiating electromagnetic energy. These orbits were characterized by the relationship

$$mvr = n\frac{h}{2\pi} \tag{1.14}$$

in which m is the mass of the electron, v is its velocity, r is the radius of the orbit, h is Planck's constant, and n is an integer, 1, 2, 3, ... As a result of n being a whole number with only specific values, it is called a *quantum number*. This enabled the problem to be solved, but it was not understood why this worked. Bohr also assumed that the emitted spectral lines resulted from the electron falling from an orbital of higher n to one of lower n. The physical aspects of the electron moving in a hydrogen atom require an analysis of the forces involved.

The physics of the hydrogen atom is based on mechanics and electrostatics. Figure 1.6 shows the forces acting on the orbiting electron. The *magnitudes* of the centrifugal and centripetal forces must be equal for an electron to be moving in a stable orbit, so

$$\frac{mv^2}{r} = \frac{e^2}{r^2} \tag{1.15}$$

Therefore, solving this equation for v gives

$$v = \sqrt{\frac{e^2}{mr}} \tag{1.16}$$

From the Bohr assumption regarding angular momentum represented by the equation

$$mvr = n\frac{h}{2\pi} \tag{1.17}$$

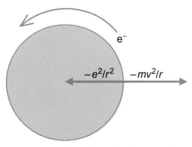

Fig. 1.6 The forces acting on an electron in a hydrogen atom.

it can be seen that solving for v gives

$$v = \frac{nh}{2\pi mr} \tag{1.18}$$

Therefore, equating the two expressions for velocity gives

$$\sqrt{\frac{e^2}{mr}} = \frac{nh}{2\pi mr} \tag{1.19}$$

Solving this equation for r gives the result

$$r = \frac{n^2 h^2}{4\pi^2 me^2} \tag{1.20}$$

This relationship shows that the radii of the allowed orbits increase as n^2 (h, m, and e are, of course, constants). Therefore, the orbit with $n=2$ is four times as large as the one with $n=1$; the one with $n=3$ is nine times as large as the one with $n=1$, etc. Figure 1.7 shows the first few allowed orbits drawn approximately to scale.

The units on r can be found from the units on the constants, because e is measured in electrostatic units (esu) and an esu is a $g^{1/2}\, cm^{3/2}\, s^{-1}$. Therefore, from Eq. (1.20), r has units that can be represented as

$$\frac{\left[(gcm^2/s^2)s \right]^2}{g \left(g^{1/2} cm^{3/2}/s \right)^2} = cm$$

The total energy is the sum of the electrostatic energy (potential) and the kinetic energy of the moving electron (total energy = kinetic + potential). This can be expressed by the equation

$$E = \frac{1}{2}mv^2 - \frac{e^2}{r} \tag{1.21}$$

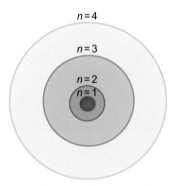

Fig. 1.7 The first four allowed orbits according to the Bohr model.

By equating the magnitudes given by the expressions for centripetal and centrifugal forces, it can be seen that

$$\frac{mv^2}{r} = \frac{e^2}{r^2}$$
(1.22)

and multiplying both sides of the equation by r gives

$$mv^2 = \frac{e^2}{r}$$
(1.23)

If both sides of this equation are multiplied by 1/2, the result is

$$\frac{1}{2}mv^2 = \frac{e^2}{2r}$$
(1.24)

The left hand side of Eq. (1.24) is simply the kinetic energy of the electron, and substituting this into Eq. (1.21) yields

$$E = \frac{1}{2}mv^2 - \frac{e^2}{r} = \frac{e^2}{2r} - \frac{e^2}{r} = -\frac{e^2}{2r}$$
(1.25)

It was shown earlier in Eq. (1.20) that

$$r = \frac{n^2 h^2}{4\pi^2 m e^2}$$

and if this result is substituted for r in Eq. (1.25), the result is

$$E = -\frac{e^2}{2r} = -\frac{e^2}{2\left(\dfrac{n^2 h^2}{4\pi^2 m e}\right)} = -\frac{2\pi^2 m e^4}{n^2 h^2}$$
(1.26)

From this equation, it can be seen that the energy of the electrons in the allowed orbits varies inversely as n^2. Note also that the energy is *negative* and gets less negative as n increases. At $n = \infty$ (i.e., the complete separation of the proton and electron), $E = 0$, and there is no binding energy of the electron to the nucleus. The units for E in the above equation depend on the units used for the constants. If h is in erg s, the mass of the electron is in grams and the charge on the electron e is in esu, so the result is

$$E = \frac{g\left(g^{1/2}\,cm^{3/2}/s\right)^4}{\left[(g cm^2/s^2)s\right]^2} = erg$$
(1.27)

If the energy is obtained in ergs, the use of conversion factors makes it possible to obtain the energy in any other desired units (J, cal, etc.).

If the expression for energy of the electron is written in the form

$$E = -\frac{1}{n^2}\frac{2\pi^2 m e^4}{h^2} \tag{1.28}$$

the resulting equation can be used to evaluate the collection of constants, and when $n=1$, the result is -2.17×10^{-11} erg. Assigning various values for n makes it possible to evaluate the energies of the allowed orbits. The results for several values of n are shown as follows:

$$
\begin{aligned}
n=1, \quad & E=-21.7 \times 10^{-12}\,\text{erg} \\
n=2, \quad & E=-5.43 \times 10^{-12}\,\text{erg} \\
n=3, \quad & E=-2.41 \times 10^{-12}\,\text{erg} \\
n=4, \quad & E=-1.36 \times 10^{-12}\,\text{erg} \\
n=5, \quad & E=-0.87 \times 10^{-12}\,\text{erg} \\
n=6, \quad & E=-0.63 \times 10^{-12}\,\text{erg} \\
n=\infty, \quad & E=0
\end{aligned}
$$

Figure 1.8 shows an energy level diagram in which the energies are shown graphically to scale for these values of n. Note that the energy levels get closer together (converge) as the value of n increases.

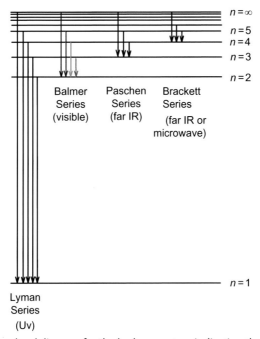

Fig. 1.8 An energy level diagram for the hydrogen atom indicating the various series in the line spectrum.

It requires energy for an electron to be moved to a higher energy level, because the positive and negative charges are held together by a strong electrostatic force. The amount of energy required for complete removal of the electron is known as the *ionization potential* (or ionization energy) and corresponds to moving the electron to the orbital where $n = \infty$. The electron in the lowest energy state is held with an energy of -21.7×10^{-12} erg, and at $n = \infty$ the energy is 0. Therefore, the ionization potential for the hydrogen atom is 21.7×10^{-12} erg.

By considering the energy difference between the $n = 2$ and $n = 3$ orbits, it can be seen that the difference is 3.02×10^{-12} erg. The calculation of the wavelength of light having this energy is achieved by the equation

$$E = h\nu = \frac{hc}{\lambda} \tag{1.29}$$

Solving for λ, it can be seen that

$$\lambda = \frac{hc}{E} = \frac{(6.63 \times 10^{-27}\,\text{erg s}) \times (3.00 \times 10^{10}\,\text{cm/s})}{3.02 \times 10^{-12}\,\text{erg}} = 6.59 \times 10^{-5}\,\text{cm} \tag{1.30}$$

which matches the wavelength of one of the lines in the Balmer Series. Using the energy difference between the $n = 2$ and $n = 4$ leads to a wavelength of 4.89×10^{-5} cm, which matches the wavelength of another line in the Balmer Series. Finally, the energy difference between the orbits for which $n = 2$ and $n = \infty$ corresponds to a wavelength of 3.66×10^{-5} cm, and this is the wavelength of the *series limit* of the Balmer Series. It should be readily apparent that Balmer's Series corresponds to light *emitted* as the electron falls from initial states with $n = 3, 4, 5, \ldots$, to the orbit with $n = 2$.

It is a simple matter to calculate the energies of lines emitted as the electron falls from orbits with $n = 2, 3, \ldots$, to the orbital with $n = 1$. When that is done, it is found that these energies match the lines in another observed spectral series, the *Lyman Series*. In that case, the wavelengths of the spectral lines are so short (i.e., they have higher energy) that the lines are no longer in the visible region of the spectrum but rather they are in the ultraviolet region. Other series of lines correspond to the transitions from higher n values to $n = 3$ (Paschen Series, infrared), $n = 4$ (Brackett Series, infrared), and $n = 5$ (Pfund Series, far infrared) as the lower n values.

The fact that the series limit for the Lyman Series represents the quantity of energy that would be required to remove the electron ($n = 1$ to $n = \infty$) suggests that this is one way to obtain the *ionization potential* for the hydrogen atom. Note that energy is *released* (negative sign) when the electron falls from the orbital with $n=\infty$ to the one with $n=1$ and that energy is *absorbed* (positive sign) when the electron is excited from the orbital with $n=1$ to the one corresponding to $n=\infty$. Ionization energies represent the energies *required* to remove electrons from atoms, and they are always *positive*.

1.5 THE PHOTOELECTRIC EFFECT

In 1887, H. R. Hertz observed that the gap between metal electrodes became a better conductor when ultraviolet light was shined on the apparatus. Soon after, W. Hallwachs observed that a negatively charged zinc surface lost its negative charge when ultraviolet light was shined on it. The negative charges that were lost were identified as electrons from their behavior in a magnetic field. The phenomenon of an electric current flowing when light was involved came to be known as the *photoelectric effect*.

The study of the photoelectric effect is made possible with an apparatus like that shown schematically in Fig. 1.9. An evacuated tube is arranged so that the highly polished metal that is to be illuminated, such as sodium, potassium, or zinc, is made the cathode. When light shines on the metal plate, electrons flow to the collecting plate (anode), and the ammeter placed in the circuit indicates the amount of current flowing. Several observations can be made as the frequency and intensity of the light varies:

1. The light must have some minimum or threshold frequency, ν_0, in order for the current to flow.
2. Different metals have different threshold frequencies.
3. If the light striking the metal surface has a frequency greater than ν_0, the electrons are ejected with a kinetic energy that increases with the frequency of the light.

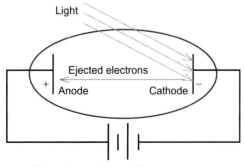

Fig. 1.9 An illustration of the photoelectric effect.

4. The number of electrons ejected depends on the intensity of the light, but their kinetic energy depends only on the frequency of the light.

An electron traveling toward the collector can be stopped if a negative voltage is applied to the collector. The voltage required to stop the motion of the electrons, which is known as the *stopping potential, V*, and causes the current to cease, depends on the frequency of the light that caused the electrons to be ejected. In fact, it is the *electrostatic* energy of the repulsion between an electron and the collector that exactly equals the *kinetic* energy of the electron. Therefore, the two energies can be equated to yield the relationship

$$Ve = \frac{1}{2}mv^2 \qquad (1.31)$$

In 1905, Albert Einstein explained the major aspects of the photoelectric effect. Einstein based his analysis on the relationship between the energy of light and its frequency that Planck established in 1900. It was assumed that the light behaved as a collection of particles (called *photons*) and the energy of a particle of light was totally absorbed by its collision with an electron on the metal surface. Electrons are bound to the surface of a metal with an energy called the *work function, w*, which is different for each type of metal. When the electron is ejected from the surface of the metal, it will have a kinetic energy that represents the difference between the energy of the incident photon and the work function of the metal. Therefore, the energies are related by the equation

$$\frac{1}{2}mv^2 = h\nu - w \qquad (1.32)$$

It can be seen that this is the equation of a straight line when the kinetic energy of the electron is plotted against the frequency of the light. By varying the frequency of the light and determining the kinetic energy of the electrons (from the stopping potential), a graph such as that shown in Fig. 1.10 can be prepared to show this relationship.

The intercept is ν_0, which is the threshold frequency, and the slope is Planck's constant, h. One of the significant points in the interpretation of the photoelectric effect is that light is considered to be particulate in nature. In other experiments, such as the diffraction experiment of T. Young, it was necessary to assume that light behaved like a wave. Many photovoltaic devices in common use today (e.g., light meters, optical counters, etc.) are based on the photoelectric effect.

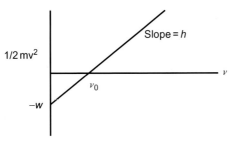

Fig. 1.10 The relationship between the kinetic energy of electrons ejected in the photoelectric effect and frequency of the light.

1.6 PARTICLE-WAVE DUALITY

Because light can behave as both waves (i.e., diffraction, as proved by Young in 1803) and particles (i.e., the photoelectric effect shown by Einstein in 1905), the nature of light was debated for many years. Of course, light has characteristics of both a wave and a particle, which is the so-called *particle-wave duality*. In 1924, Louis de Broglie, a young French doctoral student, investigated some of the consequences of relativity theory. For electromagnetic radiation,

$$E = h\nu = \frac{hc}{\lambda} \tag{1.33}$$

in which c, ν, and λ are the velocity, frequency, and wavelength, respectively, for the radiation. The photon also has an energy given by the relationship from relativity theory, which is

$$E = mc^2 \tag{1.34}$$

A particular photon can have only one energy, so the energies expressed in Eqs. (1.33) and (1.34) must be equal.

$$mc^2 = \frac{hc}{\lambda} \tag{1.35}$$

Solving this equation for λ gives the relationship

$$\lambda = \frac{h}{mc} \tag{1.36}$$

This does *not* mean that light has a mass; however, because mass and energy can be interconverted, light has an energy that is *equivalent* to some mass. The quantity represented as mass multiplied by velocity is the momentum, so

Eq. (1.36) predicts a wavelength that is Planck's constant divided by the momentum of the photon.

De Broglie reasoned that if a particle had a wave character, the wavelength would be given by

$$\lambda = \frac{h}{mv} \tag{1.37}$$

where the velocity is written as v rather than c because the particle will not be traveling at the speed of light. In 1927, C. J. Davisson and L. H. Germer, who were then working at Bell Laboratories in Murray Hill, New Jersey, verified this relationship experimentally. In their experiment, an electron beam was directed at a metal crystal, and a diffraction pattern was observed. As a result of diffraction being a property of *waves*, it was concluded that the moving electrons were behaving as waves. The reason for using a metal crystal is that in order to observe a diffraction pattern, the waves must pass through openings about the same size as the wavelength, and that distance corresponds to the distance separating atoms in a metal.

The de Broglie wavelength of moving particles (electrons in particular) has been verified experimentally. That is, of course, important, but the real value is that electron diffraction has now become a standard technique for determining molecular structure.

In developing a model for the hydrogen atom, Bohr had to assume that the stable orbits were those in which angular momentum was quantized:

$$mvr = n\frac{h}{2\pi} \tag{1.38}$$

Because de Broglie showed that the moving electron should be considered as a wave, that wave will be a stable one only if the wave joins smoothly onto itself. This means that the circular orbit must contain a whole number of wavelengths as illustrated in Fig. 1.11. This generates a standing wave that does not undergo destructive interference.

The circumference of a circle, in terms of the radius r, is $C = 2\pi r$. Therefore, a whole number of wavelengths n must be equal to the circumference of the orbit, so that

$$2\pi r = n\lambda \tag{1.39}$$

However, the de Broglie wavelength, λ, is given by

$$\lambda = \frac{h}{mv} \tag{1.40}$$

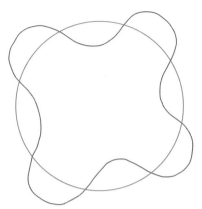

Fig. 1.11 A standing wave that matches the circumference of a Bohr orbit.

Therefore,

$$2\pi r = n\left(\frac{h}{mv}\right) \tag{1.41}$$

and this equation can be rearranged to give

$$mvr = n\left(\frac{h}{2\pi}\right) \tag{1.42}$$

This is exactly the same as the equation that Bohr assumed in order to predict which orbits were stable! We now see the connection between the wave character of a particle and the Bohr model. Only two years later, Erwin Schrödinger used the model of a standing three-dimensional wave to represent the electron in the hydrogen atom and solved the resulting wave equation. This highly significant beginning of wave mechanics will be described later. Although the Bohr model explained the spectral properties of the hydrogen atom, it did not do so for any other atoms. However, He^+, Li^{2+}, and similar species containing one electron could be treated by the same model by using an appropriate nuclear charge. Also, the model treated the atom almost as if it were a mechanical device, but because the atom did not continuously radiate energy, it violated laws of classical electricity and magnetism.

1.7 THE HEISENBERG UNCERTAINTY PRINCIPLE

A serious problem with the Bohr model stems from the fact that it is impossible to know simultaneously the position and momentum (or energy) of a particle.

A rationale for this can be given as follows: Suppose you observe a ship and determine its position. The visible light waves have a wavelength of about 3×10^{-5} to 8×10^{-5} cm (3×10^{-7} to 8×10^{-7} m) and very low energy. The light strikes the ship and is reflected to your eyes, the detector. Because of the very low energy of the light, the ship, weighing several thousand tons, does not move as a result of the light striking it. Now, suppose you wish to "see" a very small particle of perhaps 10^{-8} cm (10^{-10} m) diameter. In order to locate the particle, you must use "light" having a wavelength about the same length as the size of the particle. Radiation of a 10^{-8} cm (i.e., very short) wavelength has very high energy, as shown by the relationship

$$E = \frac{hc}{\lambda} \tag{1.43}$$

Therefore, in the process of locating (observing) the particle with the use of high-energy radiation, the particle has undergone a change in momentum and energy. Therefore, it is impossible to determine both the position and momentum of a particle simultaneously to greater accuracy than some fundamental quantity. That quantity is h and the relationship between the uncertainty in position (distance) and in momentum (mass × distance/time) is

$$\Delta x \cdot \Delta(mv) \geq h \tag{1.44}$$

This relationship, which is one form of the Heisenberg uncertainty principle, indicates that h is the fundamental quantum of action. It can be seen that this equation is dimensionally correct, because the uncertainty in position multiplied by the uncertainty in momentum has the dimensions of

$$\text{Distance} \times \left(\text{mass} \times \frac{\text{distance}}{\text{time}} \right)$$

In cgs units,

$$\text{cm} \times \left(\text{grams} \times \frac{\text{cm}}{\text{s}} \right) = \text{erg s}$$

and the units of erg s match the units on h.

If the uncertainty in time is expressed in seconds and uncertainty in energy is in ergs,

$$\Delta t \cdot \Delta E \geq h$$

so this equation is also dimensionally correct. Therefore, an equation of this form can be written between any two variables for which the units reduce to erg s or g $cm^2 s^{-1}$.

It is implied by the Bohr model that it is possible to know the details of the orbital motion of the electron and its energy at the same time. Having now shown that is not true, our attention will now focus on the wave model of the hydrogen atom.

PROBLEMS

1. A shortwave radio station in Lake Okeechobee, Florida, broadcasts on a frequency of 6.065 megahertz (MHz). What is the wavelength of the radio waves?

2. What would be the de Broglie wavelength of an electron moving at 2.00% of the speed of light?

3. An electron in the ground state of a hydrogen atom is struck by an X-ray photon with a wavelength of 50.0 nm. A scattered photon with a wavelength of 200 nm is observed after the collision. What will be the velocity and de Broglie wavelength of the ejected electron?

4. The work function of a metal is the energy required to remove an electron from the metal. What wavelength of light will eject an electron from a metal that has a work function of 2.60 electron volts (eV) ($1 \text{ eV} = 1.60 \times 10^{-12}$ erg)?

5. For Be^{3+}, calculate the wavelength of the photons emitted as the electron falls from $n=3$ to $n=2$ and from $n=4$ to $n=3$.

6. Lithium compounds containing Li^+ ions impart a red color to a flame due to light emitted that has a wavelength of 670.8 nm.
   (a) What is the frequency of this spectral line?
   (b) What is the wave number for the radiation?
   (c) In kcal mol^{-1}, what energy is associated with this spectral line?

7. The ionization potential for a certain atom is 350 kJ mol^{-1}. If the electron is in the first excited state, the ionization potential is 105 kJ mol^{-1}. If the atom undergoes deexcitation, what would be the wavelength of the photon emitted?

8. The work function for barium is 2.48 eV. If light of 400 nm is shined on a barium cathode, what is the maximum velocity of the ejected electrons?

9. Creation of matter from electromagnetic radiation can occur if the radiation has sufficient energy (pair production). What is the minimum energy of a photon that can produce an electron-positron (i.e., a positive electron) pair?

10. For a proton and an electron having kinetic energies of 2.0 eV (1 eV $= 1.6 \times 10^{-12}$ erg), what would be the ratio of the de Broglie wavelengths?

11. Neutrons having energies equivalent to the kinetic energy of gaseous molecules at room temperature (kT) are called *thermal neutrons*. What is the wavelength of a thermal neutron at 27 °C?

12. Suppose an electron remains in an excited state of an atom for 10^{-8} s. What would be the minimum uncertainty of the energy of the photon emitted as the electron falls to the ground state? To what uncertainty in the wavelength of the photon does this correspond?

13. The ionization potential for unshared electrons in the water molecule is 12.6 eV. If a single X-ray photon having a wavelength of 0.300 nm passes through water and is completely absorbed by ionizing water molecules, how many molecules could the photon ionize? Explain why this would not likely occur in practice.

14. For a gas, the root-mean-square velocity is given by

$$v = \left(\frac{3RT}{M}\right)^{1/2}$$

where M is the molecular weight, T is the temperature in Kelvin, and R is the molar gas constant. Derive an expression for the de Broglie wavelength of gaseous molecules at a temperature T. Use the expression to determine the wavelength of moving helium atoms at a temperature of 300 K.

15. Repeat the procedure of Problem 14, but use the average velocity for gaseous molecules:

$$v_a = \left(\frac{8RT}{\pi M}\right)^{1/2}$$

After deriving the relationship, determine the de Broglie wavelength for hydrogen molecules at 400 K.

16. Radon-212 emits an alpha particle (helium nucleus) having an energy of 6.26 MeV. Determine the wavelength of the alpha particle. To a good approximation, the radius of a nucleus (in centimeters) can be expressed as $R = r_0 A^{1/3}$, where r_0 is a constant with a value of 1.3×10^{-13} and A is the mass number. Compare the wavelength of the alpha particle emitted from ^{212}Rn with the diameter of the nucleus.

17. Show from the Bohr model that the ratio of the kinetic energy to the potential energy is $-\frac{1}{2}$.
18. From the relationships used in the Bohr model, show that the velocity of the electron in the first Bohr orbit is $1/137$ of the velocity of light.
19. Show that the difference in energy between any two spectral lines in the hydrogen atom is the energy corresponding to a third spectral line. This phenomenon is known as the Ritz Principle.
20. One form of the Heisenberg Uncertainty Principle is $\Delta E \times \Delta t \geq h$, where ΔE is the uncertainty in energy and Δt is the uncertainty in the time. If it requires 10^{-8} s for an electron to fall from a higher orbital to a lower one, what will be the width of the spectral line emitted?

CHAPTER 2

The Methods of Quantum Mechanics

There are several areas of chemistry that require an understanding of quantum mechanics. Therefore, quantum mechanics at an elementary level is covered in several physics and chemistry courses taken by undergraduates. The discussion of quantum mechanics presented here will begin by introducing some of the procedures and terminology by stating the postulates of quantum mechanics and showing some of their applications. The coverage here is meant to be an introduction to the field and is in no way adequate for a complete understanding of this important field. For more complete coverage of quantum mechanics and its applications, see the suggested references listed at the end of the book. The presentation of quantum mechanics begins with stating the four postulates and elucidating their meanings.

2.1 THE POSTULATES

POSTULATE I. For any possible state of a system, there is a function Ψ of the coordinates of the parts of the system and time that completely describes the system.

For a single particle whose position is described by Cartesian coordinates, the wave function can be written as

$$\Psi = \Psi(x, y, z, t) \tag{2.1}$$

For two particles, the coordinates of each particle must be specified so the wave function requires two sets of coordinates. It can be written as

$$\Psi = \Psi(x_1, y_1, z_1, x_2, y_2, z_2, t) \tag{2.2}$$

For a general system consisting of multiple particles, the wave function is written in terms of the generalized coordinates q_i:

$$\Psi = \Psi(q_i, t) \tag{2.3}$$

Fundamentals of Quantum Mechanics
http://dx.doi.org/10.1016/B978-0-12-809242-2.00002-4

Because the model is that of a *wave*, Ψ is called a *wave function*. The state of the system that i describes is called the *quantum state*.

The wave function squared, Ψ^2, is proportional to a probability. Because Ψ may be complex, the quantity that is of interest is $\Psi^*\Psi$, where Ψ^* is the *complex conjugate* of Ψ. The complex conjugate of a function is the function with i replaced by $-i$, where $i = (-1)^{1/2}$. For example, if one squares the function $(x + ib)$, the result is

$$(x + ib)(x + ib) = x^2 + 2ibx + i^2 b^2 = x^2 + 2ibx - b^2 \qquad (2.4)$$

and the resulting function is still complex. However, if the function $(x + ib)$ is multiplied by its complex conjugate $(x - ib)$, the result is

$$(x + ib)(x - ib) = x^2 - i^2 b^2 = x^2 + b^2 \qquad (2.5)$$

and the product is a real function.

The quantity $\Psi^*\Psi \; d\tau$ is proportional to the probability of finding the particles of the system in the volume element $d\tau = dx \; dy \; dz$. It is a requirement that the *total* probability be unity (1) so that the particle must be *somewhere*. That can be represented as

$$\int_{\text{all space}} \Psi^*\Psi \; d\tau = 1 \qquad (2.6)$$

If this condition is met, then Ψ is *normalized*. In addition, Ψ must be *finite, single valued*, and *continuous*. These conditions describe a "well-behaved" wave function. The reasons for these requirements are as follows:

Finite: A probability of unity (exactly 1.00) denotes a "sure thing." A probability of 0 means that a certain event cannot happen. Therefore, probability varies from 0 to unity. If Ψ could be infinite, the probability would not be limited to a value of 1.

Single valued: In a given region of space, there is only one probability of finding a particle. When this is interpreted in terms of a hydrogen atom, it means there is a single probability of finding the electron at some specified distance from the nucleus. Consequently, there are not two different probabilities of finding the electron at some given distance from the nucleus.

Continuous: If there is a certain probability of finding an electron at a given distance from the nucleus in a hydrogen atom, there will be a slightly different probability if the distance from the nucleus is changed slightly. The probability does not suddenly double if the distance is changed by 0.01%. The probability function does not have discontinuities, so the wave function must be continuous.

If two functions φ_1 and φ_2 have the property that

$$\int \varphi_1^* \varphi_2 \, d\tau = 0 \ \text{ or } \ \int \varphi_1 \varphi_2^* \, d\tau = 0 \tag{2.7}$$

the functions are said to be *orthogonal*. Whether or not the integral is equal to 0 may depend on the limits of integration, and hence one speaks of *orthogonality* within a certain interval. Therefore, the limits of integration must be clear. In the previous case, the integration is carried out over the possible range of the coordinates included in the volume element $d\tau$. If the coordinates are x, y, and z, the limits are from $-\infty$ to $+\infty$ for each variable (all space). If the coordinates are r, θ, and φ, the limits of integration are 0 to ∞, 0 to π, and 0 to 2π, respectively, for these coordinates. A more complete discussion of orthogonal wave functions will be presented in later chapters.

2.2 THE WAVE EQUATION

In 1924, it was shown by de Broglie that a moving particle has a wave character. In 1927, Davisson and Germer verified this conclusion experimentally when an electron beam was diffracted by a nickel crystal. Even before that experimental verification of de Broglie's hypothesis, Erwin Schrödinger adapted the *wave model* to the problem of electron motion in the hydrogen atom. In that case, the model needs to describe a three-dimensional wave. A problem in classical physics had dealt with such models in a conjecture known as the *flooded planet problem*. This model considers the waveforms that would result from a disturbance of a sphere that is covered with water. The classical three-dimensional wave equation is

$$\frac{\partial^2 \varphi}{\partial x^2} + \frac{\partial^2 \varphi}{\partial y^2} + \frac{\partial^2 \varphi}{\partial z^2} = \frac{1}{v^2} \frac{\partial^2 \varphi}{\partial t^2} \tag{2.8}$$

in which φ is the amplitude function and v is the phase velocity of the wave. For harmonic motion (as in the case of a sine wave), the amplitude as a function of time can be expressed by the equation

$$\frac{\partial^2 \varphi}{\partial t^2} = -4\pi^2 v^2 \varphi \tag{2.9}$$

in which v is the frequency. The de Broglie relationship

$$\lambda = \frac{h}{mv} \tag{2.10}$$

and the relationship between the frequency and the wavelength of a transverse wave,

$$\lambda(\text{distance}) \times \nu(\text{time}^{-1}) = \nu(\text{distance/time}),$$

give rise to the relationship for the frequency of the wave,

$$\nu^2 = \frac{m^2 v^4}{h^2} \tag{2.11}$$

Therefore, substituting this result in Eq. (2.9)

$$\frac{\partial^2 \varphi}{\partial t^2} = -4\pi^2 \left(\frac{m^2 v^4}{h^2} \right) \varphi \tag{2.12}$$

Substituting this in the general three-dimensional wave Eq. (2.8), it becomes

$$\frac{\partial^2 \varphi}{\partial x^2} + \frac{\partial^2 \varphi}{\partial y^2} + \frac{\partial^2 \varphi}{\partial z^2} = \left(\frac{1}{v^2} \right) \left(\frac{-4\pi^2 v^4 m^2}{h^2} \right) \varphi \tag{2.13}$$

The total energy will now be represented as E, the kinetic energy as T (the energy of translation), and the potential energy as V. Therefore, the kinetic energy is given by

$$T = \frac{mv^2}{2} = E - V \tag{2.14}$$

Solving this equation for v^2 and substituting this result into Eq. (2.13) yields the wave equation in the form

$$\frac{\partial^2 \varphi}{\partial x^2} + \frac{\partial^2 \varphi}{\partial y^2} + \frac{\partial^2 \varphi}{\partial z^2} = \left(\frac{-4\pi^2 m^2}{h^2} \right) \left[\frac{2(E-V)}{m} \right] \varphi \tag{2.15}$$

This equation, which describes a three-dimensional wave, can be rearranged and written in the form

$$\frac{\partial^2 \varphi}{\partial x^2} + \frac{\partial^2 \varphi}{\partial y^2} + \frac{\partial^2 \varphi}{\partial z^2} = \left(-\frac{8\pi^2 m}{h^2} \right) (E-V)\varphi \tag{2.16}$$

This is one form of the *Schrödinger wave equation*. Solutions to this wave equation are known as *wave functions*. Solving wave equations involves the branch of science known as *wave mechanics*, also known as quantum mechanics because of the energies of such systems are quantized.

The preceding presentation is not a *derivation* of the Schrödinger wave equation in the usual sense. Rather, it is an *adaptation* of a classical wave equation to a different type of system by making use of the de Broglie hypothesis. It is interesting to note that Schrödinger's treatment of the hydrogen atom started with an equation that was already known. As will be shown in Chapter 4, the solution of the wave equation for the hydrogen atom also makes use of mathematical techniques that were already in existence at the time. Whereas Schrödinger's work was revolutionary for its time, it was carried out with the understanding of what had been done before. In fact, his incorporation of de Broglie's hypothesis came only a couple of years after that idea became known but before it had been experimentally verified.

2.3 OPERATORS

POSTULATE II. For every dynamical variable (classical observable) there is a corresponding operator.

This postulate provides the connection between quantities that are classical observables (quantities that can be measured) and the quantum mechanical techniques for doing things. Dynamical variables are such quantities as energy, momentum, angular momentum, and position coordinates. In quantum mechanics, these are replaced by *operators*, which are symbols that indicate that some mathematical operation is to be performed. Such symbols include $(\ )^2$, d/dx, and $\int$. Coordinates are identical in operator and classical forms. For example, the coordinate x is simply used in operator form as x. (This will be illustrated later.) Other classical observables are replaced by their corresponding operators, as shown in Table 2.1. Additionally, as will be illustrated later, other operators can be formed by combining those in the table. For example, because the kinetic energy is $mv^2/2$, it can be written in terms of the momentum, p, as $p^2/2m$.

The operators that are important in quantum mechanics have two important characteristics: First, the operators are *linear*, meaning

$$\alpha(\varphi_1 + \varphi_2) = \alpha\varphi_1 + \alpha\varphi_2 \qquad (2.17)$$

in which α is the operator and φ_1 and φ_2 are the functions being operated on. Also, when C is a constant, then

$$\alpha(C\varphi) = C(\alpha\varphi) \qquad (2.18)$$

The *linear* character of the operator is related to the superposition of states and waves reinforcing each other in the process. The second property of

Table 2.1 Some common operators in quantum mechanics

Quantity	Symbol used	Operator form
Coordinates	x, y, z, r	x, y, z, r
Momentum		
x	p_x	$\dfrac{\hbar}{i}\dfrac{\partial}{\partial x}$
y	p_y	$\dfrac{\hbar}{i}\dfrac{\partial}{\partial y}$
z	p_z	$\dfrac{\hbar}{i}\dfrac{\partial}{\partial z}$
Kinetic energy	$\dfrac{p^2}{2m}$	$-\dfrac{\hbar^2}{2m}\left(\dfrac{\partial^2}{\partial x^2}+\dfrac{\partial^2}{\partial y^2}+\dfrac{\partial^2}{\partial z^2}\right)$
Kinetic energy	T	$-\dfrac{\hbar}{i}\dfrac{\partial}{\partial t}$
Potential energy	V	$V(q_i)$
Angular momentum	L_z (Cartesian)	$\dfrac{\hbar}{i}\left(x\dfrac{\partial}{\partial y}-y\dfrac{\partial}{\partial x}\right)$
	L_z (polar)	$\dfrac{\hbar}{i}\dfrac{\partial}{\partial \varphi}$

the operators is that they are *Hermitian*. This property can be illustrated as follows: If two functions φ_1 and φ_2 are considered, the operator α is said to be Hermitian if

$$\int \varphi_1^* \alpha \varphi_2\, d\tau = \int \varphi_1 \alpha^* \varphi_2^*\, d\tau \tag{2.19}$$

This requirement is necessary to ensure that the calculated quantities for physical systems are real. In the following chapters, there will be several opportunities to observe these types of behavior in the operators used.

2.4 EIGENVALUES

POSTULATE III. The permissible values that a dynamical variable may have are those given by $\alpha\, \varphi = a\, \varphi$, where φ is the *eigenfunction* of the operator α that corresponds to the observable whose permissible values are *a*.

This postulate can be stated in the form of an equation as

$$\underset{\text{operator}}{\alpha} \quad \underset{\text{wave function}}{\varphi} \quad = \quad \underset{\text{constant (eigenvalue)}}{a} \quad \underset{\text{wave function}}{\varphi} \tag{2.20}$$

If performing the operation α on the wave function yields the original function multiplied by a constant, then φ is said to be an *eigenfunction* of the operator α. The constant is called an *eigenvalue*. This can be illustrated by letting $\varphi = e^{2x}$ and the operator $\alpha = d/dx$. In this case, operating on the function with the operator gives

$$\frac{d\varphi}{dx} = 2e^{2x} = \text{constant} \cdot e^{2x} \tag{2.21}$$

Therefore, e^{2x} is an eigenfunction of the operator α with an eigenvalue of 2.

If the function used is $\varphi = e^{2x}$ and the operator is $(\ )^2$, then performing the operation gives

$$\left(e^{2x}\right) = e^{4x}$$

which is *not* a constant times the original function. Therefore, e^{2x} is not an eigenfunction of the operator $(\ )^2$.

The operator for the z component of angular momentum can be written as

$$\hat{L}_z = \frac{\hbar}{i}\frac{\partial}{\partial \varphi} \tag{2.22}$$

where $\hbar = h/2\pi$. Operating on the function $e^{in\varphi}$ (where n is a constant) with this operator gives

$$\frac{\hbar}{i}\frac{\partial}{\partial \varphi}\left(e^{in\varphi}\right) = in\frac{\hbar}{i}e^{in\varphi} = n\hbar \cdot e^{in\varphi} \tag{2.23}$$

which shows that the result is a constant $(n\hbar)$ times the original function. Therefore, the eigenvalue is $n\hbar$.

For a given system, there may be multiple values of a parameter to be calculated. Because most properties vary, such as the distance of an electron from the nucleus in a hydrogen atom, it is often desirable to determine an *average* or *expectation* value. Using the operator equation $\alpha\varphi = a\varphi$, where φ is a wave function, both sides of the equation are multiplied by the complex conjugate φ^*:

$$\varphi^*\alpha\varphi = \varphi^*a\varphi \tag{2.24}$$

Note, however, that $\varphi^* \alpha \varphi$ is not necessarily the same as $\varphi \alpha \varphi^*$. To obtain the sum of the probability over all space, this is written in the form of the integral equation

$$\int_{\text{all space}} \varphi^* \alpha \varphi \, d\tau = \int_{\text{all space}} \varphi^* a \varphi \, d\tau \tag{2.25}$$

However, a is a constant and is not affected by the order of operations. Removing it from the integral and solving for a yields

$$a = \frac{\int \varphi^* \alpha \varphi \, d\tau}{\int \varphi^* \varphi \, d\tau} \tag{2.26}$$

It should be remembered that because α is an *operator*, $\varphi^* \alpha \varphi$ is not necessarily the same as $\alpha \varphi \varphi^*$, so that the order $\varphi^* \alpha \varphi$ must be preserved, and α cannot be removed from the integral.

If φ is a normalized function, then by definition $\int \phi^* \phi \, d\tau = 1$ and

$$\bar{a} = \langle a \rangle = \int \varphi^* \alpha \varphi \, d\tau \tag{2.27}$$

In this case, $\bar{a}$ and $<a>$ are the usual ways of indicating the average or expectation value. If the explicit form of the wave function is known, then theoretically an expectation or average value can be calculated for a given parameter by using the appropriate operator and the procedure shown earlier.

Consider the following example, which illustrates the application of these ideas. Suppose a problem requires one to calculate the radius of the hydrogen atom in the 1s state. The normalized wave function ψ_{1s} can be written as

$$\psi_{1s} = \left(\frac{1}{\sqrt{\pi}} \right) \left(\frac{1}{a_0} \right)^{3/2} e^{-r/a_0} = \psi_{1s}^* \tag{2.28}$$

in which a_0 is the first Bohr radius. The equation to calculate the average radius becomes

$$\langle r \rangle = \int \psi^* (\text{operator}) \, \psi \, d\tau \tag{2.29}$$

The operator in this case is simply r because position coordinates have the same form in operator and classical form. When expressed in terms of polar coordinates, the volume element $d\tau = r^2 \sin\theta\, dr\, d\theta\, d\varphi$. Therefore, the problem now becomes

$$\langle r \rangle = \int_0^\infty \int_0^\pi \int_0^{2\pi} \frac{1}{\sqrt{\pi}}\left(\frac{1}{a_0}\right)^{3/2} e^{-r/a_0}\,(r)\,\frac{1}{\sqrt{\pi}}\left(\frac{1}{a_0}\right)^{3/2} e^{-r/a_0}\, r^2 \sin\theta\, dr\, d\theta\, d\varphi \quad (2.30)$$

Although this may look rather formidable, it simplifies greatly because the operator r becomes a multiplier and the functions involving r can be multiplied and collected. When this is done, the result is

$$\langle r \rangle = \int_0^\infty \int_0^\pi \int_0^{2\pi} \left(\frac{1}{\pi a_0^3}\right) e^{-2r/a_0}\, r^3 \sin\theta\, dr\, d\theta\, d\varphi \quad (2.31)$$

Using the technique from calculus that makes it possible to separate multiple integrals of the type

$$\int f(x)g(y)\, dx\, dy = \int f(x)\, dx \int g(y)\, dy \quad (2.32)$$

Eq. (2.31) can be written as

$$\langle r \rangle = \frac{1}{\pi a_0^3}\int_0^\infty r^3 e^{-2r/a_0}\, dr \int_0^\pi \int_0^{2\pi} \sin\theta\, d\theta\, d\varphi \quad (2.33)$$

By making use of a table of integrals, it is easily verified that

$$\int_0^\pi \int_0^{2\pi} \sin\theta\, d\theta\, d\varphi = 4\pi \quad (2.34)$$

Moreover, an exponential integral of this form is a common occurrence in quantum mechanics. It is easily evaluated using the formula

$$\int_0^\infty x^n e^{-bx}\, dx = \frac{n!}{b^{n+1}} \quad (2.35)$$

In this case, $n = 3$ and $b = 2/a_0$. Therefore, the exponential integral can be expressed as

$$\int_0^\infty r^3 e^{-2r/a_0}\, dr = \frac{3!}{(2/a_0)^4} \quad (2.36)$$

When this is simplified, the result is

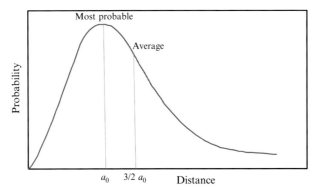

Fig. 2.1 The probability of finding the electron in the 1s state as a function of distance from the nucleus.

$$\langle r \rangle = \frac{4\pi}{\pi a_0^3} \frac{3!}{(2/a_0)^4} = \frac{3}{2} a_0 \tag{2.37}$$

Thus, $<r>_{1s} = (3/2)a_0$, where $a_0 = 0.529$ Å.

The *average* distance of the electron from the nucleus in the 1s state of hydrogen is 3/2 the radius of the first Bohr radius. However, the *most probable* distance is that where the probability is highest and the same as the radius of the first Bohr orbit. *Average* and *most probable* are *not* the same. The reason for this is that the probability distribution is not symmetric, as is shown in Fig. 2.1.

2.5 WAVE FUNCTIONS

POSTULATE IV. The state function ψ is given as a solution of

$$\hat{H}\psi = E\psi \tag{2.38}$$

where $\hat{H}$ is the operator for total energy, the *Hamiltonian operator*.

This postulate provides a starting point for formulating a problem in quantum mechanical terms because the process is to determine a wave function to describe the system being studied. The *Hamiltonian function* in classical physics is the total energy $T + V$, where T is the translational (kinetic) energy and V is the potential energy. In operator form, this can be written as

$$\hat{H} = \hat{T} + \hat{V} \tag{2.39}$$

in which $\hat{T}$ is the operator for kinetic energy and $\hat{V}$ is the operator for potential energy. Written in terms of the generalized coordinates q_i, and time, the general wave equation becomes

$$\hat{H}\Psi(q_i, t) = -\frac{\hbar}{i}\frac{\partial\Psi(q_i, t)}{\partial t} \tag{2.40}$$

When written in terms of the momentum, the kinetic energy can be expressed as

$$T = \frac{mv^2}{2} = \frac{p^2}{2m} \tag{2.41}$$

For the total kinetic energy, T can be written as the sum of the x, y, and z components:

$$T = \frac{p_x^2}{2m} + \frac{p_y^2}{2m} + \frac{p_z^2}{2m} \tag{2.42}$$

It is possible to put this in operator form by using the momentum operators given earlier in Table 2.1. When this is done, the result is

$$\hat{T} = \frac{1}{2m}\left(\frac{\hbar}{i}\frac{\partial}{\partial x}\right)^2 + \frac{1}{2m}\left(\frac{\hbar}{i}\frac{\partial}{\partial y}\right)^2 + \frac{1}{2m}\left(\frac{\hbar}{i}\frac{\partial}{\partial z}\right)^2 \tag{2.43}$$

However, the square of the momentum operator for the x-direction can be written as

$$\left(\frac{\hbar}{i}\frac{\partial}{\partial x}\right)^2 = \left(\frac{\hbar}{i}\frac{\partial}{\partial x}\right)\left(\frac{\hbar}{i}\frac{\partial}{\partial x}\right) = \left(\frac{\hbar^2}{i^2}\frac{\partial^2}{\partial x^2}\right) = -\hbar^2\frac{\partial^2}{\partial x^2} \tag{2.44}$$

When this is carried out for all three coordinates the result is

$$\hat{T} = -\frac{\hbar^2}{2m}\left(\frac{\partial^2}{\partial x^2} + \frac{\partial^2}{\partial y^2} + \frac{\partial^2}{\partial z^2}\right) = -\frac{\hbar^2}{2m}\nabla^2 \tag{2.45}$$

in which ∇^2 is known as the *Laplacian operator*, or simply the *Laplacian*. If the potential energy can be expressed in terms of the general coordinates and time, then the result is

$$V = V(q_i, t) \tag{2.46}$$

Therefore, the operator equation for total energy becomes

$$\left(-\frac{\hbar^2}{2m}\nabla^2 + V(q_i, t)\right)\psi(q_i, t) = -\frac{\hbar}{i}\frac{\Psi(q_i, t)}{\partial t}$$

(2.47)

This equation is known as Schrödinger's *time-dependent* equation or Schrödinger's *second* equation.

In most problems, the classical observables have values that do not change with time, or at least their average values do not change with time. In most cases, it would therefore be advantageous to simplify the problem by removal of the dependence on time. The separation of variables technique is now applied to see whether that dependence can be separated.

The separation of variables as a technique in solving differential equations will be discussed in Chapters 3 and 4, and so it will be used it here with very little explanation. First, it is assumed that ψ (q_i, t) is the product of two functions, one a function that contains only q_i and another that contains only the time, t. If this is done the result is

$$\Psi(q_i, t) = \psi(q_i)\tau(t)$$

(2.48)

Note that Ψ is used to denote the complete state (wave) function and lower case ψ is used to represent the state function with the time dependence removed. Because the problems that will considered in this book are time-independent ones, ψ will be used throughout. The Hamiltonian can now be written in terms of the two functions ψ and τ as

$$\hat{H}\Psi(q_i, t) = \hat{H}\psi(q_i)\tau(t)$$

(2.49)

Therefore, because $\psi(q_i)$ is not a function of t, Eq. (2.49) can be written as

$$\hat{H}\psi(q_i)\tau(t) = -\frac{\hbar}{i}\frac{\partial}{\partial t}\psi(q_i)\tau(t) = -\frac{\hbar}{i}\psi(q_i)\frac{\partial\tau(t)}{\partial t}$$

(2.50)

Dividing Eq. (2.50) by the product $\psi(q_i)\tau(t)$,

$$\frac{\hat{H}\psi(q_i)\tau(t)}{\psi(q_i)\tau(t)} = \frac{-\frac{\hbar}{i}\psi(q_i)\left(\frac{\partial\tau(t)}{\partial t}\right)}{\psi(q_i)\tau(t)}$$

(2.51)

and the result can be written as

$$\frac{1}{\psi(q_i)}\hat{H}\psi(q_i) = -\frac{\hbar}{i}\frac{1}{\tau(t)}\frac{\partial\tau(t)}{\partial t}$$

(2.52)

Note that the factor $\psi(q_i)$ cannot be canceled on the left-hand side of Eq. (2.52) because $\hat{H}\,\psi(q_i)$ does not represent $\hat{H}$ *times* $\psi(q_i)$ but rather $\hat{H}$ *operating on* ψ (q_i). The left-hand side of the equation is a function of q_i and the right-hand side is a function of t, so each can be considered equal to a constant with respect to changes in the values of the other variable. Both sides can be set equal to some new parameter W so that

$$\frac{1}{\psi(q_i)}\hat{H}\psi(q_i) = W \quad \text{and} \quad -\frac{\hbar}{i}\frac{1}{\tau(t)}\frac{\partial\tau(t)}{\partial t} = W \tag{2.53}$$

Simplifying the first of these equations leads to

$$\hat{H}\psi(q_i) = W\,\psi(q_i) \tag{2.54}$$

and the second gives the relationship

$$\frac{d\tau(t)}{dt} = -\frac{i}{\hbar}W\tau(t) \tag{2.55}$$

The differential equation involving the time is of a common form and can be solved readily to give

$$\tau(t) = e^{-(i/\hbar)Wt} \tag{2.56}$$

Substituting this result into Eq. (2.48), it is found that the total state function, Ψ, is

$$\Psi(q_i, t) = \psi(q_i)\,e^{-(i/\hbar)Wt} \tag{2.57}$$

Therefore, Eq. (2.50) can be written as

$$e^{-(i/\hbar)Wt}\hat{H}\psi(q_i) = \frac{\hbar}{i}\frac{i}{\hbar}W\psi(q_i)\,e^{-(i/\hbar)Wt} \tag{2.58}$$

or

$$e^{-(i/\hbar)Wt}\hat{H}\psi(q_i) = W\psi(q_i)\,e^{-(i/\hbar)Wt} \tag{2.59}$$

The factor $e^{-(i/\hbar)Wt}$ can be dropped from both sides of Eq. (2.59), which results in

$$\hat{H}\psi(q_i) = W\psi(q_i) \tag{2.60}$$

This equation shows that the time dependence has been separated from the general equation.

In Eq. (2.60), neither the Hamiltonian operator nor the wave function is time dependent. It is this form of the equation that will be used to solve problems discussed in this book. Therefore, the time-independent wave function ψ will be indicated any time the general wave equation is written as $\hat{H}\psi = E\psi$.

For the hydrogen atom, the electrostatic potential energy is $V = -e^2/r$, which remains unchanged in operator form because e is a constant (the charge on the electron) and r represents a coordinate. Therefore, the Hamiltonian operator can be written as

$$\hat{H} = -\frac{\hbar^2}{2m}\nabla^2 - \frac{e^2}{r} \tag{2.61}$$

Using this form of the operator, the equation $\hat{H}\psi = E\psi$ leads to the wave equation

$$\hat{H}\psi = -\frac{\hbar^2}{2m}\nabla^2\psi - \frac{e^2}{r}\psi \tag{2.62}$$

Rearrangement of Eq. (2.62) gives a familiar form of the Schrödinger wave equation for the hydrogen atom most often written as

$$\nabla^2\psi + \frac{2m}{\hbar^2}(E - V)\psi = 0 \tag{2.63}$$

Several relatively simple models are capable of being treated by the methods of quantum mechanics, and much of the remainder of this introductory book is the discussion of those important models. To treat these models, the four postulates can be applied in a relatively straightforward manner. However, for any of these models the starting point will be

$$\hat{H}\psi = E\psi \tag{2.64}$$

In applying this equation to problems of interest, it will be necessary to use the appropriate expressions for the operators corresponding to the potential and kinetic energies. In practice, it will be found that a rather limited number of potential functions are applicable, with the most common being a Coulombic (electrostatic) potential.

In the following chapters, quantum mechanical models are presented because they can be applied to several systems of chemical and physical interest. For example, the barrier penetration phenomenon has application as a model for nuclear decay and transition state theory in chemical kinetics. The rigid rotor and harmonic oscillator models are useful as models in applying rotational and vibrational spectroscopy. The particle in the box model has some utility as an approximation in the treatment of electrons in metals or conjugated molecules. Given the utility of these models, some familiarity with each of them is essential for all who would understand the application of quantum mechanics to the problems of relevance to a wide range of sciences. The next several chapters will deal with the basic models and their applications.

PROBLEMS

1. The operator for the z component of angular momentum $\hat{L}_z$ in polar coordinates is $(\hbar/i)\,(\partial/\partial\varphi)$. Determine which of the following functions are eigenfunctions of this operator and determine the eigenvalues for those that are:
   (a) $\sin\varphi\, e^{i\varphi}$,
   (b) $\sin^l\varphi\, e^{il\varphi}$ (where l is an integer constant),
   (c) $\sin\varphi\, e^{-i\varphi}$.

2. Calculate the expectation value for the z component of angular momentum for functions (a) and (b) in Problem 1.

3. Normalize the following functions in the interval 0 to ∞
   (a) e^{-5x}, and
   (b) e^{-bx} (where b is a constant).

4. Show that the function $(x+iy)/r$ is an eigenfunction of the operator for the z component of angular momentum (see Table 2.1).

5. Show that the $1s$ wave function for hydrogen,

$$\psi_{1s} = \frac{1}{\sqrt{\pi}}\left(\frac{1}{a_0}\right)^{3/2} e^{-r/a_0}$$

   is normalized.

6. Show that $\psi = a\, e^{-bx}$ (where a and b are constants) is an eigenfunction of the operator d^2/dx^2.

7. Normalize the function $\varphi = a\, e^{-bx}$ in the interval 0 to ∞.

8. Determine whether the function $\varphi = \sin x\, e^{ax}$ (where a is a constant) is an eigenfunction of the operators d/dx and d^2/dx^2. If it is, determine any eigenvalue(s).

9. Normalize the function $\psi = \sin(\pi x/L) + i\sin(2\pi x/L)$ in the interval 0 to L. Determine the expectation values for the momentum, p, and the kinetic energy, T.

10. Functions and operators are said to be symmetric if $f(x) = f(-x)$. Determine whether the operator for kinetic energy is symmetric or antisymmetric.

CHAPTER 3

Particles in Boxes

As one begins the study of quantum mechanics, it is desirable to consider some simple problems that can be solved exactly whether they represent important models of nature or not. Such is the case with the models of particles in boxes. Although they have some applicability as approximations to real problems, such models are most useful in illustrating the methods of formulating problems and applying quantum mechanical procedures. As a result, almost every introductory book on quantum mechanics includes a discussion on particles in boxes.

3.1 THE PARTICLE IN A ONE-DIMENSIONAL BOX

In this model, a particle is considered to have motion that is confined within a box. In the first case, the box is assumed to be one dimensional for simplicity, although a three-dimensional problem is not much more difficult, and that problem will be considered next. To confine the particle absolutely to the box, the walls of the box are considered to be infinitely high. Otherwise, there is a small but finite probability that the particle can "leak" out of the box by *tunneling*. The problem of tunneling through a potential energy barrier will be discussed in a later chapter. The coordinate system and energy parameters for this problem are shown in Fig. 3.1.

The Hamiltonian, H, is

$$H = T + V = \frac{p^2}{2m} + V \tag{3.1}$$

where p is the momentum, m is the mass of the particle, and V is the potential energy. Outside the box, $V = \infty$, so

$$H = \frac{p^2}{2m} + \infty \tag{3.2}$$

which means that the Hamiltonian operator $\hat{H}$ is

$$\hat{H} = -\frac{\hbar^2}{2m}\frac{d^2}{dx^2} + \infty \tag{3.3}$$

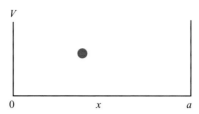

Fig. 3.1 A particle in a one-dimensional box.

As a result, the wave equation can be written as

$$\hat{H}\psi = E\psi = -\frac{\hbar^2}{2m}\frac{d^2}{dx^2}\psi + \infty\psi \tag{3.4}$$

Therefore, for the equation to be valid, ψ must be 0, which leads to the conclusion that the probability of finding the particle outside the box is zero. Inside the box, the potential energy is zero ($V = 0$), so the wave equation can be written as

$$\hat{H}\psi = E\psi \tag{3.5}$$

or

$$-\frac{\hbar^2}{2m}\frac{d^2\psi}{dx^2} = E\psi \tag{3.6}$$

This equation is of the form

$$\frac{d^2\psi}{dx^2} + k^2\psi = 0 \tag{3.7}$$

where $k^2 = 2\,mE/\hbar^2$. This is a linear differential equation with constant coefficients, which is a standard type of equation having a solution of the form

$$\psi = A\cos kx + B\sin kx \tag{3.8}$$

Actually, it is possible to guess a form of the solution in this case, because Eq. (3.7) shows that the original function times a constant must be equal and opposite in sign to the second derivative of the function. Very few common functions meet this requirement, but $\sin bx$, $\cos bx$, and an exponential function e^{ibx} do have this property. This can be shown as follows:

$$\frac{d^2}{dx^2}(\sin bx) = -b^2\sin bx$$

$$\frac{d^2}{dx^2}e^{ibx} = -b^2 e^{ibx}$$

In the solution shown in Eq. (3.8), A and B are constants and

$$k = \frac{(2mE)^{1/2}}{\hbar}$$

The appearance of two constants in the solution is normal for a second order differential equation. These constants must be evaluated using *boundary conditions*, which are those requirements that must be met because of the physical limits of the system. In order for the probability of finding the particle to vanish at the walls of the box, it is necessary that ψ be zero at the boundaries. That is, $\psi = 0$ for $x = 0$ or $x = a$. At $x = 0$ the solution can be written as

$$\psi = 0 = A\cos\left[\frac{(2mE)^{1/2}}{\hbar} \times 0\right] + B\sin\left[\frac{(2mE)^{1/2}}{\hbar} \times 0\right] \qquad (3.9)$$

Now, $\sin 0 = 0$, so the last term is 0, and $\cos 0 = 1$, so $\psi = 0 = A \times 1$. This can be true only for $A = 0$. Therefore, the constant A in the solution must be 0 and the wave function reduces to

$$\psi = B\sin\left[\frac{(2mE)^{1/2}}{\hbar} x\right] \qquad (3.10)$$

However, the constant B must now be evaluated by making use of the boundary conditions. Using the requirement that the wave function must vanish at the boundary a so that $\psi = 0$ for $x = a$, it can be seen that

$$\psi = 0 = B\sin\left[\frac{(2mE)^{1/2}}{\hbar} a\right] \qquad (3.11)$$

Now $\sin\theta = 0$ for $\theta = 0°, 180°, 360°, \ldots$, which represents $\theta = n\pi$ rad and where n is an integer. Consequently, the condition

$$\frac{(2mE)^{1/2}}{\hbar} a = n\pi \qquad (3.12)$$

must be met in which $n = 1, 2, 3, \ldots$. The value $n = 0$ would lead to $\psi = 0$, which would give a probability of 0 for finding the particle inside the box. Because the particle is required to be somewhere inside the box, the value of $n = 0$ is rejected. Eq. (3.12) can now be solved to obtain the allowed energy levels in terms of the quantum number n,

$$E = \frac{n^2\hbar^2\pi^2}{2ma^2} = \frac{n^2h^2}{8ma^2} \qquad (3.13)$$

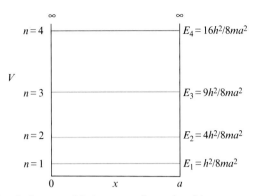

Fig. 3.2 Energy levels for a particle in a one-dimensional box.

where $n = 1, 2, 3, \ldots$, a *quantum number*. Note the requirement that the wave function must vanish at the boundaries of the box (0 and a) leads to the quantization of *energy*. *This occurs because the trigonometric functions vanish only for certain values of θ.* Therefore, for a *free* particle, the energy levels are not quantized, but rather they form a continuum. It is only for the bound (confined or constrained) system that the energy levels are quantized. It would therefore be expected that an electron bound in a hydrogen atom should exhibit discrete energy levels. The *diverging* energy levels for the particle in a one-dimensional box are shown in a graphical way in Fig. 3.2. In most systems of a chemical nature (recall the energy level diagram for the hydrogen atom), the energy levels *converge*. Note also the energy of the lowest state is not zero and the energy of the lowest state is $h^2/8ma^2$:

Although it has been possible to obtain values for the allowed energy levels for the particle and the general form of the wave function, the wave function has not been normalized. The wave function is normalized when

$$\int_{\text{all space}} \psi^* \psi \, dt = 1 \tag{3.14}$$

Therefore, for this problem the integration is over the interval in which x can vary from 0 to a. If we let B be the *normalization constant* that is multiplied by the wave function, the result is

$$\int_0^a B^* \psi^* B\psi \, d\tau = 1 \tag{3.15}$$

As has been shown, ψ is given as

$$\psi = B \sin \left(\frac{n\pi x}{a} \right)$$

and $B^* = B$, so Eq. (3.15) can be written as

$$\int_0^a B^2 \sin^2 (n\pi/a) \, x \, dx = 1 \qquad (3.16)$$

This integral (where a is a constant) is a standard form given in tables of integrals as

$$\int (\sin^2 ax) \, dx = \frac{1}{2} x - \frac{1}{4a} \sin 2ax \qquad (3.17)$$

After solving for B^2 and evaluating the integral, the result is

$$B^2 = \frac{1}{\dfrac{x}{2} \left. \dfrac{(\sin 2\pi x/a)}{4n\pi/a} \right|_{x=0}^{x=a}} \qquad (3.18)$$

The denominator evaluates to $a/2$ so that $B = (2/a)^{1/2}$. The complete normalized wave function can be written as

$$\psi = \left(\frac{2}{a}\right)^{1/2} \sin \left(\frac{n\pi}{a}\right) x \qquad (3.19)$$

With this wave function, the average or expectation value of position or momentum of the particle can be calculated using the results of Postulate III shown in Chapter 2. Figure 3.3 shows the plots of ψ and ψ^2 for the first few values of n.

If a carbon chain such as

$$C = C - C = C - C \qquad (3.20)$$

is considered to be a "box" in which the π electrons can move along the chain, the results of the particle in a one-dimensional box can be considered as a model. If the average C—C bond length is 1.40 Å (140 pm), the entire chain would be 5.60 Å (560 pm) in length. Therefore, the difference between the $n=1$ and the $n=2$ states would be

$$E = \frac{2^2 h^2}{8ma^2} - \frac{1^2 h^2}{8ma^2} = \frac{3h^2}{8ma^2} \qquad (3.21)$$

$$E = \frac{3(6.63 \times 10^{-27} \text{ erg s})^2}{8(9.10 \times 10^{-28} \text{ g}) \times (5.60 \times 10^{-8} \text{ cm})^2} = 5.78 \times 10^{-12} \text{ erg} \qquad (3.22)$$

This corresponds to a photon having wavelength of 344 nm, and the actual maximum in the absorption spectrum of 1,3-pentadiene is found at 224 nm.

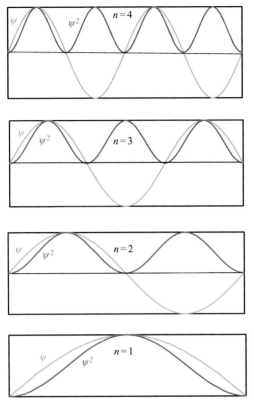

Fig. 3.3 Plots of the wave function *(blue line)* and the square of the wave function *(red line)* for the first four states of the particle in a box.

Although it is not in close agreement, the simple model does predict absorption in the ultraviolet region of the spectrum.

Although this model is of limited usefulness for representing physical systems, the methodology that it shows is valuable for illustrating the quantum mechanical way of doing things. A few observations are in order here: First, the energy of a confined particle is quantized. The application of the boundary conditions leads to the quantization of energy, and the energy increases as the square of the quantum number describing the state. Moreover, the energy also increases as the mass of the particle decreases. This has implications for the confinement of a particle having the mass of an electron to a region the size of an atomic nucleus ($\sim 10^{-13}$ cm). Therefore, the β^- particles (electrons) emitted during beta decay could not exist in the nucleus prior to the actual decay. For example, confining an electron this way would require that the lowest state have an energy of

$$E = \frac{1^2 h^2}{8ma^2} = \frac{1(6.63 \times 10^{-27} \text{ erg s})^2}{8(9.10 \times 10^{-28} \text{ g}) \times (10^{-13} \text{ cm})^2} = 0.604 \text{ erg}$$

Using the conversion factor $1 \text{ erg} = 6.242 \times 10^{11} \text{ eV}$, this amounts to 3.77×10^{11} eV or 3.77×10^5 MeV! For a reference, this energy can be compared to the value of 13.6 eV for the ionization potential of the hydrogen atom.

It is interesting to note that the energy, in addition to being quantized, depends on the quantum number n, which cannot be zero. Therefore, there is *some* energy for the particle $(E = h^2/8ma^2$ when $n = 1)$ even in the lowest state. This is known as *zero-point energy*. As will be shown later, other systems (the hydrogen atom, the harmonic oscillator, etc.) have a zero-point energy as well.

The second important result that can be seen from the solution of the particle in a box problem is that *one quantum number arises from the solution of an equation for a one-dimensional system*. This quantum number arises as a mathematical restriction or condition rather than as an *assumption,* as it did in the case of the Bohr treatment of the hydrogen atom. It should not be surprising when it turns out that a two–dimensional system gives rise to two quantum numbers, three dimensions to three quantum numbers, and so on. Also, it is apparent that the particle in the one-dimensional box can serve as a useful first approximation for electrons moving along a conjugated hydrocarbon chain. An extension of this problem is the "particle on a ring" problem, in which the particle moves along a closed path.

3.2 SEPARATION OF VARIABLES

In this section. one of the techniques used to solve certain differential equations that arise in quantum mechanics will be illustrated. Suppose a differential equation can be written as

$$\frac{\partial^2 U}{\partial x^2} - \frac{\partial U}{\partial y} = 0 \tag{3.23}$$

The solution of the equation requires finding a solution that is a function of x and y, which can be represented as $U = U(x, y)$. Let us now *assume* that a solution exists such that

$$U(x, y) = X(x)Y(y)$$

in which X and Y are functions of x and y, respectively, so that $U=XY$. Substituting the product for U in Eq. (3.23) leads to

$$\frac{\partial XY}{\partial y} = \frac{\partial^2 XY}{\partial x^2} \tag{3.24}$$

Now X is *not* a function of y, and Y is *not* a function of x, so it is possible to treat X and Y as constants to give

$$\frac{X \partial Y}{\partial y} = \frac{Y \partial^2 X}{\partial x^2} \tag{3.25}$$

This equation can be rearranged to give

$$\frac{X''}{X} = \frac{Y'}{Y} \tag{3.26}$$

Each side of the equation is a constant with respect to the other, because one is a function of x and the other is a function of y. Therefore, representing the two sides as C gives two equations

$$\frac{X''}{X} = C \text{ and } \frac{Y'}{Y} = C \tag{3.27}$$

Each differential equation can now be solved independently of the other to obtain $X(x)$ and $Y(y)$. The desired solution is $U(x, y) = X(x) Y(y)$. The separation of the variables technique is commonly used in solving partial differential equations. In the next section the model known as a particle in a three-dimensional box involves this technique. Solution of the wave mechanical equation for the hydrogen atom also requires this technique, as will be shown in the next chapter.

3.3 THE PARTICLE IN A THREE-DIMENSIONAL BOX

The particle in a three-dimensional box model illustrates additional aspects of elementary quantum mechanical methods. In this problem, a particle is contained in a box of dimensions a, b, and c in the x, y, and z directions, respectively, as shown in Fig. 3.4.

As in the case of a particle in a one-dimensional box, the potential energy inside the box is assumed to be zero, but outside the box it will be assumed that $V=\infty$. Therefore, the total potential energy can be expressed in terms of that for the three dimensions,

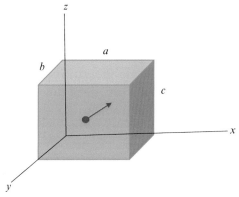

Fig. 3.4 A particle in a three-dimensional box.

$$V_{\text{total}} = V_x + V_y + V_z \tag{3.28}$$

The boundary conditions are expressed as follows:

$$V = 0 \quad \text{for} \quad 0 < x < a; \;\; 0 < y < b; \;\; 0 < z < c$$

and

$$V = \infty \quad \text{for} \quad 0 > x > a; \;\; 0 > y > b; \;\; 0 > z > c$$

For the particle inside the box, the Hamiltonian can now be written as

$$H = T + V = \frac{p^2}{2m} + 0 \tag{3.29}$$

The Hamiltonian operator can be written using the kinetic energy expressed in terms of the momentum, as shown in Chapter 2. The result is

$$-\frac{\hbar^2}{2m}\left(\frac{\partial^2}{\partial x^2} + \frac{\partial^2}{\partial y^2} + \frac{\partial^2}{\partial z^2}\right) = -\frac{\hbar^2}{2m}\nabla^2 \tag{3.30}$$

Therefore, the general equation

$$\hat{H}\psi = E\psi$$

can be written as

$$-\frac{\hbar^2}{2m}\nabla^2\psi = E\psi \tag{3.31}$$

Rearranging this equation gives

$$\nabla^2\psi + \frac{2m}{\hbar^2}E\psi = 0 \tag{3.32}$$

This is a partial differential equation in three variables (x, y, and z). The simplest method to solve such an equation and the one that should be tried first is the *separation of variables* method described in Section 3.2. In the next chapter, it will be shown that this is also the case in solving the wave equation for the hydrogen atom. To separate the variables, it is assumed that the desired solution ψ can be factored into three functions, each of which is a function of one variable only. In mathematical form, it is assumed that

$$\psi(x, y, z) = X(x)Y(y)Z(z) \tag{3.33}$$

This product of three functions is now written in place of ψ, but simplified notation will be used so that $X = X(x)$, etc. The resulting equation is

$$\frac{\partial^2 XYZ}{\partial x^2} + \frac{\partial^2 XYZ}{\partial y^2} + \frac{\partial^2 XYZ}{\partial z^2} + \frac{2m}{\hbar^2}E(XYZ) = 0 \tag{3.34}$$

Because YZ is not a function of x, XZ is not a function of y, and XY is not a function of z, they can be removed from the derivatives of the other variables to give

$$YZ\frac{\partial^2 X}{\partial x^2} + XZ\frac{\partial^2 Y}{\partial y^2} + XY\frac{\partial^2 Z}{\partial z^2} + \frac{2m}{\hbar^2}E(XYZ) = 0 \tag{3.35}$$

If Eq. (3.35) is divided by XYZ, the result is

$$\frac{1}{X}\frac{\partial^2 X}{\partial x^2} + \frac{1}{Y}\frac{\partial^2 Y}{\partial y^2} + \frac{1}{Z}\frac{\partial^2 Z}{\partial z^2} = -\frac{2m}{\hbar^2}E \tag{3.36}$$

Because each term on the left-hand side of Eq. (3.36) is a function of only one variable, each will be independent of any change in the other two variables. Therefore, each term can be considered to be equal to a constant, which will be represented as $-k^2$. Because there must be three such constants (i.e., one for each variable), they will be represented as $-k_x^{\,2}$, $-k_y^{\,2}$, and $-k_z^{\,2}$ for the x, y, and z directions, respectively. The results are three equations that can be written using ordinary derivatives as

$$\frac{1}{X}\frac{\partial^2 X}{\partial x^2} = -k_x^2 \tag{3.37}$$

$$\frac{1}{Y}\frac{\partial^2 Y}{\partial y^2} = -k_y^2 \tag{3.38}$$

$$\frac{1}{Z}\frac{\partial^2 Z}{\partial z^2} = -k_z^2 \tag{3.39}$$

The sum of the three constants must be equal to the right-hand side of Eq. (3.36), which is $-2mE/\hbar^2$. Therefore,

$$k_x^2 + k_y^2 + k_z^2 = \frac{2mE}{\hbar^2} \qquad (3.40)$$

The total energy E is the sum of the contributions from each degree of freedom in the x, y, and z coordinates, so it can be expressed as

$$E = E_x + E_y + E_z \qquad (3.41)$$

Therefore, the following equation is obtained by substitution:

$$k_x^2 + k_y^2 + k_z^2 = \frac{2m}{\hbar^2}\left(E_x + E_y + E_z\right) \qquad (3.42)$$

As a result of the energy associated with the degree of freedom in the x direction being independent of the y and z coordinates, it is possible to separate Eq. (3.42) to give

$$k_x^2 = \frac{2mE_x}{\hbar^2}; \quad k_y^2 = \frac{2mE_y}{\hbar^2}; \quad k_z^2 = \frac{2mE_z}{\hbar^2} \qquad (3.43)$$

The first of the equations, Eq. (3.37), can be written as

$$\frac{\partial^2 X(x)}{\partial x^2} + k_x^2 X(x) = 0 \qquad (3.44)$$

This equation is of the same form as Eq. (3.7), so the solution can be written directly as

$$X(x) = \sqrt{\frac{2}{a}}\sin\frac{n_x\pi}{a}x \qquad (3.45)$$

Similarly, the other two equations yield solutions that can be written as

$$Y(y) = \sqrt{\frac{2}{b}}\sin\frac{n_y\pi}{b}y \qquad (3.46)$$

and

$$Z(z) = \sqrt{\frac{2}{c}}\sin\frac{n_z\pi}{c}z \qquad (3.47)$$

The general solution can be written as the product of the three partial solutions (the assumption made earlier that permitted the separation of variables)

$$\psi(x, y, z) = X(x)Y(y)Z(z) \tag{3.48}$$

and

$$\psi(x, y, z) = \sqrt{\frac{2}{a}}\sin\frac{n_x\pi}{a}x \cdot \sqrt{\frac{2}{b}}\sin\frac{n_y\pi}{b}y \cdot \sqrt{\frac{2}{c}}\sin\frac{n_z\pi}{c}z \tag{3.49}$$

in which n_x, n_y, and n_z are the quantum numbers for the x, y, and z components of energy, respectively. The general solution can be simplified somewhat by combining constants to give

$$\psi(x, y, z) = \sqrt{\frac{8}{abc}}\sin\frac{n_x\pi}{a}x\sin\frac{n_y\pi}{b}y\sin\frac{n_z\pi}{c}z \tag{3.50}$$

It is now possible to draw some analogies to the particle in the one-dimensional box. First, the results will be used to find expressions for the energies using Eq. (3.43):

$$\frac{n_x\pi}{a} = k_x = \sqrt{\frac{2mE_x}{\hbar^2}} \tag{3.51}$$

Solving for E_x gives an expression for the energy levels based on the x component that can be written as

$$E_x = \frac{\hbar^2 n_x^2 \pi^2}{2ma^2} = \frac{h^2 n_x^2}{8ma^2} \tag{3.52}$$

The equivalent expressions for the energy in terms of the y and z directions are

$$E_y = \frac{\hbar^2 n_y^2 \pi^2}{2mb^2} \quad \text{and} \quad E_z = \frac{\hbar^2 n_z^2 \pi^2}{2mc^2} \tag{3.53}$$

The total energy E is the sum of the three components and is represented by the equation

$$E = E_x + E_y + E_z = \frac{\hbar^2 \pi^2}{2m}\left(\frac{n_x^2}{a^2} + \frac{n_y^2}{b^2} + \frac{n_z^2}{c^2}\right) \tag{3.54}$$

where $n_x = 1, 2, 3, \ldots$; $n_y = 1, 2, 3, \ldots$; and $n_z = 1, 2, 3 \ldots$.

It should be noted that one quantum number has been introduced for each degree of freedom of the system, corresponding to the three coordinates of the particle. Therefore, the energy is dependent on the quantum numbers n_x, n_y, and n_z. If it is assumed that the box is cubic, then $a = b = c$. Therefore, the denominators of the fractions in Eq. (3.54) would be

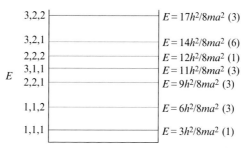

Fig. 3.5 Energy levels for a particle in a three-dimensional cubic box.

identical, and the lowest energy will occur when the numerators have the smallest values, which occurs when all of the quantum numbers are 1. This state can be designated as the 111 state, where the digits indicate the values of the quantum numbers n_x, n_y, and n_z, respectively. The state of next lowest energy would be with two of the quantum numbers being 1 and the other being 2. One way that this could occur would be with $n_x = 2$ and $n_y = n_z = 1$. In this case, the energy would be equal to 6 $\hbar^2\pi^2/2ma^2$ or 6 $h^2/8ma^2$ when $\hbar = h/2\pi$ is substituted. It should be apparent that this state, designated as the 211 state, has the same energy as the 121 and 112 states. Therefore, these states are *degenerate* in the case where $a = b = c$. However, if the dimensions of the box are not equal, then the 211, 121, and 112 states are not degenerate. It is easy to see that if one of the dimensions is twice another (or in some other appropriate relationship), the energies might still happen to be degenerate simply because of the relationship of a, b, and c, as the quantum numbers are assigned different values. Such a situation is known as *accidental degeneracy*.

The energy level diagram that results when $a = b = c$ is shown in Fig. 3.5. The states are indicated in terms of the quantum numbers n_x, n_y, and n_z (e.g., 112, 123, and 322), and the degeneracy is given after the combination of quantum numbers.

When $a \neq b \neq c$, the energies must be calculated using the actual dimensions of the box. Choosing unequal values for a, b, and c and then using several integers for n_x, n_y, and n_z will quickly enable one to see the nondegeneracy of the energy states.

3.4 F-CENTERS IN CRYSTALS

When potassium vapor is passed over a crystal of KCl, the crystal takes on a color. It can be shown that as a result of the reaction

$$K(vap) \rightarrow K^+(crystal) + e^-(anion\ site) \tag{3.55}$$

the electrons occupy anion sites in the KCl lattice. In reality, the electrons are distributed over the cations that surround the lattice site. The centers where the electrons reside are responsible for the absorption of light, which results in the crystal being colored. Such centers are called *f-centers* because of the German word *farbe*, which means "color." When other alkali metals are added in the same way to the corresponding alkali halides, color centers are also produced. As a very crude approximation, the electrons in anion sites can be treated as particles in three-dimensional boxes. It is interesting to note that the wavelength of the light absorbed depends on the nature of the crystal lattice, and the maxima in the absorptions for several crystals are as follows:

Crystal	Absorption maximum (erg)
LiCl	4.96×10^{-12}
NaCl	4.32×10^{-12}
KCl	3.52×10^{-12}
RbCl	3.20×10^{-12}
LiBr	4.32×10^{-12}
NaBr	3.68×10^{-12}

For a given chloride compound, the size of the anion site (where the electron resides) is dependent on the size of the cation. Because the energies for a particle in a three-dimensional box are inversely related to the size of the box, it is to be expected that the greatest difference between energy levels would be for LiCl. Accordingly, the absorption energy is highest for LiCl, for which the anion site is smallest. In fact, the series from LiCl to RbCl shows this trend clearly based on the size of the anion site. Therefore, this phenomenon shows a correlation that would be predicted when the particle in a three-dimensional box model is employed.

3.5 SOLVATED ELECTRONS

When alkali metals are dissolved in liquid ammonia, the dilute solutions have a blue color. There have been many studies conducted on these solutions and how their properties vary with concentration. Conductivity studies have shown that the dilute solutions behave as 1:1 electrolytes with the charged species consisting of Na^+ and electrons. As a result of ammonia molecules being polar, it is believed that the electrons are solvated, probably residing in some sort of solvent cage. It is of interest to note the same blue

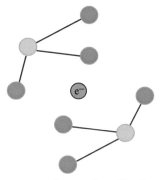

Fig. 3.6 A model to represent an electron in a "box" consisting of two ammonia molecules.

color results when any of the alkali metals are dissolved in liquid ammonia, with the absorption reaching its maximum at ~400 nm.

If the electron is presumed to reside in a cage formed by the bases of two pyramidal ammonia molecules, the result might be represented as shown in Fig. 3.6. For purposes of illustration, it will be assumed that the "box" in which the electron resides is cubic with an edge length of 6 Å ($a=b=c=6 \times 10^{-8}$ cm). Given the size of ammonia molecules and the fact that the hydrogen atoms have partial positive charges, this distance of approach may not be exactly correct, but it gives a starting point for the calculation. The difference between the first two energy levels are given by

$$\Delta E = \frac{h^2}{8m} \left(\frac{2^2}{a^2} + \frac{1^2}{a^2} + \frac{1^2}{a^2} \right) - \frac{h^2}{8m} \left(\frac{1^2}{a^2} + \frac{1^2}{a^2} + \frac{1^2}{a^2} \right) = \frac{3h^2}{8ma^2} \qquad (3.56)$$

When values are substituted in this equation for the constants, ΔE is 5.04×10^{-12} erg. This is equivalent to a photon with a wavelength of 395 nm or 3950 Å. Of course, this means that the dimension chosen for the box was a very lucky guess, but the point is that even if the solvent cage has somewhat different dimensions, treating the solvated electron as a particle in a three-dimensional box is not totally unrealistic. Even very simple quantum mechanical models are sometimes useful for approximating real systems.

In this chapter, two simple quantum mechanical models have been considered. In the process of solving the wave equations for these models, several of the important aspects of quantum mechanics have been illustrated. Moreover, in each case, an application of the model to a physical system has been made. Therefore, the particle in one- and three-dimensional box models serve as useful ways to illustrate quantum mechanical principles and practices.

PROBLEMS

1. Solve the equation $y + ay = 0$ subject to the boundary conditions $y(0) = y(\pi) = 0$.

2. If a hexatriene molecule absorbs light of 2500 Å (250 nm) to move a π electron from $n = 1$ to $n = 2$, what is the length of the molecule?

3. What would be the translational energies of the first two levels for a hydrogen molecule confined to a length of 10 cm?

4. What would be the length of a one-dimensional box necessary for the separation between the first two energy levels for a proton to be 2.00 eV?

5. Calculate the probability of finding the particle in a one-dimensional box of length a in the interval $0.100a$ to $0.250a$.

6. Planck's constant is the fundamental quantum of action (energy × time). Explain how the behavior of a particle in a box becomes classical as the fundamental quantum of action approaches zero.

7. Calculate the average value of the x coordinate of a particle in a one-dimensional box.

8. Consider an atomic nucleus to be a potential box 10^{-13} cm in diameter. If a neutron falls from $n = 2$ to $n = 1$, what energy is released? If this energy is emitted as a photon, what will be its wavelength? In what region of the electromagnetic spectrum will it be observed?

9. Consider a particle of mass m moving in a planar circular path of length l. Assume that the potential for the particle along the path is zero, whereas the potential for the particle to be out of the path is infinite to confine the particle to the path
   (a) Set up the wave equation for this model.
   (b) Solve the wave equation to get a general form of the solution.
   (c) Use the fact that the solution for any points x and $(l + x)$ must be equal to simplify the solution.

10. When sodium dissolves in liquid ammonia, some dissociation occurs:

$$Na \rightarrow Na^+(\text{solvated}) + e^-(\text{solvated})$$

The solvated electron can be treated as a particle in a three dimensional box. Assume that the box is cubic with an edge length of

1.55×10^{-7} cm and suppose that excitation occurs in all directions simultaneously from the lowest state to the first excited state. What wavelength of radiation would the electron absorb? Would the solution be colored?

11. What size would a one-dimensional box holding an electron have to be in order for it to have the same energy as a hydrogen molecule would have in a box of length 10 Å?

12. Suppose a helium atom is in a box of length 50 Å. Calculate the energy at a few distances and sketch the energy as a function of box length.

13. Suppose a particle in a three-dimensional box has an energy of $14h^2/8ma^2$. If $a=b=c$ for the box, what is the degeneracy of this state?

14. Show that the wave functions for the first two energy levels of a particle in a one-dimensional box are orthogonal.

15. Consider an electron in the π bond in ethylene as a particle in a one-dimensional box of length 133 pm. What is the energy difference between the first two energy levels? In what region of the electromagnetic spectrum would a photon emitted as the electron falls from the first excited state to the ground state be observed?

16. An electron trapped in a three-dimensional lattice defect (vacant anion site) of a crystal can be considered as a particle in a three-dimensional box. If the length of the box in each dimension is 200 nm, what would be the difference between the first two allowed states? What would be the effect on the respective energies of the first two allowed states if the defect site were 200 nm in length in the x and y directions but 250 nm in the z direction?

CHAPTER 4

The Hydrogen Atom

The work of de Broglie showed that a moving particle has a wave character. In Chapter 2, it was shown that this could lead to an adaptation of an equation that is known to apply to vibrations in three dimensions, resulting in an equation that describes the electron in a hydrogen atom as a three-dimensional wave. Although this was illustrated in the last chapter, we did not solve the resulting equation. We now address that problem for which Erwin Schrödinger received the Nobel Prize: the solution of the wave equation for the hydrogen atom. Therefore, it should be anticipated that the solution is not a trivial problem!

4.1 SCHRÖDINGER'S SOLUTION TO THE HYDROGEN ATOM PROBLEM

A hydrogen atom can be described in terms of polar coordinates, as shown in Fig. 4.1. As the electron circles around the nucleus, the system (i.e., the proton and the electron) rotates around the center of gravity. For the rotating system, we can write the reduced mass μ as

$$\frac{1}{\mu} = \frac{1}{m_e} + \frac{1}{m_p} \tag{4.1}$$

or, in solving for μ,

$$\mu = \frac{m_e m_p}{m_e + m_p} \tag{4.2}$$

Because the mass of the electron is so much less than that of the proton, $m_e + m_p \approx m_p$ and $\mu = m_e$. When we assume that the nucleus is stationary and that the electron does all of the moving (known as the Born-Oppenheimer approximation), it leads to the same result. Therefore, we will assume that this approximation can be used, although μ is indicated. As we saw in the last chapter, the Hamiltonian can be written as the sum of the potential and kinetic energies

$$H = T + V \tag{4.3}$$

Fundamentals of Quantum Mechanics
http://dx.doi.org/10.1016/B978-0-12-809242-2.00004-8

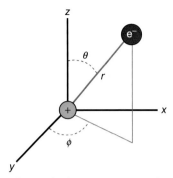

Fig. 4.1 Coordinate system for the hydrogen atom in polar coordinates.

and $T = p^2/2\mu$. The potential energy V for the interaction of the electron with the proton is $-e^2/r$. Therefore, the Hamiltonian, H, for the hydrogen atom is

$$H = \frac{p^2}{2\mu} - \frac{e^2}{r} \qquad (4.4)$$

In operator form, $-e^2/r$ is unchanged because e is a *constant* and r is a *coordinate*. It has already been shown that the kinetic energy in operator form (where $\hbar = h/2\pi$, usually called h-bar) can be written as

$$\hat{T} = -\frac{\hbar^2}{2\mu} \nabla^2 \qquad (4.5)$$

Therefore, the Hamiltonian operator $\hat{H}$ is

$$\hat{H} = -\frac{\hbar^2}{2\mu} \nabla^2 - \frac{e^2}{r}$$

The general form of the wave equation is

$$\hat{H}\psi = E\psi \qquad (4.6)$$

After substituting in this equation for the Hamiltonian operator $\hat{H}$, the wave equation is

$$-\frac{\hbar^2}{2\mu} \nabla^2 \psi - \frac{e^2}{r} \psi = E\psi \qquad (4.8)$$

When the Laplacian operator ∇^2 is written out in explicit form and the equation is rearranged, the result is

$$\left(\frac{\partial^2}{\partial x^2} + \frac{\partial^2}{\partial y^2} + \frac{\partial^2}{\partial z^2}\right)\psi + \frac{2\mu}{\hbar^2}(E - V)\psi = 0 \qquad (4.9)$$

This is a partial differential equation in three variables, and one should expect that the separation of variables technique must be applied to solve it. However, the distance of the electron from the nucleus when expressed in Cartesian coordinates is

$$r = \left(x^2 + y^2 + z^2\right)^{1/2}$$

Therefore, the variables cannot be separated. The difficult part now is to change coordinate systems to get variables (coordinates) that can be separated. The transformation made is to describe the hydrogen atom in terms of *polar* coordinates. Figure 4.1 shows the coordinate system and the relationship between the Cartesian and the polar coordinates.

A tedious aspect of this problem now consists of transforming

$$\frac{\partial^2}{\partial x^2} + \frac{\partial^2}{\partial y^2} + \frac{\partial^2}{\partial z^2} \qquad (4.10)$$

into a function of r, θ, and φ. That transformation is laborious, but the result is

$$\nabla^2 = \frac{1}{r^2}\frac{\partial}{\partial r}\left(r^2\frac{\partial}{\partial r}\right) + \frac{1}{r^2\sin\theta}\frac{\partial}{\partial\theta}\left(\sin\theta\frac{\partial}{\partial\theta}\right) + \frac{1}{r^2\sin^2\theta}\frac{\partial^2}{\partial\varphi^2} \qquad (4.11)$$

Now, the Schrödinger equation becomes

$$\frac{1}{r^2}\frac{\partial}{\partial r}\left(r^2\frac{\partial}{\partial r}\psi\right) + \frac{1}{r^2\sin\theta}\frac{\partial}{\partial\theta}\left(\sin\theta\frac{\partial}{\partial\theta}\psi\right) + \frac{1}{r^2\sin^2\theta}\frac{\partial^2}{\partial\varphi^2}\psi + \frac{2\mu}{\hbar^2}\left(\frac{e^2}{r} + E\right)\psi = 0 \qquad (4.12)$$

This equation contains only the variables r, θ, and φ.

In this case, it is assumed there is a solution of the form $\psi(r, \theta, \varphi) = R(r)\,\Theta(\theta)\,\Phi(\varphi)$, which is substituted in Eq. (4.12). For simplicity, the partial solutions will be written as R, Θ, and Φ without showing the functionality, $R(r)$, etc. The Schrödinger equation can then be written as

$$\frac{1}{r^2}\frac{\partial}{\partial r}\left(r^2\frac{\partial R\Theta\Phi}{\partial r}\right) + \frac{1}{r^2\sin\theta}\frac{\partial}{\partial\theta}\left(\sin\theta\frac{\partial R\Theta\Phi}{\partial\theta}\right)$$
$$+ \frac{1}{r^2\sin^2\theta}\frac{\partial^2 R\Theta\Phi}{\partial\varphi^2} + \frac{2\mu}{\hbar^2}\left(\frac{e^2}{r} + E\right)R\Theta\Phi = 0 \qquad (4.13)$$

However, because Θ and Φ are not functions of r, they can be removed (as constants) from the differentiation with respect to r. The same action is possible in other terms of the equation involving the other variables. This allows the wave equation to be written as

$$\frac{\Theta\Phi}{r^2}\frac{\partial}{\partial r}\left(r^2\frac{\partial R}{\partial r}\right) + \frac{R\Phi}{r^2\sin\theta\,\partial\theta}\frac{\partial}{\partial\theta}\left(\sin\theta\frac{\partial\Theta}{\partial\theta}\right)$$

$$+\frac{R\Theta}{r^2\sin^2\theta}\frac{\partial^2\Phi}{\partial\varphi^2}+\frac{2\mu}{\hbar^2}\left(\frac{e^2}{r}+E\right)R\Theta\Phi=0$$

(4.14)

When both sides of Eq. (4.14) are divided by $R\Theta\Phi$ and multiplied by $r^2\sin^2\theta$, the result is

$$\frac{\sin^2\theta}{R}\frac{\partial}{\partial r}\left(r^2\frac{\partial R}{\partial r}\right) + \frac{\sin\theta}{\Theta}\frac{\partial}{\partial\theta}\left(\sin\theta\frac{\partial\Theta}{\partial\theta}\right)+\frac{1}{\Phi}\frac{\partial^2\Phi}{\partial\varphi^2}+\frac{2\mu r^2\sin^2\theta}{\hbar^2}\left(\frac{e^2}{r}+E\right)=0$$

(4.15)

Inspection shows that for the four terms on the left-hand side of the equation, there is no functional dependence on φ except in the third term. Therefore, with respect to the other variables, the third term can be treated as a constant. For convenience, that constant will be represented as equal to $-m^2$. This m is *not* the same as the electron mass used in the expression $2m/\hbar^2$. Therefore, the term involving Φ leads to the equation

$$\frac{1}{\Phi}\frac{\partial^2\Phi}{\partial\varphi^2}=-m^2$$

(4.16)

which can be rearranged to give

$$\frac{\partial^2\Phi}{\partial\varphi^2}+m^2\Phi=0$$

(4.17)

This equation is sometimes referred to as the "φ equation," and this represents the first equation obtained from the separation of variables.

By making use of $-m^2$ as a constant that represents the third term of Eq. (4.15), the wave equation can now be written as

$$\frac{\sin^2\theta}{R}\frac{\partial}{\partial r}\left(r^2\frac{\partial R}{\partial r}\right) + \frac{\sin\theta}{\Theta}\frac{\partial}{\partial\theta}\left(\sin\theta\frac{\partial\Theta}{\partial\theta}\right)-m^2+\frac{2\mu r^2\sin^2\theta}{\hbar^2}\left(\frac{e^2}{r}+E\right)=0$$

(4.18)

If this equation is divided by $\sin^2\theta$ and rearranged, the result can be written as

$$\frac{1}{R}\frac{\partial}{\partial r}\left(r^2\frac{\partial R}{\partial r}\right) + \frac{2\mu r^2}{\hbar^2}\left(\frac{e^2}{r} + E\right) + \frac{1}{\Theta\sin\theta}\frac{\partial}{\partial\theta}\left(\sin\theta\frac{\partial\Theta}{\partial\theta}\right) - \frac{m^2}{\sin^2\theta} = 0$$

$$(4.19)$$

Inspection of this equation shows that the first two terms contain the functional dependence on r and the last two terms reflect the dependence on θ. As was done earlier, the sum of two terms is set equal to a constant β, such that

$$\frac{1}{R}\frac{\partial}{\partial r}\left(r^2\frac{\partial R}{\partial r}\right) + \frac{2\mu r^2}{\hbar^2}\left(\frac{e^2}{r} + E\right) = \beta \tag{4.20}$$

and

$$\frac{1}{\Theta\sin\theta}\frac{\partial}{\partial\theta}\left(\sin\theta\frac{\partial\Theta}{\partial\theta}\right) - \frac{m^2}{\sin^2\theta} = -\beta \tag{4.21}$$

If Eq. (4.20) is multiplied by R and Eq. (4.21) is multiplied by Θ, the results are

$$\frac{\partial}{\partial r}\left(r^2\frac{\partial R}{\partial r}\right) + \frac{2\mu r^2}{\hbar^2}\left(\frac{e^2}{r} + E\right)R - R\beta = 0 \tag{4.22}$$

and

$$\frac{1}{\sin\theta}\frac{\partial}{\partial\theta}\left(\sin\theta\frac{\partial\Theta}{\partial\theta}\right) - \frac{m^2}{\sin^2\theta}\Theta + \beta\Theta = 0 \tag{4.23}$$

The equation containing the variables r, θ, and φ has now been separated. The result is that the second-order partial differential equation in three variables has been transformed into three second-order differential equations, each containing only one variable. Solving the three equations is now the task. Only the equation involving φ is simple in its solution because it is of the same form as the equation for the particle in a box problem. The solution of the "φ equation" can be written as

$$\Phi(\varphi) = \frac{1}{\sqrt{2}}e^{im\varphi} \tag{4.24}$$

Therefore, the solution of the overall equation can be written as

$$\psi(r, \theta, \varphi) = \frac{1}{\sqrt{2\pi}}e^{im\varphi}R(r)\Theta(\theta) \tag{4.25}$$

Having solved the first of the three separated equations, attention will now be turned to the equation involving θ:

$$\frac{1}{\sin\theta}\frac{\partial}{\partial\theta}\left(\sin\theta\frac{\partial\Theta}{\partial\theta}\right) - \frac{m^2}{\sin^2\theta}\Theta + \beta\Theta = 0 \qquad (4.26)$$

This equation can be put in the form

$$\frac{d}{\sin\theta\,d\theta}\left(\frac{\sin^2\theta\,d\Theta}{\sin\theta\,d\theta}\right) - \frac{m^2\Theta}{\sin^2\theta} + \beta\Theta = 0 \qquad (4.27)$$

The standard method for solving this equation is to make the transformations

$$u = \cos\theta \text{ so that } du = -\sin\theta\,d\theta$$
$$\cos^2\theta = u^2 = 1 - \sin^2\theta$$
$$\sin^2\theta = 1 - u^2$$

Substituting for $\sin^2\theta$ and $\sin\theta\,d\theta$ in Eq. (4.27) leads to the equation

$$\frac{d}{du}\left(\frac{(1-u^2)}{du}d\Theta\right) - \frac{m^2\Theta}{1-u^2} + \beta\Theta = 0 \qquad (4.28)$$

By rearrangement, this equation can be transformed into the form

$$(1-u^2)\frac{d^2\Theta}{du^2} - 2u\frac{d\Theta}{du} + \left(\beta - \frac{m^2}{(1-u^2)}\right)\Theta = 0 \qquad (4.29)$$

This equation is similar in form to a well-known differential equation encountered in advanced mathematics. That equation,

$$(1-z^2)\frac{d^2 P_l^{|m|}(z)}{dz^2} - 2z\frac{dP_l^{|m|}(z)}{dz} + l(l+1) - \frac{m^2}{1-z^2}P_l^{|m|}(z) = 0 \qquad (4.30)$$

is known as Legendre's equation, in which β in Eq. (4.29) is equivalent to $l(l+1)$ in Eq. (4.30). Solving equations of this type requires the use of series, but rather than getting too involved with mathematics at this point, the discussion of series solutions of differential equations will be delayed until Chapter 6. The series solutions of Legendre's equation are known as the *associated Legendre polynomials,* and they can be written as

$$P_l^{|m|}(\cos\theta), \text{ where } l = 0,\ 1,\ 2, ..., \text{ and } m = 0,\ \pm 1,\ \pm 2, ..., \pm l$$

The first few associated Legendre polynomials are as follows:

$$l = 0,\ m = 0: \quad \Theta(\theta) = 1/\sqrt{2} = \Theta_{0,0}$$
$$l = 1,\ m = 0: \quad \Theta(\theta) = \sqrt{3/2}\,\cos\theta = \Theta_{1,0}$$
$$l = 1,\ m = \pm 1: \quad \Theta(\theta) = \sqrt{3/4}\,\sin\theta = \Theta_{1,\pm 1}$$
$$l = 2,\ m = 0: \quad \Theta(\theta) = \sqrt{5/8}\left(3\cos^2\theta - 1\right) = \Theta_{2,0}$$

The equation that is a function of r is known as the *radial* equation, and it can be put in the form

$$\frac{1}{r^2}\frac{d}{dr}r^2\frac{dR}{dr} + \frac{2\mu}{\hbar^2}\left(\frac{e^2}{r} + E\right)R - \frac{l(l+1)}{r^2}R = 0 \qquad (4.31)$$

This equation can be written in a general form that is shown as

$$xu'' + u'(2l+2) + (-l-1+n)u = 0 \qquad (4.32)$$

This equation can be solved only when $n \geq l+1$. This equation is known as *Laguerre's equation,* and the solutions are the *Laguerre polynomials.* It was shown earlier that $l = 0$, 1, 2, ..., so it is apparent that $n = 1, 2, 3, ...$. For example, if $n = 3$, l can take on the values 0, 1, and 2. This gives rise to the familiar restrictions on the quantum numbers that students learn in general chemistry (see also Chapter 5):

$n =$ principal quantum number $= 1, 2, 3,...$
$l =$ orbital angular momentum quantum number $= 0, 1, 2, ..., (n-1)$
$m =$ magnetic quantum number $= 0, \pm 1, \pm 2, ..., \pm l$.

The spin quantum number, s, (equal to $\pm(1/2)\hbar$) is a property of the electron because it is a particle that has an intrinsic spin. It should be recalled from previous chemistry courses that each value of m defines an orbital, so there are $2l+1$ orbitals for each value of l. It should also be recalled that the designation of the types of orbitals is related to the l value by

$$l = 0 \quad 1 \quad 2 \quad 3$$
$$s \quad p \quad d \quad f$$

The solutions of the equations involving φ and θ are combined by multiplication to give the complete angular dependence of the wave functions. These angular functions are known as the *spherical harmonics*, which are written as $Y_{l,m}(\theta, \varphi)$. Solutions of the equation involving r are called the *radial wave functions*, $R_{n,l}(r)$, and the overall solutions are $R_{n,l}(r) Y_{l,m}(\theta, \varphi)$. Table 4.1 gives the wave functions for the hydrogen-like species. The hydrogen wave functions are indicated when $Z = 1$.

4.2 INTERPRETING THE SOLUTIONS

In Chapter 2, the idea that ψ^2 is related to the probability of finding a particle described by the wave function ψ was discussed briefly. In classical physics, the square of the amplitude gives the total energy of a vibrating system (e.g., a vibrating object on a spring or a vibrating string). Similarly, the square of

Table 4.1 Complete wave functions for hydrogen-like species[a]

$$\psi_{1s} = \frac{1}{\pi^{1/2}}\left(\frac{Z}{a}\right)^{3/2} e^{-Zr/a}$$

$$\psi_{2s} = \frac{1}{4(2\pi)^{1/2}}\left(\frac{Z}{a}\right)^{3/2}\left(2 - \frac{Zr}{a}\right)e^{-Zr/2a}$$

$$\psi_{2p_z} = \frac{1}{4(2\pi)^{1/2}}\left(\frac{Z}{a}\right)^{5/2} r\,e^{-Zr/2a}\cos\theta$$

$$\psi_{2p_x} = \frac{1}{4(2\pi)^{1/2}}\left(\frac{Z}{a}\right)^{5/2} r\,e^{-Zr/2a}\sin\theta\cos\varphi$$

$$\psi_{2p_y} = \frac{1}{4(2\pi)^{1/2}}\left(\frac{Z}{a}\right)^{5/2} r\,e^{-Zr/2a}\sin\theta\sin\varphi$$

$$\psi_{3s} = \frac{1}{81(3\pi)^{1/2}}\left(\frac{Z}{a}\right)^{3/2}\left(27 - 18\frac{Zr}{a} + 2\frac{Z^2r^2}{a^2}\right)e^{-Zr/3a}$$

$$\psi_{3p_z} = \frac{2^{1/2}}{81\pi^{1/2}}\left(\frac{Z}{a}\right)^{5/2}\left(6 - \frac{Zr}{a}\right)r\,e^{-Zr/3a}\cos\theta$$

$$\psi_{3p_x} = \frac{2^{1/2}}{81\pi^{1/2}}\left(\frac{Z}{a}\right)^{5/2}\left(6 - \frac{Zr}{a}\right)r\,e^{-Zr/3a}\sin\theta\cos\varphi$$

$$\psi_{3p_y} = \frac{2^{1/2}}{81\pi^{1/2}}\left(\frac{Z}{a}\right)^{5/2}\left(6 - \frac{Zr}{a}\right)r\,e^{-Zr/3a}\sin\theta\sin\varphi$$

$$\psi_{3d_{xy}} = \frac{1}{81(2\pi)^{1/2}}\left(\frac{Z}{a}\right)^{7/2} r^2\,e^{-Zr/3a}\sin^2\theta\sin 2\varphi$$

$$\psi_{3d_{xz}} = \frac{2^{1/2}}{81\pi^{1/2}}\left(\frac{Z}{a}\right)^{7/2} r^2\,e^{-Zr/3a}\sin\theta\cos\theta\cos\varphi$$

$$\psi_{3d_{yz}} = \frac{2^{1/2}}{81\pi^{1/2}}\left(\frac{Z}{a}\right)^{7/2} r^2\,e^{-Zr/3a}\sin\theta\cos\theta\sin\varphi$$

$$\psi_{3d_{x^2-y^2}} = \frac{1}{81(2\pi)^{1/2}}\left(\frac{Z}{a}\right)^{7/2} r^2\,e^{-Zr/3a}\sin\theta\cos 2\theta$$

$$\psi_{3d_{z^2}} = \frac{1}{81(6\pi)^{1/2}}\left(\frac{Z}{a}\right)^{7/2} r^2\,e^{-Zr/3a}\left(3\cos^2\theta - 1\right)$$

[a]The nuclear charge is given by Z, and a is the first Bohr radius 0.529 Å.

the wave function for an electron is proportional to the amplitude function squared, because in reality the electron has been represented as a de Broglie wave by means of the Schrödinger equation.

Solving a differential equation to obtain ψ does not uniquely determine a probability, because solving such an equation leads to arbitrary constants.

In the case of the electron, the particle must be *somewhere*, so the probability integral is written as

$$\int \psi^2 \, d\tau = 1 \tag{4.33}$$

If this relationship does not hold, then the constants included in the wave function are "adjusted" by introducing a constant N to make the integral evaluate to 1:

$$\int N\psi^* \cdot N\psi \, d\tau = 1 \tag{4.34}$$

In this case, N is called a *normalization constant*. The value of ψ^2 is positive regardless of whether ψ is positive or negative, so the probability ranges from 0 to 1.

Another interpretation of the square of the wave function is not only possible, but it also provides a very useful concept for describing certain properties of electrons in atoms. The concept being described is that of the *density* of the electron "cloud." If flash photography could show the location of an electron, and if the process could be repeated an enormous number of times, then a plot could be made to show the position of the electron at the instant it appeared in a photograph. The result would appear as shown in Fig. 4.2A.

The area in which the dots have the highest density represents the regions in which the electron is found most of the time. If the dots represent particles of a cloud, then the cloud has its highest density where the dots are closest together. Obviously, a particle cannot be "smeared out" over space, but it is a useful concept nonetheless. In fact, one qualitative definition of a covalent bond is the increased probability for finding electrons between two

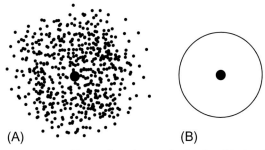

Fig. 4.2 (A) Instantaneous positions of an electron in the 1s orbital and (B) the contour surrounding the electron 95% of the time.

atoms or the increased density of the electron cloud between two atoms. Although the charge cloud does not represent the nature of *particles*, it provides a pictorial way to describe a *probability*.

Having plotted the position of the electron in the ground state of a hydrogen atom, it is now possible to draw a surface to encompass the positions where the electron was found a specified percentage (perhaps 95 percent) of the time. The surface is a sphere in the case of the electron in a 1s orbital. Therefore, the 1s orbital is described as *spherical* or with *spherical symmetry* (as shown in Fig. 4.2B), and the probability of finding the electron depends on r. The quantum state in which the electron resides is referred to as an *orbital*, but this word in no way indicates a *path* of the electron.

Finally, as it has already been discussed, the probability of finding the electron in terms of three dimensions (its radial density) is given by

$$P(r)\, dr = [R(r)]^2\, 4\pi r^2\, dr \tag{4.35}$$

Figure 2.1 shows a plot of the radial density that indicates the distance of highest probability (most probable distance) is a_0. This can be shown mathematically as follows: After squaring the radial wave function, the result is

$$P(r)\, dr = 4\pi \left(\frac{1}{a_0}\right)^3 r^2\, e^{-2r/a_0}\, dr \tag{4.36}$$

Differentiating this equation with respect to r and setting the derivative equal to 0 gives

$$\frac{d\,P(r)}{dr} = 0 = (4\pi)2r \left(\frac{1}{a_0}\right)^3 e^{-2r/a_0} - (4\pi)\frac{2}{a_0}r^2 \left(\frac{1}{a_0}\right)^3 e^{-2r/a_0} \tag{4.37}$$

Therefore,

$$(4\pi)2r \left(\frac{1}{a_0}\right)^3 e^{-2r/a_0} = (4\pi)\frac{2}{a_0}r^2 \left(\frac{1}{a_0}\right)^3 e^{-2r/a_0} \tag{4.38}$$

After canceling like terms from both sides and simplifying, it leads to

$$1 = \frac{r}{a_0} \tag{4.39}$$

Therefore, the *most probable radius* is a_0.

To this point, the discussion has dealt with properties of the 1s wave function. If the 2s wave function is considered, it is found that the radial density plot has a very different appearance. The 2s wave function (where $Z=1$) for the hydrogen atom is

$$\psi_{2s} = \frac{1}{4(2\pi)^{1/2}} \left(\frac{Z}{a}\right)^{3/2} \left(2 - \frac{Zr}{a}\right) e^{-Zr/2a} \tag{4.40}$$

From this equation, it can be seen that the probability of finding the electron has a node where the probability goes to 0 at $r=2a_0$, which can be seen from the $2-(r/a_0)$ part of the wave function when $Z=1$. The 3s wave function is

$$\psi_{3s} = \frac{1}{81(3\pi)^{1/2}} \left(\frac{Z}{a}\right)^{3/2} \left(27 - 18\frac{Zr}{a} + 2\frac{Z^2 r^2}{a^2}\right) e^{-Zr/3a} \tag{4.41}$$

Because the only part of this wave function (and hence the square of ψ) that can go to 0 is the polynomial, it should be clear that at *some* value of r (in units of a_0),

$$27 - 18x + 2x^2 = 0 \tag{4.42}$$

where $x=r/a_0$. When this quadratic equation is solved to determine x, it is found that

$$x = \frac{18 \pm \sqrt{18^2 - 4(2)(27)}}{4} = 7.10 \text{ and } 1.90 \tag{4.43}$$

Therefore, the probability of finding the electron as a function of distance in the 3s state goes to 0 at $r=1.90\, a_0$ and $r=7.10\, a_0$, so the 3s wave function has two nodes. In fact, it is easy to see that the probability has $n-1$ nodes, where n is the principal quantum number. Figure 4.3 shows the probability density as a function of r for the 2s and 3s states.

4.3 *p* AND *d* WAVE FUNCTIONS AND ORBITALS

It has already been explained that the *s* wave functions give rise to a spherical surface that encompasses the electron an arbitrary percentage of the time (perhaps 90% or 95%). The surfaces that depict the probabilities of finding the electron in the 2s and 3s states are also spherical, although they are larger than the 1s surface. Consequently, the electron density is more diffuse when the electron is in one of these states. This has implications for the strengths of bonds formed using such orbitals. This point will be revisited in Chapter 8 in a discussion of bonds formed by the overlap of *s* orbitals. Surfaces that correspond to the electron density in other states must also be shown.

The surfaces for the set of three *p* orbitals, for which $l=1$, are shown in Fig. 4.4. The appropriate mathematical signs of the wave functions are

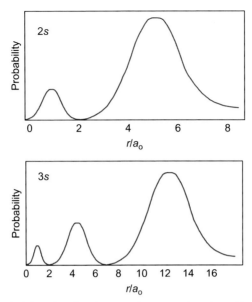

Fig. 4.3 Radial probability plots for the 2s and 3s wave functions. Note the positions of the minima, which indicate nodes.

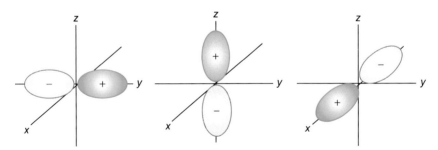

Fig. 4.4 A set of three p orbitals.

shown in the figures, and each p orbital has two lobes separated by a nodal plane.

When considering the cases that arise when $l=2$, it is found that there are five orbitals of the d type, shown in Fig. 4.5. However, if the d_{xy}, d_{yz}, and d_{xz}, orbitals are rotated by 45°, this generates three new orbitals, $d_{x^2-y^2}$, $d_{y^2-z^2}$, and $d_{z^2-x^2}$, which have lobes lying along the axes. It can be shown that combining these wave functions leads to the relationship

$$\psi_{d_{x^2-y^2}} + \psi_{d_{y^2-z^2}} + \psi_{d_{z^2-x^2}} = 0 \tag{4.44}$$

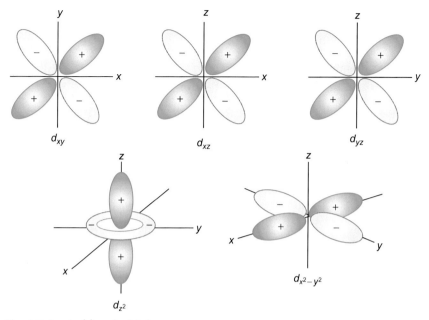

Fig. 4.5 A set of five *d* orbitals.

Therefore, only two of the three orbitals are independent. In the usual case, the $d_{x^2-y^2}$ is chosen to represent one of the independent orbitals, which leaves the other two to be combined to give the d_{z^2} orbital, as shown in Fig. 4.5. Thus, the d_{z^2} orbital is usually shown as a combination of the other two functions, $d_{y^2-z^2}$ and $d_{z^2-x^2}$. For our purposes, the usual diagrams shown in Fig. 4.5 will suffice.

4.4 ORTHOGONALITY

Our ideas about bonding between atoms focus on the combination of atomic orbitals (i.e., wave functions). One of the more important aspects of orbital combination is that of *orthogonality*. For certain combinations of wave functions, it is found that

$$\int \psi_1 \cdot \psi_2 \, d\tau = 0 \tag{4.45}$$

This type of integral, known as an *overlap integral*, gives a measure of the extent to which orbitals overlap in molecules (see Chapter 8). If the condition shown in Eq. (4.45) is met, then the wave functions are said to be *orthogonal*. This relationship shows that there is no effective overlap or congruency

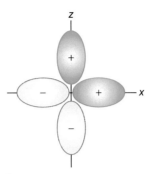

Fig. 4.6 The p_x and p_z orbitals.

of the wave functions. For example, the p_z and p_x orbitals are perpendicular to each other, and the overlap is 0. Although this is shown pictorially in Fig. 4.6, it can also be shown mathematically as follows:

The wave functions corresponding to the p_x and p_z wave functions are

$$p_z = \frac{1}{4\sqrt{2\pi}}\left(\frac{1}{a_0}\right)^{5/2} r\, e^{-r/2a_0} \cos\theta \tag{4.46}$$

and

$$p_x = \frac{1}{4\sqrt{2\pi}}\left(\frac{1}{a_0}\right)^{5/2} r\, e^{-r/2a_0} \sin\theta \cos\varphi \tag{4.47}$$

Therefore, the integral

$$\int \psi^* \psi \, d\tau$$

can be written as

$$\int \psi_{p_z} \psi_{p_z} \, d\tau = \frac{1}{32\pi a_0^5} \int_0^\infty \int_0^\pi \int_0^{2\pi} r^4 e^{-r/a_0} \sin^2\theta \, \cos\theta \, \cos\varphi \, dr \, d\theta \, d\varphi \tag{4.48}$$

$$= \frac{1}{32\pi\, a_0^5} \int_0^\infty r^4 e^{-r/a_0} \, dr \int_0^\pi \sin^2\theta \, d\theta \int_0^{2\pi} \cos\varphi \, d\varphi$$

The exponential integral is of a form discussed earlier, and it can be evaluated immediately to give

$$\int_0^\infty r^4 e^{-r/a_0} \, dr = \frac{4!}{(1/a_0)^5} \tag{4.49}$$

The two integrals giving the angular dependence can be looked up in a table of integrals because they are standard forms. It is found that

$$\int \cos ax \, dx = \frac{1}{a}\sin ax \tag{4.50}$$

$$\int \sin^n ax \cos ax \, dx = \frac{1}{a(n+1)}\sin^{n+1} ax \tag{4.51}$$

For the integrals being considered, $a=1$ and $n=2$, so the evaluation is as follows:

$$\int_0^{2\pi} \cos\varphi \, d\varphi = \sin\varphi \Big|_0^{2\pi} = \sin 2\pi - \sin 0 = 0 \tag{4.52}$$

$$\int_0^{\pi} \sin^2\theta \cos\theta \, d\theta = \frac{1}{3}\sin^3\theta \Big|_0^{\pi} = \frac{1}{3}\left[\sin^3\pi - \sin^3 0\right] = 0 \tag{4.53}$$

Therefore, it has been shown that

$$\int \psi_{p_z}\psi_{p_x} \, d\tau = 0$$

The orbitals are shown to be orthogonal, as expected. Note that it is the *angular* dependence that is different for the two orbitals, which results in their being orthogonal.

The implications of the orthogonality of orbitals are of great importance. For example, it can be seen immediately that any overlap between certain types of orbitals will not occur; two of these are shown in Fig. 4.7. For a given atom, it can be shown that if ψ_1 and ψ_2 are orbitals with different symmetry types, they must be orthogonal. However, the orbitals are orthogonal

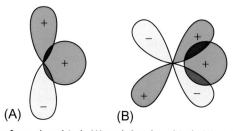

(A)   (B)

Fig. 4.7 Interaction of *p* and *s* orbitals (A) and *d* and *s* orbitals (B) to give no net overlap.

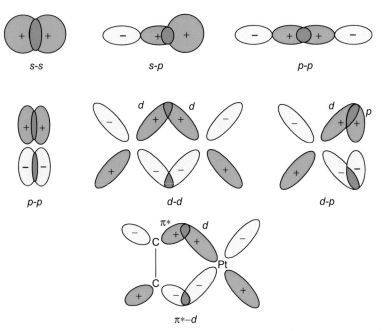

Fig. 4.8 Favorable overlap of orbitals leading to $S > 0$. *(Reprinted with permission from House, J. E. Inorganic Chemistry, 2nd ed.; Academic Press/Elsevier: Amsterdam, 2013.)*

(e.g. p_x, p_y, and p_z) within the same *type* of orbital, as are the five orbitals in a set of d orbitals. However, it can also be seen that numerous types of overlaps, some of which are shown in Fig. 4.8, lead to favorable situations (overlap integral > 0). The discussion of the overlap of atomic orbitals will be presented in more detail in Chapters 8 and 10.

4.5 APPROXIMATE WAVE FUNCTIONS AND THE VARIATION METHOD

For many problems, it is not practical to obtain a wave function by the exact solution of a wave equation that describes the system. It is still possible to perform many types of calculations. One of the most useful techniques is that known as the *variation method*. The procedure will be illustrated by applying it to the solution of a simple problem. The variation method will also be used in the next chapter in dealing with the problem of the helium atom, as well as in Chapter 8 when dealing with diatomic molecules.

By using the principles introduced in Section 2.4, the expectation value for the energy of a system is given by

$$E = \frac{\int \psi^* \hat{H} \psi \, d\tau}{\int \psi^* \psi \, d\tau} \qquad (4.54)$$

If the exact form of the wave function is unknown, then the first step is to assume some form of the wave function. It is frequently described as "guessing" a *trial wave function*, but this is hardly the case. The general form of the wave functions found from the solution of the hydrogen atom problem shows that they are exponential functions. Therefore, a function of that general type is a good starting point (i.e., the initial "guess").

The variation *theorem* provides the basis for the variation *method*. This theorem states that the correct energy is obtained from the use of Eq. (4.54) only when the correct wave function is used. Any "incorrect" wave function will give an energy that is higher than the actual energy. This premise is stated without providing a proof at this point. If a trial wave function, ψ_i, is chosen, then the energy E_i calculated using it is greater than the correct energy E_0, so that for any incorrect wave function, $E_i > E_0$. Generally, a trial wave function is chosen that contains some adjustable parameter so that its value can be varied to "improve" the wave function. This "improved" wave function can then be used to calculate an "improved" value for the energy, etc., and the process repeated until a constant value is obtained.

A simple illustration of the variation method is provided by considering the hydrogen atom in the $1s$ state. For this state, the trial wave function chosen will be one having the form

$$\psi = e^{-br} \qquad (4.55)$$

In this case, b is a parameter whose actual value can be changed as information about it is gained. For the hydrogen atom, the potential energy is $V = -e^2/r$, so the Hamiltonian operator is

$$\hat{H} = -\frac{\hbar^2}{8\pi^2 m} \nabla^2 - \frac{e^2}{r} \qquad (4.56)$$

The energy depends only on r for the $1s$ state of the hydrogen atom, so the angular portion of the Laplacian can be omitted and replaced by the factor 4π after integration. Therefore, it is not necessary to use the radial portion of ∇^2, and the required function is

$$\nabla^2 = \frac{1}{r^2}\frac{\partial}{\partial r}\left(r^2\frac{\partial}{\partial r}\right) \tag{4.57}$$

When this operator is used with the trial wave function ψ, the result is

$$\nabla^2\psi = \frac{1}{r^2}\frac{\partial}{\partial r}\left(r^2\frac{\partial}{\partial r}\right)e^{-br} \tag{4.58}$$

By taking the derivatives indicated and simplifying, the result is

$$\nabla^2\psi = \left(b^2 - \frac{2b}{r}\right)e^{-br} \tag{4.59}$$

When this value for ∇^2 is substituted in the Hamiltonian operator shown in Eq. (4.56) and the result is substituted into Eq. (4.54), the expression for the energy is

$$E = \frac{\displaystyle\int_0^\infty e^{-br}\left(-\frac{h^2}{8\pi^2 m}\nabla^2 - \frac{e^2}{r}\right)e^{-br}\,4\pi\,r^2\,dr}{\displaystyle\int_0^\infty \left(e^{-br}\right)\left(e^{-br}\right)4\pi\,r^2\,dr} \tag{4.60}$$

The value of ∇^2 shown in Eq. (4.59) is substituted in Eq. (4.60), and the factor of 4π is canceled. The resulting equation can be written as

$$E = \frac{\displaystyle\int_0^\infty e^{-br}\left(-\frac{h^2}{8\pi^2 m}\left(b^2 - \frac{2b}{r}\right)e^{-br} - \frac{e^2}{r}e^{-br}\right)r^2\,dr}{\displaystyle\int_0^\infty \left(e^{-2br}\right)r^2\,dr} \tag{4.61}$$

After simplifying this equation by expanding the terms in the numerator by multiplication, the result is

$$E = \frac{\displaystyle\int_0^\infty -\frac{h^2 b^2}{8\pi^2 m}r^2 e^{-br}\,dr + \int_0^\infty -\frac{2h^2 b}{8\pi^2 m}r\,e^{-2br}\,dr - \int_0^\infty e^2\,re^{-2br}\,dr}{\displaystyle\int_0^\infty \left(e^{-2br}\right)r^2\,dr} \tag{4.62}$$

Fortunately, each of the exponential integrals is of the easily recognized form

$$\int_0^\infty x^n e^{-ax}\, dx = \frac{n!}{a^{n+1}} \qquad (4.63)$$

Therefore, evaluating the integrals gives

$$E = \frac{-\dfrac{h^2 b^2}{8\pi^2 m}\dfrac{2}{8b^3} + \dfrac{h^2 b}{4\pi^2 m}\dfrac{1}{b^2} - e^2\dfrac{1}{4b^2}}{\dfrac{2}{8b^3}} \qquad (4.64)$$

Finally, after simplifying this expression, the energy is represented as

$$E = \frac{h^2 b^2}{8\pi^2 m} - be^2 \qquad (4.65)$$

This equation gives the energy in terms of fundamental constants and the adjustable parameter b. It is necessary to find the value of b that will give the minimum energy, and this is done by taking the derivative of E with respect to b and setting the derivative equal to 0. The relationship that is obtained can be written as

$$\frac{\partial E}{\partial b} = \frac{2h^2 b}{8\pi^2 m} - e^2 = 0 \qquad (4.66)$$

Solving this equation for b gives

$$b = \frac{4\pi^2 m e^2}{h^2} \qquad (4.67)$$

When this value for b is substituted into Eq. (4.65) for the energy, the result after simplification is

$$E = -\frac{2\pi^2 m e^4}{h^2} \qquad (4.68)$$

This expression for the energy of a hydrogen atom in the $1s$ state is exactly the same as that found using the Bohr model (see Chapter 1). How was it possible to obtain the correct energy in a single "improvement" of the wave function? The answer is that the correct form of the wave function to use as the trial was "guessed." It was known in advance that an exponential function involving r was the form of the actual wave function, so the variation method enabled us to evaluate the constant in a single calculation cycle.

Of course, this is not always the case. The variation method will be used as the basis for other types of calculations in later chapters.

PROBLEMS

1. Calculate the velocity of an electron in the $n=1$ state of a hydrogen atom.

2. Show that the de Broglie wavelength of an electron moving at the velocity found in Problem 1 corresponds to the circumference of the first Bohr orbit.

3. What is the total number of electrons that can be accommodated if $n=5$?

4. The potential energy for an electron attracted to a +1 nucleus is $V=-e^2/r$. Using the variation method, determine $<V>$ in this case.

5. Determine the value for $<r>$ for an electron in the $2p_z$ state of the hydrogen atom.

6. Use the procedure described in the text to determine the probability that the electron in the $1s$ state of hydrogen will be found outside a_0.

7. Use the procedure described in the text to determine the probability that the electron in a hydrogen atom will be found between a_0 and $2 a_0$.

8. Calculate the average or expectation energy $<E>$ for the electron in the $1s$ state of the hydrogen atom.

9. It has been shown that the most probable radius of the hydrogen atom is $(3/2)a_0$. Consider the hydrogen atom as a particle in a one-dimensional box (the electron can travel on either side of the nucleus) and calculate the energy (in eV) of the electron in the state with $n=1$.

10. If a sphere is to be drawn with the nucleus at the center, how large must the sphere be to encompass the electron in a hydrogen atom 90 percent of the time for the $n=1$ state?

11. Show that the wave functions for the $1s$ and $2s$ states in the hydrogen atom are orthogonal.

CHAPTER 5

Structure and Properties of More Complex Atoms

As will become apparent, wave equations cannot be solved exactly for complex atoms. Although a rigorous quantum mechanical treatment cannot result in a closed form solution to the wave equation for a complex atom, such solutions are usually not necessary for an understanding of most aspects of chemical bonding. At this point the approaches used to describe the helium atom, as well as some of the empirical and experimental properties of atoms, will be presented.

5.1 THE HELIUM ATOM

Although it is easy to formulate the wave equation for atoms that are more complex than hydrogen, such equations cannot be solved exactly. A consideration of the helium atom will show this is so. The helium atom can be represented as shown in Fig. 5.1. From the figure, it can be seen that the Hamiltonian must include the kinetic energy of each electron, the electrostatic attraction of the nucleus for each electron, and the repulsion of the two electrons. Taking all of these energies into account with the distances shown in Fig. 5.1, the Hamiltonian operator can be written as

$$\hat{H} = -\frac{\hbar^2}{2m}\nabla_1^2 - \frac{\hbar^2}{2m}\nabla_2^2 - \frac{2e^2}{r_1} - \frac{2e^2}{r_2} + \frac{e^2}{r_{12}} \qquad (5.1)$$

From the general equation

$$\hat{H}\psi = E\psi$$

this leads directly to the wave equation

$$\left(-\frac{\hbar^2}{2m}\nabla_1^2 - \frac{\hbar^2}{2m}\nabla_2^2 - \frac{2e^2}{r_1} - \frac{2e^2}{r_2} + \frac{e^2}{r_{12}}\right)\psi = E\psi \qquad (5.2)$$

For the hydrogen atom, the Hamiltonian operator has only one term involving $1/r$ (where r is the distance of the electron from the nucleus), and the difficulty this causes is avoided by the use of polar coordinates. In the case of

Fundamentals of Quantum Mechanics
http://dx.doi.org/10.1016/B978-0-12-809242-2.00005-X

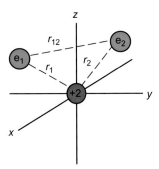

Fig. 5.1 A coordinate system for the helium atom.

the helium atom, even a change to polar coordinates does not help because of the term containing $1/r_{12}$. Thus, when the term involving $1/r_{12}$ is included, the variables cannot be separated. One approach to solving the wave equation for the helium atom is by constructing a *trial* wave function and using the variation method to optimize it. A trial wave function of the following form can be assumed:

$$\psi = \varphi_1 \varphi_2 = \frac{Z'^3}{\pi a_0^3} e^{-Z' r_1/a_0} e^{-Z' r_2/a_0} \qquad (5.3)$$

in which φ_1 and φ_2 are hydrogen-like wave functions. Z' is an *effective* nuclear charge that is less than the actual value of 2 because each electron does not experience the full effect of a +2 nucleus due to the other electron. The Hamiltonian operator can be written as

$$\hat{H} = -\frac{\hbar^2}{2m}(\nabla_1^2 + \nabla_2^2) - Z' e^2 \left(\frac{1}{r_1} + \frac{1}{r_2}\right) + \frac{e^2}{r_{12}} \qquad (5.4)$$

Therefore, the wave equation can be written as

$$\hat{H}\psi = E\psi = \hat{H}\varphi_1\varphi_2 = E_H\varphi_1\varphi_2 \qquad (5.5)$$

In this equation, E_H is the energy of the hydrogen atom in the $1s$ state. The quantity $Z' E_H$ is the energy that results when the Bohr model is applied to a hydrogen-like species where the nuclear charge is not equal to 1. Therefore,

$$\hat{H}\varphi_1 = -\frac{\hbar^2}{2m}\nabla_1^2\varphi_1 = \frac{Z' e^2 \varphi_1}{r_1} - Z'^2 E_H\varphi_1 \qquad (5.6)$$

A similar equation can be written using the second atomic wave function φ_2. Therefore, the combination can be written as

$$\hat{H}\varphi_1\varphi_2 = \frac{Z'\,e^2}{r^2}\varphi_1\varphi_2 - Z'^2 E_H\varphi_1\varphi_2 + \frac{Z'\,e^2}{r_1}\varphi_1\varphi_2 - Z'^2 E_H\varphi_1\varphi_2$$
$$- Z'e^2\left(\frac{1}{r_1}+\frac{1}{r_2}\right)\varphi_1\varphi_2 + \frac{e^2}{r_{12}}\varphi_1\varphi_2 \tag{5.7}$$

By multiplying by $\varphi_1^*\varphi_2^*$, the result is

$$\varphi_1^*\varphi_2^*\hat{H}\varphi_1\varphi_2 = \varphi_1^*\varphi_2^*\frac{Z'\,e^2}{r^2}\varphi_1\varphi_2 - \varphi_1^*\varphi_2^* Z'^2 E_H\varphi_1\varphi_2$$
$$+ \varphi_1^*\varphi_2^*\frac{Z'\,e^2}{r_1}\varphi_1\varphi_2 - \varphi_1^*\varphi_2^* Z'^2 E_H\varphi_1\varphi_2 \tag{5.8}$$
$$- \varphi_1^*\varphi_2^* Z'e^2\left(\frac{1}{r_1}+\frac{1}{r_2}\right)\varphi_1\varphi_2 + \varphi_1^*\varphi_2^*\frac{e^2}{r_{12}}\varphi_1\varphi_2$$

After writing the terms in integral form, the total energy can be expressed as

$$E = -2Z'^2 E_H + (Z'-Z)e^2\int \varphi_1^*\varphi_2^*\left(\frac{1}{r_1}+\frac{1}{r_2}\right)\varphi_1\varphi_2\,d\tau + \int \varphi_1^*\varphi_2^*\frac{e^2}{r_{12}}\varphi_1\varphi_2\,d\tau \tag{5.9}$$

The integral in middle term on the right-hand side of this equation can be expanded to give

$$\int \varphi_1^*\left(\frac{1}{r_1}\right)\varphi_1\,d\tau_1\int \varphi_2^*\varphi_2\,d\tau_2 + \int \varphi_2^*\left(\frac{1}{r_2}\right)\varphi_2\,d\tau_2\int \varphi_1^*\varphi_1\,d\tau_1 \tag{5.10}$$

When normalized atomic wave functions are used, two of the integrals are equal to 1:

$$\int \varphi_2^*\varphi_2\,d\tau_2 = \int \varphi_1^*\varphi_1\,d\tau_1 = 1 \tag{5.11}$$

Therefore, the middle term on the right-hand side of Eq. (5.9) reduces to

$$(Z'-Z)e^2 2\int \varphi_1^*\left(\frac{1}{r_1}\right)\varphi_1\,d\tau_1 = 4(Z'-Z)Z'\,E_H \tag{5.12}$$

The energy can now be written from Eqs. (5.9) and (5.12) as

$$E = -2Z'^2 E_H + 4(Z'-Z)Z'\,E_H + \frac{5}{4}Z'\,E_H \tag{5.13}$$

$$E = -2Z'^2 E_H + 4Z'^2 E_H - 4ZZ'E_H + \frac{5}{4}Z'E_H \tag{5.14}$$

Having obtained an expression for the energy in terms of the effective nuclear charge, Z', it is necessary to determine the value of Z' that results in the minimum energy, which is done by taking the derivative and setting it equal to 0:

$$\left(\frac{\partial E}{\partial Z'}\right) = 0 = -4Z'E_H + 8Z'E_H - 4ZE_H + \frac{5}{4}E_H \tag{5.15}$$

Solving this expression for Z', we find

$$Z' = Z - \frac{5}{16} \tag{5.16}$$

Thus, each electron experiences an *effective* nuclear charge of 27/16 instead of the nuclear charge of +2 as a result of screening by the other electron.

Having calculated that the effective nuclear charge in the helium atom is 27/16, it is possible to use that value in Eq. (5.14) to determine the total binding energy for the two electrons. When the value for Z' is substituted, we find $E = +5.696$ eV. As a result of the ionization potential for the hydrogen atom, E_H, being 13.6 eV, the total ionization potential for the two electrons in helium is calculated to be 77.5 eV. The experimental value is approximately 79.0 eV, so the calculation using the variation method yields a value that is in reasonable agreement with the observed value.

5.2 PERTURBATION METHOD

The discussion of the variation method presented in Section 4.5 and its application to the calculation of the total electron binding energy in the helium atom have shown that the variation method is an important tool in quantum mechanics. Another such tool is the *perturbation method*. The basic idea behind perturbation theory is that the system does not behave perfectly because of some "slight" deviation from a system that can be treated exactly. In the orbits of planets, deviation from perfect orbits results from gravitational forces, which become more important as the planets get closer together in their orbits. In the harmonic oscillator model, the perturbation might be a potential that is not expressed exactly by $\frac{1}{2}kx^2$ (see Section 6.4). In Eq. (5.2), it can be seen that the helium atom could be treated as the sum of two hydrogen atomic problems, if the e^2/r_{12} term that arises from repulsion of the electrons was not present in the Hamiltonian. Therefore, the repulsion term is treated as a "slight" perturbation of an otherwise "perfect" system that could be solved exactly. An overview of the general principles of the perturbation method will be presented to show how the method is approached and then show how it applies to the helium atom case.

For an unperturbed system that can be treated exactly, the wave equation can be written as

$$\hat{H}^\circ \psi^\circ = E^\circ \psi^\circ \qquad (5.17)$$

The solutions are the wave functions ψ_0°, ψ_1°, ψ_2°, ... and the energy eigenvalues are E_0°, E_1°, E_2°, ..., etc. The Hamiltonian is that which is appropriate for the particular system, but it has the form

$$\hat{H} = -\frac{\hbar^2}{2m}\nabla^2 + V \qquad (5.18)$$

If the system becomes slightly perturbed, that perturbation is represented by a slight alteration in the potential term of the Hamiltonian. Therefore, the Hamiltonian can be written as

$$\hat{H} = \hat{H}^\circ + \lambda \hat{H}' + \lambda^2 \hat{H}' + \cdots \qquad (5.19)$$

in which λ is a parameter that expresses the extent of perturbation and $\hat{H}'$ is the adjustment to the Hamiltonian. In performing the calculation of the first-order perturbation, the terms beyond $\lambda\hat{H}'$ in the series are ignored. Because the perturbation energy is presumed to be small compared to the total energy, the wave function for the kth state of the system is written as

$$\psi_k = \psi_k^\circ + \lambda \psi_k' \qquad (5.20)$$

and the corresponding energy of the state is given by

$$E_k = E_k^\circ + \lambda E_k' \qquad (5.21)$$

Using the expressions above and the fundamental relationship that

$$\hat{H}\psi = E\psi$$

the first-order correction is obtained by omitting terms in λ^2 (because λ is small). Therefore, for the unperturbed kth state, the resulting relationship can be shown as

$$\hat{H}^\circ \psi_k^\circ = E\psi_k^\circ \qquad (5.22)$$

For the first-order perturbation, the equation becomes

$$\hat{H}^\circ \psi_k' + \hat{H}^\circ \psi_k^\circ - E_k^\circ \psi_k' - E_k' \psi_k^\circ = 0 \qquad (5.23)$$

This equation can also be written as

$$\left(\hat{H}^\circ - E_k^\circ\right)\psi_k' = \left(E_k' - \hat{H}'\right)\psi_k^\circ = 0 \qquad (5.24)$$

A linear combination of solutions is used to represent the wave function of the perturbed kth state in terms of the wave functions for all of the i unperturbed states. This series is written as

$$\psi'_k = \sum_i a_i \psi^o_i \tag{5.25}$$

From a combination of Eqs. (5.22) and (5.25), it is possible to obtain

$$\hat{H}^o \psi'_k = \sum_i a_i \hat{H}^o \psi^o_i = \sum_i a_i E^o_i \psi^o_i \tag{5.26}$$

Equation (5.26) can also be written as

$$\sum_i \left(a_i E^o_i \psi^o_i - a_i E^o_k \psi^o_i \right) = E'_k \psi^o_k - \hat{H}' \psi^o_k \tag{5.27}$$

Simplifying this equation gives

$$\sum_i a_i \left(E^o_i - E^o_k \right) \psi^o_i = E'_k \psi^o_k - \hat{H}' \psi^o_k \tag{5.28}$$

By multiplying both sides of this equation by ψ_k^{o*} (see Section 2.4) and integrating over all space, the left-hand side of Eq. (5.28) can be written as

$$\int \psi^o_k \sum_i a_i \left(E^o_i - E^o_k \right) \psi^o_i \, d\tau \tag{5.29}$$

This expression can be greatly simplified because for orthogonal wave functions, the integral

$$\int \psi^{o*}_i \psi^o_k \, d\tau = 0 \tag{5.30}$$

gives a value of 0 when $i \neq k$, and a value of 1 when $i = k$. When $i = k$, it can be seen that $E^o_k = E^o_i$ so the entire integral vanishes, and the left-hand side of Eq. (5.28) must equal 0. Therefore, because the right-hand side of Eq. (5.28) must equal 0

$$\int \psi^{o*}_k \left(E' - \hat{H}' \right) \psi^o_k \, d\tau = 0 \tag{5.31}$$

Separation of the integral can be performed, and because the perturbation energy E' is a constant and is removed from the integral, it is possible to write

$$E' \int \psi^{o*}_k \psi^o_k \, d\tau - \int \psi^{o*}_k \hat{H}' \psi^o_k \, d\tau = 0 \tag{5.32}$$

For normalized wave functions, the integral multiplied by E' is equal to 1. Therefore,

$$E' = \int \psi_k^{o*} \hat{H}' \psi_k^o d\tau \tag{5.33}$$

This result shows that the perturbation energy correction to the kth state is the familiar expectation value, with only the perturbation Hamiltonian being used in the integration. After E' is calculated, the total energy of the kth energy level (the $1s$ state in helium is considered in this case) will be given as $E^o + E'$. The application of the perturbation method to the helium atom involves treating the term e^2/r_{12} as the perturbation to an otherwise exactly solvable system consisting of two hydrogen atoms. It is already known that the ionization potential of the hydrogen atom is 13.6 eV. From the exact solution of the approximate equation obtained by neglecting the e^2/r_{12} term in the Hamiltonian, the energy for the $1s$ state in helium is

$$E^o = -Z^2 E_H - Z^2 E_H = -2Z^2 E_H. \tag{5.34}$$

The perturbation term involving e^2/r_{12} gives an energy E', which can be expressed as

$$E' = \int\int \psi^*_{(1)} \psi^*_{(2)} \left(e^2/r_{12}\right) \psi_{(1)} \psi_{(2)} d\tau_1 d\tau_2 \tag{5.35}$$

where e^2/r_{12} is the perturbation operator $\hat{H}$. Of course, for the $1s$ wave function, $\psi^* = \psi$. This integral must be evaluated to give the perturbation correction to the energy $-2Z^2 E_H$, which was also obtained by neglecting the repulsion between the two electrons. Each electron is represented as a spherically symmetric charge field, and the integral representing their interaction can be transformed to give

$$\frac{Z^6}{\pi^2 a_o^6} \int \frac{\exp(-2Zr_1/a_o) \exp(-2Zr_2/a_o)}{r_{12}} dV \tag{5.36}$$

In this integral, the exponential functions are charge distributions of two spherically symmetric electrostatic fields produced by the two electrons. An evaluation of the integral (Pauling and Wilson, 1935) leads to a perturbation energy of

$$E' = \frac{5}{4} Z E_H \tag{5.37}$$

Therefore, the total energy for the $1s$ level in the helium atom is

$$E = -2Z^2 E_H + \frac{5}{4} Z E_H = -\left[2Z^2 - \frac{5}{4}Z\right] E_H \tag{5.38}$$

Note that the perturbation caused by the repulsion of the two electrons raises the energy (destabilization) of the $1s$ level so that the total binding energy of the two electrons is *not* as great as it would be for a +2 nucleus with no repulsion between the two electrons. (See Chapter 6 of the book by Pauling and Wilson (1935) for details.) By substituting $Z=2$ and $E_H=13.6$ eV, it is found that the *binding* energy is

$$E = -\left[2(2)^2 - 2\left(\frac{5}{4}\right)\right](13.6) = -5.50(13.6) = -74.8 \text{ eV} \qquad (5.39)$$

Therefore, the total *ionization* energy is 74.8 eV. As mentioned earlier, the experimental value is approximately 79.0 eV. In this section, the applications of two of the very important approximation methods widely used in quantum mechanical calculations have been shown.

Although two of the important approximation methods have been illustrated, other applications of the variation method to molecules will be shown in Section 8.3. From the discussion presented in this section, it should be clear why variation and perturbation methods are among the most important techniques used in quantum mechanical calculations.

Although the wave equation cannot be solved exactly in most cases (even for the helium atom), it is possible to arrive at approximate solutions. In this case, evaluating the total energy by the variation and perturbation methods leads to values that are quite close to the actual binding energy of the two electrons. It is a simple matter to write the wave equation for complex atoms by writing the Hamiltonian in terms of the various attraction and repulsion energies, but various approximation methods must be used to solve the equations. In the next section, a different approach to obtaining wave functions for complex atoms will be discussed.

5.3 SLATER WAVE FUNCTIONS

The exact solution of the Schrödinger wave equation for complex atoms is not possible. However, an examination of the form of the wave functions obtained for the hydrogen atom suggests that approximate wave functions might be obtained if we were to take into account the mutual electron repulsion. J. C. Slater devised such a procedure, and the resulting approximate wave functions are known as *Slater wave functions* [or Slater-type orbitals (STO)]. The wave functions are written in the form

$$\psi_{n,l,m} = R_{n,l}(r)e^{-Zr/a_0 n^*} Y_{l,m}(\theta, \varphi) \qquad (5.40)$$

Specifically, the wave functions have the form

$$\psi_{n,l,m} = r^{n^*-1} e^{-(Z-s)r/a_o n^*} Y_{l,m}(\theta, \varphi) \tag{5.41}$$

in which s is the screening constant and n^* is a parameter that varies with the principal quantum number, n. For a rigorous presentation of the theory of perturbation and variation methods, see the book by Pauling and Wilson (1935).

The screening constant, s, for a given electron is determined by considering all of the contributions from all the populated orbitals in the atom. The electrons are grouped according to the procedure that follows, and the weightings from each group are determined according to the other rules:

1. The electrons are grouped in this manner:

$$1s|2s|2p|3s3p|3d|4s4p|4d|4f|5s5p|5d|\cdots$$

2. No contribution to the screening of an electron is considered to arise from electrons in orbitals *outside* the orbital holding the electron for which the wave function is being written.
3. For the $1s$ level, the contribution is 0.30, but for other groups, 0.35 is added for each electron in that group.
4. For an electron in an s or p orbital, 0.85 is added for each other electron when the principal quantum number is one less than that for the orbital being written. For still lower levels, 1.00 is added for each electron.
5. For electrons in d and f orbitals, 1.00 is added for each electron residing below the one for which the wave function is being written.
6. The value of n^* varies with n as follows:

$n=1$	2	3	4	5	6
$n^*=1$	2	3	3.7	4.0	4.2.

Suppose one needs to write the Slater wave function for an electron in a $2p$ orbital of oxygen ($Z=8$). For that electron, $n=2$ so $n^*=2$ also. The screening constant for the fourth electron in the $2p$ level is determined as follows: For the two electrons in the $1s$ level, $2(0.85)=1.70$. For the five electrons in the $2s$ and $2p$ levels, $5(0.35)=1.75$. Summing these contributions to the screening constant for the electron in question, it is found that $s=3.45$ and the effective nuclear charge is $8-3.45=4.55$, so that $(Z-s)/n^*=2.28$. Therefore, the Slater wave function for an electron in the $2p$ level of oxygen can be written as

$$\psi = r^1 e^{-2.28r/a_o} Y_{2,m}(\theta, \varphi) \tag{5.42}$$

The variation method was used earlier to determine the optimum value of the nuclear charge for helium. Although the *actual* nuclear charge is 2, the variation method predicts a value of $27/16 = 1.6875$ as the *effective* nuclear charge that each electron experiences. This difference is an obvious result of screening by the other electron. It is now possible to compare the result from the variation method to the effect of screening in the helium atom obtained using Slater's rules. For an electron in the $1s$ level, the only screening is that of the other electron, for which the value 0.30 is used. The effective nuclear charge is $(Z-s)/n^* = (2-0.30)/1 = 1.70$, which is in good agreement with the result obtained by the variation method (1.6875).

Slater-type orbitals are frequently useful as a starting point in other calculations. As has been shown earlier, an approximate wave function can give useful results when used in the variation method. Also, *some* atomic wave function must be used in the construction of wave functions for molecules. Consequently, these approximate, semiempirical wave functions are especially useful, and molecular orbital calculations are frequently carried out using a STO basis set. More discussion of this topic will appear in Chapter 14.

5.4 ELECTRON CONFIGURATIONS

As has been shown, four quantum numbers are required to completely describe an electron in an atom. There are certain restrictions on the values that these quantum numbers can have. Thus, for $n = 1, 2, 3,...,$ the values for l are $l = 0, 1, 2,..., (n-1)$. For a given value of n, the quantum number l can have all integer values from 0 to $(n-1)$. The quantum number m can have the series of values $+l, +(l-1),..., 0,..., -(l-1), -l$. Thus, there are $(2l+1)$ values for m. The fourth quantum number, s, can have values of $\pm(\frac{1}{2})$, with this being the spin angular momentum in units of $h/2\pi$.

It is possible to write a set of four quantum numbers to describe each electron in an atom. It is necessary to use the *Pauli Exclusion Principle*, which states that no two electrons in the same atom can have the same set of four quantum numbers. In the case of the hydrogen atom, states characterized by lower n values represent those of lower energy. For the single electron in a hydrogen atom, the four quantum numbers to describe the electron can be written as $n = 1$, $l = 0$, $m = 0$, $s = +\frac{1}{2}$, or $s = -\frac{1}{2}$. The value chosen for s is arbitrary. For helium, which has two electrons, the two sets of quantum numbers are

Electron 1	Electron 2
$n=1$	$n=1$
$l=0$	$l=0$
$m=0$	$m=0$
$s=+\frac{1}{2}$	$s=-\frac{1}{2}$

An atomic energy level is denoted by the n value followed by a letter (s, p, d, or f to denote $l=0$, 1, 2, or 3, respectively), and the ground state for hydrogen is $1s^1$, whereas that for helium is $1s^2$. The two sets of quantum numbers written above complete the sets that can be written for the first shell with $n=1$.

For $n=2$, l can have the values of 0 and 1. In general, the levels increase in energy as the sum $(n+l)$ increases. By taking the value of $l=0$ first, the state with $n=2$ and $l=0$ is designated as the 2s state. Like the 1s state, it can hold two electrons:

Electron 1	Electron 2
$n=2$	$n=2$
$l=0$	$l=0$
$m=0$	$m=0$
$s=+\frac{1}{2}$	$s=-\frac{1}{2}$

These two sets of quantum numbers describe the electrons residing in the 2s level. Now, by taking the value $l=1$, we find that six sets of quantum numbers can be written:

Electron: 1	2	3	4	5	6
$n=2$	$n=2$	$n=2$	$n=2$	$n=2$	$n=2$
$l=1$	$l=1$	$l=1$	$l=1$	$l=1$	$l=1$
$m=+1$	$m=0$	$m=-1$	$m=+1$	$m=0$	$m=-1$
$s=+\frac{1}{2}$	$s=+\frac{1}{2}$	$s=+\frac{1}{2}$	$s=-\frac{1}{2}$	$s=-\frac{1}{2}$	$s=-\frac{1}{2}$

These six sets of quantum numbers represent six electrons residing in the 2p level, which consists of three orbitals, each holding two electrons. Each value of m denotes an orbital that can hold two electrons with $s=+\frac{1}{2}$ and $s=-\frac{1}{2}$. This was shown to be the case for the 1s and 2s orbitals, but in those cases, m was restricted to the value 0 because $l=0$ for an s orbital. Table 5.1 shows the types of atomic orbitals and the number of electrons that can populate them.

Table 5.1 Maximum numbers of electrons that energy states can hold

l Value	m Values	State	Maximum number of electrons
0	0	s	2
1	0, ±1	p	6
2	0, ±1, ±2	d	10
3	0, ±1, ±2, ±3	f	14
4	0, ±1, ±2, ±3, ±4	g	18

For a given value of l, there are always as many orbitals as there are m values, with each orbital capable of holding a pair of electrons. Thus, for $l=3$ (an f state), there are seven possible values for m (0, ±1, ±2, and ±3) so that such an f state can hold 14 electrons. For the various types of orbitals, this information is summarized in Table 5.1.

For convenience, the practice will be followed that sets of quantum numbers will be written by using the highest positive value of m first and working down. Also, the positive value of s is used first. Thus, for Al, the "last" electron is in the $3p$ level and it will be assigned the set of quantum numbers $n=3$, $l=1$, $m=1$, and $s=+\frac{1}{2}$.

Except for minor variations, the order of increasing energy levels in an atom is given by the sum $(n+l)$. The lowest value for $(n+l)$ occurs when $n=1$ and $l=0$, which are the quantum numbers for the $1s$ state. The next lowest sum of $(n+l)$ is 2, which occurs when $n=2$ and $l=0$. There will not be a $1p$ state where $n=1$ and $l=1$ because of the restrictions on n and l that arise from the solution of the wave equation. For $(n+l)=3$, the possible combinations are $n=2$ and $l=1$ ($2p$) and $n=3$ and $l=0$ ($3s$). Although the sum $(n+l)$ is the same in both cases, the level with $n=2$ is filled first. Therefore, it can be concluded that when two or more ways exist for the same sum $(n+l)$ to arise, the level with the lower n will usually fill first. Thus, the approximate order of filling the energy states in atoms is shown in Table 5.2.

The filling of energy states and the maximum occupancies of the orbitals can be described by making use of the order shown in Tables 5.1 and 5.2. The order in which orbitals are filled follows the pattern shown until Cr is reached. There, it is predicted that the outer electrons would populate the states to give $3d^4\ 4s^2$, but instead they give $3d^5\ 4s^1$. The reason for this is the more favorable coupling of spin and orbital angular momenta that results for the configuration $3d^5\ 4s^1$, which has six unpaired electrons. Coupling of angular momenta will be discussed in the next section.

Table 5.2 Energy states according to the $(n+l)$ sum[a]

n	l	$(n+l)$	State
1	0	1	$1s$
2	0	2	$2s$
2	1	3	$2p$
3	0	3	$3s$
3	1	4	$3p$
4	0	4	$4s$
3	2	5	$3d$
4	1	5	$4p$
5	0	5	$5s$
4	2	6	$4d$
5	1	6	$5p$
6	0	6	$6s$
4	3	7	$4f$
5	2	7	$5d$
6	1	7	$6p$
7	0	7	$7s$

[a]Energy of the states increases going down in the table.

The relationship of the electronic structure to the periodic table should be readily apparent. Groups IA and IIA represent the groups where an s level is being filled as the outer shell. The first, second, and third series of transition elements are the groups where the $3d$, $4d$, and $5d$ levels are being filled. As a result, such elements are frequently referred to as d group elements. The main group elements to the right in the periodic table represent the periods for which $2p$, $3p$, $4p$, $5p$, and $6p$ levels are the outside shells in the various long periods. Finally, the rare earths and the actinides represent groups of elements where the $4f$ and $5f$ levels are being filled.

The periodic table shows the similarities in electron configurations of elements in the same group. For example, the alkali metals (Group IA) all have an outside electron arrangement (valence shell) of ns^1, where $n=2$ for Li, $n=3$ for Na, etc. Because the chemical properties of elements depend on the outer shell electrons, it is apparent why elements in this group are similar chemically. By adding one electron, the halogens (Group VIIA), which have the configurations $ns^2\,np^5$, are converted to the configuration of the next noble gas, $ns^2\,np^6$. It should be emphasized, however, that although there are many *similarities*, numerous *differences* also exist between elements in the same group. Thus, it should not be inferred that the same electron configuration in the valence shell gives rise to the same chemical properties for all members of the group.

5.5 SPECTROSCOPIC STATES

After the overall electronic configuration of an atom has been determined, there are still other factors that affect the energy state. For example, the electronic configuration of carbon is $1s^2\,2s^2\,2p^2$. The fact that two electrons are indicated in the $2p$ state is insufficient for a complete description of the atom, because there are several ways in which those electrons may be arranged. However, this description does not take into account either electron repulsion or spin-orbit coupling, meaning the electronic configuration alone is based only on the n and l quantum numbers. Therefore, within the $1s^2\,2s^2\,2p^2$ electron configuration, there are several different energy states possible.

The energy states arise because the orbital and spin angular momentum vectors may couple to provide several states of different energy. Two ways in which vector coupling can occur will be discussed. These represent limiting cases, and intermediate coupling schemes are known. In the first coupling scheme, the individual orbital moments (l) couple to give a resultant orbital angular moment, L. Also, the individual spin moments (s) couple to produce a total spin moment, S. The two vector quantities, L and S, then couple to give the total angular momentum quantum number, J. This scheme is known as Russell-Saunders or $L-S$ coupling. In this case, the coupling of individual spin moments and individual orbital moments is stronger than is the coupling between individual spin and orbital moments. In the other extreme, the individual spin and orbital moments for a given electron couple to produce a resultant j for that electron. These j vectors then couple to produce the resultant J, which is the overall angular momentum. Coupling of this type is called $j-j$ coupling.

For relatively light atoms, $L-S$ coupling provides the more commonly followed model. The $j-j$ coupling scheme occurs for heavier atoms in the lower part of the periodic table. For our purposes, the $L-S$ coupling scheme will suffice. This coupling of spin and orbital momenta according to the $L-S$ scheme will now be described.

An electron in an atom is characterized by a set of four quantum numbers. The orbital angular momentum quantum number l gives the length of the orbital angular momentum vector in units of $h/2\pi$. The actual quantum mechanical result is $[l(l+1)]^{1/2}$ instead of l. This is because the quantity $l(l+1)$ is the square of the eigenvalue of the operator for the z component of angular momentum. Although the correct value is $[l(l+1)]^{1/2}$, l is commonly used because a vector of length $[l(l+1)]^{1/2}$ has exactly the same quantized projections on the z-axis as a vector of l units in length.

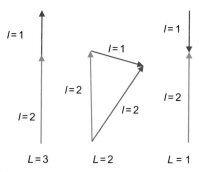

Fig. 5.2 Quantized combinations of vectors $l=2$ and $l=1$.

Because the overall angular momentum is produced by the coupling of vectors, it is necessary to see how that coupling can occur. Figure 5.2 shows the coupling of two vectors, $l=1$ and $l=2$, according to quantization. It is readily apparent that for two vectors of lengths $l=1$ and $l=2$, the resultant L can be formed according to quantum restrictions in three ways. Note that the values of L are $|l_1+l_2|$, $|l_1+l_2-1|$, and $|l_1-l_2|$. It will be necessary to make use of this quantum mechanical coupling of vectors in arriving at the overall angular momentum state for an atom.

As was mentioned earlier, the quantum number m gives the projection of the orbital angular momentum vector on the z-axis. This vector can undergo precession around the z-axis, sweeping out cones of revolution around that axis. This is shown in Fig. 5.3 for $l=2$ (for which $[l(l+1)]^{1/2}=(6)^{1/2}$), which has quantized projections on the z-axis of 0, ± 1, and ± 2. In the absence of an external magnetic field, these orientations are degenerate. If spin–orbit coupling occurs, it is not possible to say what the value of the orbital component will be merely based on the orbital angular momentum quantum number, because this number can give only the *maximum* value of the projection on the z-axis. Consequently, the *microstates* (detailed arrangements of the electrons in orbitals) must be written in order to predict the spin-orbit coupling. However, this is not necessary if only the *lowest* energy state (the *ground state*) is desired.

The essential aspect of spin–orbit coupling is that the spin angular momenta for two or more unpaired electrons will couple to form a resultant spin vector, S. For the configuration

$$\frac{\uparrow \quad \uparrow}{m = \quad +1 \quad 0 \quad -1}$$

$S=(1/2)+(1/2)=1$. The orbital angular momenta couple similarly in their z projections. It is the sum of the m values that gives the maximum length of

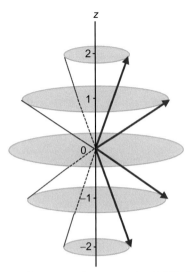

Fig. 5.3 The projections on the z-axis of a vector that is $[l(l+1)]^{1/2}$ units in length (for $l=2$). The angular momentum vector can precess around the z-axis and has projections having lengths of +2, +1, 0, −1, and −2.

the L vector. These two resultant vectors, L and S, then couple to give a third vector, J, which can have all integral values from $|L+S|$ to $|L-S|$ (see Fig. 5.2). These rules can be summarized as follows:

$$L = \Sigma l_i$$

$$S = \Sigma s_i$$

$$M = \Sigma m_i = L, L-1, L-2, \ldots, 0, \ldots, -L$$

After the values for M are obtained, it should be obvious which L projections the M values represent. The L values 0, 1, 2, and 3, respectively, correspond to the *states* designated as S, P, D, and F, in accord with the practice for electronic energy levels (s, p, d, f).

The *multiplicity* of a state is given by $(2S+1)$, where S is the sum of the spins. Thus, a single unpaired electron (as in Na, $3s^1$) gives rise to a *doublet* $[2(1/2)+1=2]$. A *term symbol* is written as $^{(2S+1)}L_J$, which summarizes this information and some examples are $^2S_0$, $^2P_2$, and $^3D_1$. There are also three rules that will make it possible to determine the relative energy of the terms once they are obtained. These rules are known as *Hund's rules*:

1. For equivalent electrons, the state with the highest multiplicity gives the lowest energy.

2. Of those states having the highest multiplicity, the one with the highest L is lowest in energy.
3. For shells that are less than half filled, the lowest J gives the state of lowest energy; for shells more than half filled, the highest J gives the lowest energy.

A few examples will be used to illustrate how the rules are applied:

Example 1

Consider first the case of a configuration ns^1: $S = \frac{1}{2}$ and $L = 0$, because the m value for an s state is 0. Therefore, the state is designated as ^2S in accord with the rules. Only a single J value is possible, that being $\frac{1}{2}$. The resulting spectroscopic state (term) is $^2S_{1/2}$. This is the only term possible for a configuration of ns^1.

Example 2

Consider next the case of ns^2. In this case, $S = 0$ because the two electrons have opposite spins, and $L = 0$ because the electrons reside in an orbital for which $l = 0$ (hence $m = 0$ and the sum of the $m_i = 0$). The only possible value for J is 0, so the ground state term is $^1S_0$. Actually, this must be the result for any filled shell, p^6, d^{10}, etc. Because filled shells contribute only 0 to the S and L values, we can ignore them when determining the spectroscopic state and consider only the partially filled outer shell.

Example 3

Consider now the np^1 configuration. In this case, there are three orbitals with m values of $+1$, 0, and -1 in which the electron may be found. These projections on the z-axis can arise only for a vector $L = 1$ because $L = \Sigma m_i = 1$. The ground state must be a P state ($L = 1$). For a single unpaired electron, $S = \frac{1}{2}$, so the multiplicity is equal to $(2S + 1) = 2$. For $L = 1$ and $S = \frac{1}{2}$, two values are possible for the vector J. These are $|1 + (\frac{1}{2})| = 3/2$ and $|1 - (\frac{1}{2})| = \frac{1}{2}$. Therefore, the two spectroscopic states that exist for the np^1 configuration are $^2P_{1/2}$ and $^2P_{3/2}$, with $^2P_{1/2}$ being lower in energy because the p shell is less than half filled. For the boron atom, the $^2P_{1/2}$ state lies 16 cm^{-1} lower than the $^2P_{3/2}$ state, whereas for the aluminum atom, the difference between the same states is 112 cm^{-1}.

Example 4

An analysis of the spectroscopic states for the np^2 configuration will now be presented. The analysis will begin with writing the

Table 5.3 The microstates arising from the np^2 configuration

$m = +1$	$m = 0$	$m = -1$	$S = \Sigma s_i$	$L = \Sigma m_l$	Label
↑	↑		1	1	*
↑		↑	1	0	*
	↑	↑	1	-1	*
↓	↓		-1	1	*
↓		↓	-1	0	*
	↓	↓	1	-1	*
↓	↑		0	1	*
↓		↑	0	0	*
	↓	↑	0	-1	*
↑↓			0	2	**
↑	↓		0	1	**
	↑↓		0	0	**
	↑	↓	0	-1	**
		↑↓	0	-2	**
↑		↓	0	0	***

15 microstates (shown in Table 5.3) that are possible for this configuration. It is apparent that the highest value of L is 2. This occurs with
$S=0$, but the vector $L=2$ can have *five* projections on the z-axis.
These will be represented by the series of M values of 2, 1, 0, −1,
and −2, which are the states designated in the table by **. These five
states constitute a ^1D term. Of the remaining microstates, the highest
M value is 1 and the highest S is 1. Therefore, the coupling of these
vectors must represent a ^3P state. Because there exists the possible
combinations of $M=1$, 0, and −1 with $S=1$, 0, and −1, there are
a total of nine microstates corresponding to the ^3P term. The resulting
J values are given by the series $|1+1|$, $|1+1-1|$, and $|1-1|$, so that
the J values are 2, 1, and 0. Thus, nine microstates designated by * in
the table have been used. One microstate remains, indicated by *** in
the table, which has $M=0$ and $S=0$, remains. This combination can
only correspond to the term $^1S_0$.
The terms associated with the np^2 configuration have been found to be $^3P_0$,
$^3P_1$, $^3P_2$, $^1D_2$, and $^1S_0$, with Hund's rules predicting exactly that order of
increasing energy. The energies of the spectroscopic states relative to the
$^3P_0$ ground state for atoms having the np^2 (where $n=2$, 3, 4, 5, or 6) configuration are shown in Table 5.4.

Table 5.4 Relative energy levels (in cm^{-1}) for terms arising from the np^2 configuration

State	Carbon	Silicon	Germanium	Tin	Lead
			Atom		
$^3P_0$	0.0	0.0	0.0	0.0	0.0
$^3P_1$	16.4	77.2	557.1	1691.8	7819.4
$^3P_2$	43.5	223.3	1409.9	3427.7	10,650.5
$^1D_2$	10,193.7	6298.8	7125.3	8613.0	21,457.9
$^1S_0$	21,648.4	15,394.2	16,367.1	17,162.6	29,466.8

In the lighter elements, such as carbon and silicon, the coupling scheme is of the L–S type. For elements as heavy as lead, the j–j coupling scheme is followed. Note that while all of the elements have a ground state of $^3P_0$, the singlet terms are much higher in energy for the heavier elements.

In many instances, only the ground state term must be determined. This can be done without going through the complete procedure just outlined for the np^2 configuration. For example, Hund's rules indicate that the state with the highest multiplicity will be lowest in energy. Consequently, to determine the ground state term, it is necessary only to consider the states where the sum of spins, S, is highest, a condition that results when the electrons are unpaired and have the same spin. For the np^2 configuration, the states are those where $S=1$ and $L=1$. This corresponds to a ^3P state, and the values permissible for J can then be worked out. This amounts to simply placing electrons in orbitals with the highest m values and working down while placing the electrons in the orbitals with the same spin. Therefore, for np^2, the microstate of lowest energy is

$$\frac{\uparrow}{+1} \quad \frac{\uparrow}{0} \quad \frac{}{-1}$$
$$m =$$

It can be seen immediately that $S=1$ and $L=1$, which leads directly to a multiplicity of 3, a P state, and J values of 2, 1, and 0. Thus, the ground state $^3P_0$ has been found for the np^2 configuration, but the states $^3P_1$ and $^3P_2$ also exist.

For the d^1 configuration, $S=\frac{1}{2}$ and the maximum M value is 2 because $l=2$ for a d state. Thus, the ground state term will be a ^2D term with a J value of 3/2. There are other terms corresponding to higher energies, but in this way the ground state term can be found easily.

Consider a d^2 configuration, where two possible arrangements for the electrons could be shown as follows:

$$\underset{m\,=\quad +2}{\uparrow} \quad \underset{+1}{\uparrow} \quad \underset{0}{} \quad \underset{-1}{} \quad \underset{-2}{} \qquad L=3,\,S=1$$

or

$$\underset{m\,=\quad +2}{\uparrow\downarrow} \quad \underset{+1}{} \quad \underset{0}{} \quad \underset{-1}{} \quad \underset{-2}{} \qquad L=4,\,S=0$$

Hund's rules indicate that the first of these will lie lower in energy (i.e., higher multiplicity). Consequently, the *maximum* value of Σm_i is 3 when the electrons are unpaired, so in the ground state the L vector is three units in length ($L=3$ corresponds to an F state). Therefore, the term lying lowest in energy is ^3F with a J value of 2, so the ground state term is $^3F_2$.

One further point should be mentioned: If a set of degenerate orbitals can hold x electrons, the terms arising from $(x-y)$ electrons in those orbitals are exactly the same as the terms arising for y electrons in those orbitals. This means that permuting a *vacancy* among the orbitals produces the same effect as permuting an electron among the orbitals. Thus, a p^4 configuration gives rise to the same spectroscopic ground state as does the p^2 configuration. Only the order of J values is inverted for the shell that is greater than half filled. Table 5.5 shows the terms arising from the various electron configurations, and the significance of the energies represented by the various microstates is shown in Fig. 5.4. The inclusion of electron repulsion separates the np^2 state into the various triplet and singlet states arising from that configuration.

Furthermore, the inclusion of spin-orbit interactions separates the ^3P state into as many components as there are J values. Finally, the application of a magnetic field removes the degeneracy within the $^3P_2$ and $^3P_1$ states to produce multiplets that are represented by the different orientations that are possible for the J vector. Figure 5.4 shows a summary of these states as separated by spin-orbit coupling and the effect of an external magnetic field.

Although no attempt was made to solve Schrödinger wave equations for complex atoms, this chapter shows the approaches taken for the helium atom.

Table 5.5 Spectroscopic terms arising for equivalent electrons[a]

Electron configuration	Spectroscopic states	Electron configuration	Spectroscopic states
s^1	^2S	d^2	$^3F, {}^3P, {}^1G,$ ${}^1D, {}^1S$
s^2	^1S	d^3	^4F (ground state)
p^1	^2P	d^4	^5D (ground state)
p^2	$^3P, {}^1D, {}^1S$	d^5	^6S (ground state)
p^3	$^4S, {}^2D, {}^2P$	d^6	^5D (ground state)
p^4	$^3P, {}^1D, {}^1S$	d^7	^4F (ground state)
p^5	^2P	d^8	^3F (ground state)
p^6	^1S	d^9	^2D
d^1	^2D	d^{10}	^1S

[a]The j values have been omitted for simplicity.

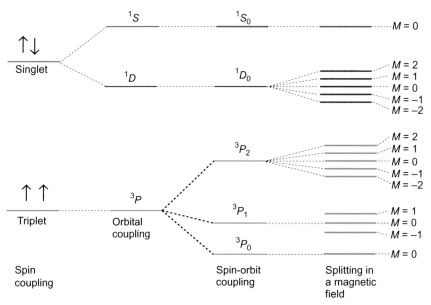

Fig. 5.4 Splitting of states due to spin-orbit coupling. There are $2J+1$ states in a magnetic field that correspond to projections of the J vector in the z direction.

Electron configurations of atoms have been shown, and the vector model of the atom that leads to coupling of angular momenta has been explained. For the reader interested in more details, this introduction should be supplemented by consultation of the references listed at the end of the book.

LITERATURE CITED

Pauling, L.; Wilson, E. B. *Introduction to Quantum Mechanics*; McGraw–Hill: New York, 1935.

PROBLEMS

1. Determine by means of vector diagrams the possible values that the total angular momentum can have for the following combinations:
   (a) $L=3$ and $S=5/2$
   (b) $L=2$ and $S=5/2$.

2. Use a vector diagram to show what total angular momenta that a single electron in an f state can have.

3. For each of the following, determine the ground state spectroscopic term and sketch the splitting pattern that would result for the atom in a magnetic field:
   (a) Al
   (b) P
   (c) Ca
   (d) Ti
   (e) Se

4. Write a set of four quantum numbers for the "last" electron in each of the following:
   (a) Sc
   (b) Cl
   (c) Sr
   (d) V
   (e) Co

5. Calculate the most probable radius of He^+ using $P(r) = 4\pi r\,[R(r)]^2 r^2$.

6. The ionization potential for the potassium atom is 4.341 eV. Estimate the effective atomic number (nuclear charge) Z' for the $4s$ electron in potassium.

7. Using Slater's rules, determine the effective nuclear charge for the fifth electron in the $2p$ level of fluorine. Write the wave function for this electron.

8. Determine the most probable distance for the electron in the $1s$ state of Be^{3+}.

9. Determine the spectroscopic ground state term for the following:
   (a) Ti^{3+}
   (b) Cr^{3+}
   (c) O^{2-}
   (d) Mn^{2+}
   (e) Ni^{2+}

10. Calculate $<r>$ for the electron in the $1s$ state of Li^{2+}. Use the wave function shown in Table 4.1.

CHAPTER 6

Vibrations and the Harmonic Oscillator

Most of what is known about the structure of atoms and molecules has been obtained by studying the interaction of electromagnetic radiation with matter or its emission from matter. The vibrations in molecular systems constitute one of the properties that provides a basis for studying molecular structure by spectroscopic techniques. *Infrared* spectroscopy provides the experimental technique for studying changes in vibrational states in molecules, a technique that is familiar to chemistry students even at a low level. However, molecular vibrations in gaseous molecules also involve changes in rotational states, so changes in these types of energy levels are sometimes considered together. In this chapter, some of the concepts required for an interpretation of molecular vibrations will be developed, and the application of spectroscopic experiments will be illustrated.

The discussion of molecular vibrations will begin by considering an object vibrating on a spring to illustrate some of the physical concepts and mathematical techniques. Solving problems related to vibrations requires the use of differential equations, but some individuals studying quantum mechanics for the first time may not yet have taken such a course or may need a review. Consequently, this chapter also includes a very limited coverage of this area of mathematics because it is so important in the physical sciences.

6.1 THE VIBRATING OBJECT

Some aspects of vibratory motion can be illustrated utilizing a simple problem in vibrational mechanics, illustrated by the arrangement shown in Fig. 6.1. For an object attached to a spring, Eq. (6.1), which is known as *Hooke's law*, describes the system in terms of the force (F) on the object and the displacement (x) from the equilibrium position:

$$F = -kx \tag{6.1}$$

In this equation, x is the distance the object is displaced from its *equilibrium* position, and k is known as the *spring constant* or *force constant*, which is

Fundamentals of Quantum Mechanics
http://dx.doi.org/10.1016/B978-0-12-809242-2.00006-1

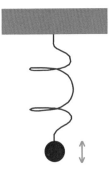

Fig. 6.1 An object vibrating on a spring.

expressed in dimensions of force/distance. Therefore, appropriate units for k are dynes cm^{-1}, $N\,m^{-1}$, or, in the case of chemical bonds, millidynes Angstrom^{-1} (mdyn $\overset{\circ}{A}^{-1}$). Both $N\,m^{-1}$ and mdyn $\overset{\circ}{A}^{-1}$ are in common use when describing force constants for chemical bonds. However, with 1 N being equal to 10^5 dyn, 1 m = 100 cm, and 1 Å being 10^{-8} cm, it works out that 1 mdyn $\overset{\circ}{A}^{-1}$ is equal to $100\,N\,m^{-1}$. Consequently, the force constant for the bond in H_2 is 5.75 mdyn $\overset{\circ}{A}^{-1}$ or $575\,N\,m^{-1}$. An empirical relationship between force constant and bond length is known as Badger's Rule (Badger, 1934).

The negative sign in Eq. (6.1) indicates that the restoring force (spring tension) is in the direction *opposite* to the displacement. The energy (or work) that is required to cause the displacement becomes the potential energy given to the object. This amount of energy is expressed by the force law integrated over the interval that the spring is stretched:

$$\int_0^x F(x)\,dx = \int_0^x -kx\,dx = \frac{1}{2}kx^2 \tag{6.2}$$

If the mass m is displaced by a distance of x and then released, the object vibrates in simple harmonic motion. The *angular frequency* of the vibration ω will be given by

$$\omega = \sqrt{\frac{k}{m}} \tag{6.3}$$

whereas the *classical* or *vibrational frequency* ν is given by

$$\nu = \frac{1}{2\pi}\sqrt{\frac{k}{m}} \tag{6.4}$$

It should be clear from Eqs. (6.3) and (6.4) that $\omega = 2\pi\nu$.

The maximum displacement from the equilibrium position is called the *amplitude,* and the variation of the displacement with time is found by making use of Newton's Second Law of Motion, $F = ma$. Velocity is the first derivative of distance with respect to time dx/dt, and acceleration is the derivative of velocity with time d^2x/dt^2. Therefore, in differential form, $F = ma$ is equivalent to

$$m\frac{d^2x}{dt^2} = -kx \qquad (6.5)$$

Rearrangement of this equation leads to

$$\frac{d^2x}{dt^2} + \frac{k}{m}x = 0 \qquad (6.6)$$

which is a linear differential equation with constant coefficients. Before progressing to the solution of this problem as it applies to vibrations, a brief discussion of the type of differential equation involved in the analysis will be presented.

6.2 LINEAR DIFFERENTIAL EQUATIONS WITH CONSTANT COEFFICIENTS

Because differential equations are required when representing motion, some familiarity with that branch of mathematics is beneficial when studying quantum mechanics. This section provides a very brief introduction to such mathematics. However, in this section, the results of several important theorems in differential equations will be presented. Because of the nature of this book, the theorems will be presented in an operational manner and used without proof. The interested reader should consult a text on differential equations for more details.

A linear differential equation with constant coefficients is an equation of the form

$$a_n(x)\frac{d^n y}{dx^n} + a_{n-1}(x)\frac{d^{n-1} y}{dx^{n-1}} + \cdots + a_1(x)\frac{dy}{dx} + a_0(x)y = F(x) \qquad (6.7)$$

in which the constants $a_0(x)$, $a_1(x)$, ..., and $F(x)$ have values that change only with changes in x. A particularly important equation of this type is the second-order case:

$$a_2(x)\frac{d^2 y}{dx^2} + a_1(x)\frac{dy}{dx} + a_0(x)y = F(x) \qquad (6.8)$$

The *differential operator D* is defined as

$$D=\frac{d}{dx}, \quad D^2 =\frac{d^2}{dx^2}, \text{ etc.} \qquad (6.9)$$

When an operator meets the conditions represented as

$$D(f+g)=Df+Dg \qquad (6.10)$$

and that which can be shown by the relationship

$$D^n(f+g)=D^nf+D^ng \qquad (6.11)$$

the operator is called a *linear operator*.

From Eq. (6.8), it can be seen that a second-order linear differential equation can be written in the operator notation as

$$a_2D^2y+a_1Dy+a_0=F(x) \qquad (6.12)$$

The solution of an equation of this form is obtained by considering an *auxiliary equation*, which is obtained by writing an equation in the form

$$f(D)y=0 \qquad (6.13)$$

when the general differential equation is written as

$$f(D)y=F(x) \qquad (6.14)$$

The auxiliary equation is called the *complementary equation*, and its solution is known as the *complementary solution*. The *complete solution* of the differential equation is the sum of the particular and general solution of the complementary equation. These principles will be illustrated by working through the following example:

Suppose it is necessary to find the general solution of

$$\frac{d^2y}{dx^2}-5\frac{dy}{dx}+4y=10x \qquad (6.15)$$

In operator form, this equation can be written as

$$\left(D^2-5D+4\right)y=10x \qquad (6.16)$$

A solution of this type of equation is frequently of the form

$$y=C_1e^{ax}+C_2e^{bx} \qquad (6.17)$$

in which a and b are to be determined by the solutions of the complementary equation,

$$m^2 - 5m + 4 = 0 \tag{6.18}$$

Therefore, factoring the polynomial gives

$$(m - 4)(m - 1) = 0 \tag{6.19}$$

from which it can be seen that $m = 4$ and $m = 1$. In this case, the general solution of Eq. (6.15) is

$$y = C_1 e^x + C_2 e^{4x} \tag{6.20}$$

It can easily be verified that this is the solution by using it in the complementary equation. If this is the solution, then

$$Dy = \frac{dy}{dx} = C_1 e^x + 4C_2 e^{4x}$$

and the second derivative can be written as

$$D^2 y = \frac{d^2 y}{dx^2} = C_1 e^x + 16 C_2 e^{4x}$$

Therefore, the auxiliary equation becomes

$$\left(D^2 - 5D + 4\right) y = D^2 y - 5Dy + 4y = 0 \tag{6.21}$$

By substituting for the first and second derivatives, it is found that

$$C_1 e^x + 16 C_2 e^{4x} - 5\left(C_1 e^x + 4C_2 e^{4x}\right) + 4\left(C_1 e^x + C_2 e^{4x}\right) = 0 \tag{6.22}$$

By simplification, it can be seen that this equation reduces to $0 = 0$. However, it can also be shown that a *particular* solution is

$$y = \frac{5}{2}x + \frac{25}{8} \tag{6.23}$$

It will now be shown that this particular solution also satisfies the general equation. From Eq. (6.23) it is found that

$$Dy = \frac{dy}{dx} = \frac{5}{2} \quad \text{and} \quad D^2 y = \frac{d^2 y}{dx^2} = 0 \tag{6.24}$$

When these values are substituted in Eq. (6.15) the result is

$$-5\left(\frac{5}{2}\right) + 4\left(\frac{5x}{2} + \frac{25}{8}\right) = 10x$$

$$10x = 10x$$

Therefore, the *complete* solution of Eq. (6.15) is the sum of the two expressions,

$$y = C_1 e^x + C_2 e^{4x} + \frac{5}{2}x + \frac{25}{8} \tag{6.25}$$

In most problems, it is sufficient to obtain a general solution, and "singular" solutions that do not describe the physical behavior of the system are ignored.

It must be mentioned that there are two arbitrary constants that characterize the solution that was obtained. Of course, the solution of an nth order equation would result in n constants. In quantum mechanics, these constants are determined by the physical constraints of the system (known as boundary conditions), as was shown in Chapter 3.

The equation

$$D^2 y + y = 0 \tag{6.26}$$

has the auxiliary equation that can be written as

$$m^2 + 1 = 0 \tag{6.27}$$

From this equation, it can be seen that $m^2 = -1$ and $m = \pm i$. Therefore, the general solution is

$$y = C_1 e^{ix} + C_2 e - e^{-ix} \tag{6.28}$$

At this point, it is useful to remember that

$$\frac{d}{dx}(\sin x) = \cos x \tag{6.29}$$

and

$$\frac{d}{dx}(\cos x) = -\sin x = \frac{d^2}{dx^2}(\sin x) \tag{6.30}$$

From these relationships, it is evident that

$$\frac{d^2}{dx}(\sin x) + \sin x = 0 \tag{6.31}$$

and the solution $y = \sin x$ satisfies the equation. In fact, if it is assumed that a solution to Eq. (6.26) is of the form

$$y = A \sin x + B \cos x \tag{6.32}$$

then

$$Dy = A \cos x - B \sin x \tag{6.33}$$

and

$$D^2 y = -A \sin x - B \cos x \tag{6.34}$$

Therefore,

$$D^2 y + y = -A \sin x - B \cos x + A \sin x + B \cos x = 0 \tag{6.35}$$

Therefore, Eq. (6.35) shows the solution represented by Eq. (6.32) satisfies Eq. (6.26). A differential equation can have only one general solution, so the solution shown in Eqs. (6.28) and (6.32) must be equal:

$$y = C_1 e^{ix} + C_2 e^{-ix} = A \sin x + B \cos x \tag{6.36}$$

It can be seen that when $x=0$, $C_1 + C_2 = B$. Differentiating Eq. (6.36) gives

$$\frac{dy}{dx} = C_1 i e^{ix} - C_2 i e^{-ix} = A \cos x - B \sin x \tag{6.37}$$

However, when $x=0$, $\sin x=0$. It is apparent, then, that

$$i(C_1 - C_2) = A \tag{6.38}$$

By substituting for A and B and simplifying, the result can be shown as

$$C_1 e^{ix} + C_2 e^{-ix} = C_1(\cos x + i \sin x) + C_2(\cos x - i \sin x) \tag{6.39}$$

If $C_2 = 0$ and $C_1 = 1$, then

$$e^{ix} = \cos x + i \sin x \tag{6.40}$$

Therefore, if $C_2 = 1$ and $C_1 = 0$, the result can be shown as

$$e^{-ix} = \cos x - i \sin x \tag{6.41}$$

The relationships shown in Eqs. (6.40) and (6.41) are known as *Euler's formulas*.

Suppose it is necessary to solve the differential equation

$$y'' + 2y' + 5y = 0 \tag{6.42}$$

When written in operator form the result is

$$\left(D^2 + 2D + 5\right)y = 0 \tag{6.43}$$

Therefore, the auxiliary equation is written as

$$m^2 + 2m + 5 = 0 \tag{6.44}$$

and its roots are found by using the quadratic formula,

$$m = \frac{-2 \pm \sqrt{4-20}}{2} = -1 \pm 2i \qquad (6.45)$$

Therefore, the solution of Eq. (6.42) is

$$y = C_1 e^{(-1+2i)x} + C_2 e^{(-1-2i)x} \qquad (6.46)$$

By expanding the exponentials, this equation can also be written as

$$y = C_1 e^{-x} e^{2ix} + C_2 e^{-x} e^{-2ix} = e^{-x}\left(C_1 e^{2ix} + C_2 e^{-2ix}\right) \qquad (6.47)$$

Using Euler's formulas to express the exponential functions in terms of sin and cos, the result is

$$y = e^{-x}\left(A \sin 2x + B \cos 2x\right) \qquad (6.48)$$

In general, if the auxiliary equation has roots $a \pm bi$, the solution of the differential equation has the form

$$y = e^{ax}\left(A \sin bx + B \cos bx\right) \qquad (6.49)$$

An equation having the form

$$y'' + a^2 y = 0 \qquad (6.50)$$

was obtained in solving the particle in the one-dimensional box problem in quantum mechanics (see Chapter 3). The auxiliary equation is

$$m^2 + a^2 = 0 \qquad (6.51)$$

and it has the solutions $m = ia$. Therefore, the solution of Eq. (6.50) can be written as

$$y = C_1 e^{aix} - C_2 e^{-aix} = A \cos ax + B \sin ax \qquad (6.52)$$

This is exactly the form of the solution found when solving the particle in a one-dimensional box. The boundary conditions of a particular system make it possible to evaluate the constants in the solution, as was shown in Chapter 3.

6.3 BACK TO THE VIBRATING OBJECT

The discussion will now return to the vibrating system described in Section 6.1. Suppose a force of 6.0 N stretches the spring 0.375 m. It is customary to take the displacement of the vibrating object be negative so that the spring constant k can be represented as

$$k = -\frac{f}{x} = -\frac{6.0\,\text{N}}{-0.375\,\text{m}} = 16\,\text{N/m} \tag{6.53}$$

For purposes of illustration, it will be assumed that the object has a mass of 4.00 kg, and it is raised 0.375 m above its equilibrium position and released, as shown in Fig. 6.2. As shown earlier, the motion of the object is described by the equation

$$\frac{d^2x}{dt^2} + \frac{k}{m}x = 0 \tag{6.54}$$

By using the data specified in this case and the condition $x(0) = 0.375$, the equation of motion can be written as

$$\frac{d^2x}{dt^2} + \frac{16}{4}x = 0 \tag{6.55}$$

Therefore, the auxiliary equation can be written as

$$m^2 + 4 = 0 \tag{6.56}$$

from which it apparent that $m^2 = -4$ and $m = 2i$. Therefore, the general solution of Eq. (6.54) can be written as

$$x = C_1 e^{2it} + C_2 e^{-2it} = A \sin 2t + B \cos 2t \tag{6.57}$$

At the beginning of the motion, $t = 0$ and $x = 0.375$ m, and the velocity of the object is 0 or $v = dx/dt = 0$. Therefore, Eq. (6.57) can be written as

$$x = A \sin 2t + B \cos 2t = 0.375\,\text{m} \tag{6.58}$$

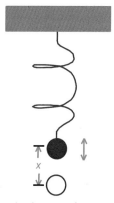

Fig. 6.2 An object vibrating after displacing it from its rest or equilibrium position (the open circle).

However, because $\sin 0 = 0$ and $\cos 0 = 1$, Eq. (6.58) reduces to

$$0.375 = B \cos 0 = B \qquad (6.59)$$

Consequently, $B = 0.375$, and the partial solution is

$$x = 0.375 \cos 2t + A \sin 2t \qquad (6.60)$$

Taking the derivative dx/dt and setting it equal to 0 yields

$$\frac{dx}{dt} = -0.375(2) \sin 2t + 2A \cos 2t = 0 \qquad (6.61)$$

At the time when $t = 0$, the $\sin \theta$ term goes to 0 ($\sin 2t = 0$), so the result is

$$2A \cos 2t = 2A \cos 0 = 2A \times 1 = 0 \qquad (6.62)$$

Therefore, A must be equal to 0. The required solution giving the displacement as a function of time is

$$x = 0.375 \cos 2t \qquad (6.63)$$

Figure 6.3 shows the graphical nature of this solution with the displacement of the object being 0.375 m at $t = 0$, then varying as a cosine function thereafter.

The *period* of the vibration is the time necessary for one complete vibration (or "cycle") whereas the frequency is given by

$$\nu = \frac{1}{2\pi} \sqrt{\frac{k}{m}} = \frac{1}{2\pi} \sqrt{\frac{16(\mathrm{kgms^{-2}})/m}{4\mathrm{kg}}} = \frac{1}{\pi} \mathrm{s}^{-1}$$

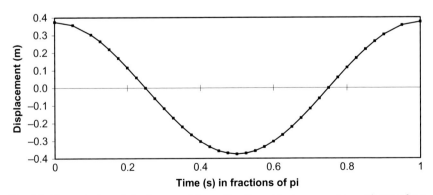

Fig. 6.3 The variation of displacement for a vibrating object on a spring with time for a complete cycle.

The vibrating object treated by classical mechanics serves to introduce the terminology and techniques used for the quantum mechanical oscillator. The latter is a much more complex problem, which will now be considered.

6.4 THE QUANTUM MECHANICAL HARMONIC OSCILLATOR

One of the very useful models in quantum mechanics is that known as the harmonic oscillator. This model provides the basis for discussing vibrating chemical bonds (discussed further in Chapter 7), so it is a necessary part of the discussion of spectroscopy. This section is devoted to discussing this important quantum mechanical model.

Earlier, it was shown that for a vibrating object, the potential energy is given by

$$V = \frac{1}{2}kx^2 \tag{6.64}$$

but it can also be written as

$$V = \frac{1}{2}mx^2\omega^2 \tag{6.65}$$

where ω is the angular frequency of vibration $(k/m)^{1/2}$. The classical vibrational frequency ν is $(1/2\pi)(k/m)^{1/2}$. From these definitions, it can be seen that $\omega = 2\pi\nu$. The total energy of the oscillator is the sum of the potential and kinetic energies. In order to write the Schrödinger equation, it is necessary to find the form of the Hamiltonian operator. The kinetic energy T can be written in operator form as

$$\hat{T} = -\frac{\hbar^2}{2m}\frac{d^2}{dx^2} \tag{6.66}$$

The potential energy is $2\pi^2\nu^2mx^2$, so the Hamiltonian operator can be written as

$$\hat{H} = -\frac{\hbar^2}{2m}\frac{d^2}{dx^2} + 2\pi^2\nu^2mx^2 \tag{6.67}$$

If a substitution is made such that $b = 2\pi\nu m/\hbar$, then the general operator equation

$$\hat{H}\psi = E\psi$$

can be written explicitly as

$$\frac{d^2\psi}{dx^2} + \left(\frac{2mE}{\hbar^2} - b^2x^2\right)\psi = 0 \tag{6.68}$$

In this equation, the potential energy varies with x^2, and because it is a nonlinear function, this equation is much more complex than those that describe the classical harmonic oscillator or the particle in the one-dimensional box. Inspection of the wave equation shows that the solution must be a function such that its second derivative contains both the original function *and* a factor of x^2. A function such as $\exp(-bx^2)$ satisfies that requirement. In fact, it will be shown later that the solution can actually be written as

$$\psi = c\left[\exp\left(-bx^2\right)\right] \tag{6.69}$$

where b and c are constants. Furthermore, it is possible to show that this solution satisfies Eq. (6.68). The solution of the equation by rigorous means requires that it be solved by a method using an infinite series. Before applying this technique to an equation of the complexity of Eq. (6.68), the solution of differential equations by this technique will be illustrated for some simple cases.

6.5 SERIES SOLUTIONS OF DIFFERENTIAL EQUATIONS

A large number of problems in science and engineering are formulated in terms of differential equations that have variable coefficients. The quantum mechanical treatment of the harmonic oscillator is one such problem, but another such equation was encountered in the solution of the wave equation for the hydrogen atom that was described in Chapter 4. At that time, the solutions were simply stated without any indication of how such equations must be solved. Some of the most famous equations of this type are shown in Table 6.1. In these equations, n or ν represents a constant.

The equations shown in Table 6.1 are differential equations for which the solutions are given as infinite series or polynomial solutions, usually bearing the name of the person who solved the problem. It is not intended to solve all of these equations, but they are given to show some of the "name" differential equations that will be encountered in a more advanced study of quantum mechanics. They constitute some of the most important

Table 6.1 Some important nonlinear differential equations that are solved by a series technique

Name	Equation	Solutions
Hermite	$y'' - 2xy' + 2ny = 0$	Hermite polynomials
Bessel	$y'' + \dfrac{1}{x}y' + \left(1 - \dfrac{v^2}{x^2}\right)y = 0$	Bessel functions
Legendre	$y'' - \dfrac{2x}{1-x^2}y' + \dfrac{v(v+1)}{1-x^2}y = 0$	Legendre polynomials
Chebyshev	$y'' - \dfrac{x}{1-x^2}y' + \dfrac{n^2}{1-x^2}y = 0$	Chebyshev polynomials
Laguerre	$x^2 y'' + (1-x)y' + ny = 0$	Laguerre polynomials

differential equations to be encountered in theoretical work. It should be mentioned that these equations can be written in other forms, so they may not be readily recognized. One of the commonly used techniques in quantum mechanics is that of manipulating an equation to get it in a recognizable form, as was done in Chapter 4.

Because the solution of the harmonic oscillator problem will be given in somewhat greater detail than was given in treating the hydrogen atom problem in Chapter 4, the discussion will begin by illustrating the solution of differential equations by means of a series. This is done for the reader whose mathematics background includes calculus, but not differential equations. For a more complete discussion of this technique, see the references at the end of the book. Some of the differential equations that have already been solved have had solutions that were trigonometric or exponential functions. Because these functions can also be written as series, the solutions of such equations could have been written in the form of series.

Consider the differential equation

$$\frac{dy}{dx} = y \tag{6.70}$$

with the boundary conditions such that $y = 1$ at $x = 0$. This equation can be solved by inspection because the only function that equals its first derivative is e^x. Also, Eq. (6.70) can be written in the form

$$\frac{dy}{dx} - y = 0 \tag{6.71}$$

From Eq. (6.71) the auxiliary equation can be written as

$$m - 1 = 0 \qquad (6.72)$$

From this equation, it can be seen that $m=1$, and by using the techniques illustrated in Section 6.2, the solution can be written directly as e^x.

Suppose the previous approaches were not taken but rather that a solution was *assumed*, such that it can be represented by the series

$$y = a_0 + a_1 x + a_2 x^2 + a_3 x^3 + a_4 x^4 + \cdots \qquad (6.73)$$

Substitution of this series for y in Eq. (6.70) gives

$$\frac{d}{dx}\left(a_0 + a_1 x + a_2 x^2 + a_3 x^3 + \cdots\right) = a_0 + a_1 x + a_2 x^2 + a_3 x^3 \cdots \qquad (6.74)$$

Therefore, by setting $dy/dx = y$, it is possible to write

$$a_1 + 2a_2 x + 3a_3 x^2 \cdots = a_0 + a_1 x + a_2 x^2 + a_3 x^3 + \cdots \qquad (6.75)$$

which is true if coefficients of like powers of x are equal. Assuming that this is so, the following relationships are generated:

$$a_0 = a_1 \qquad a_1 = 2a_2 \qquad a_2 = 3a_3 \qquad a_3 = 4a_4$$

$$a_2 = \frac{a_1}{2} = \frac{a_0}{2!} \qquad a_3 = \frac{a_2}{3} = \frac{a_0}{6} = \frac{a_0}{3!} \qquad a_4 = \frac{a_3}{4} = \frac{a_0}{4!}$$

Therefore, because the assumed solution has the form

$$y = a_0 + a_1 x + a_2 x^2 + a_3 x^3 + a_4 x^4 + \cdots \qquad (6.76)$$

substituting the coefficients found earlier and factoring out a_0 gives

$$y = a_0\left(1 + x + \frac{x^2}{2!} + \frac{x^3}{3!} + \frac{x^4}{4!} + \cdots\right) \qquad (6.77)$$

From the initial condition that $y=1$ when $x=0$, it can be seen that $y=a_0=1$. Therefore, $a_0=1$ and the required solution is

$$y = 1 + x + \frac{x^2}{2!} + \frac{x^3}{3!} + \frac{x^4}{4!} + \cdots \qquad (6.78)$$

It is interesting to note that when written in series form, e^x is given by

$$e^x = 1 + x + \frac{x^2}{2!} + \frac{x^3}{3!} + \frac{x^4}{4!} + \cdots \qquad (6.79)$$

and this agrees with the solution given earlier. Of course, it is not always this simple, but it is reassuring to see that the solution obtained from the series is exactly the same as that already known from other methods.

In order to provide another less obvious illustration of the method, the equation

$$\frac{dy}{dx} = xy \text{ with } y(0) = 2 \tag{6.80}$$

will be solved by the series approach. Assuming as before that y can be expressed by the series, the result is

$$y = a_0 + a_1 x + a_2 x^2 + a_3 x^3 + a_4 x^4 + \cdots \tag{6.81}$$

Taking the derivative dy/dx gives

$$\frac{dy}{dx} = a_1 + 2a_2 x + 3a_3 x^2 + 4a_4 x^3 + 5a_5 x^4 + 6a_6 x^5 + 7a_7 x^6 + \cdots \tag{6.82}$$

The product xy can be written as

$$xy = a_0 x + a_1 x^2 + a_2 x^3 + a_3 x^4 + a_4 x^5 + a_5 x^6 + 6a_6 x^7 + \cdots \tag{6.83}$$

Therefore, from the original equation, Eq. (6.80), it can be seen that

$$a_1 + 2a_2 x + 3a_3 x^2 + 4a_4 x^3 + \cdots = a_0 x + a_1 x^2 + a_2 x^3 + a_3 x^4 + a_4 x^5 + \cdots \tag{6.84}$$

By equating coefficients of like powers of x, the following relationships are obtained:

$$2a_2 = a_0 \quad 3a_3 = a_1 \quad 4a_4 = a_2 \qquad 5a_5 = a_3 \quad 6a_6 = a_4$$

$$a_2 = \frac{a_0}{2} \quad a_3 = 0 \quad a_4 = \frac{a_2}{4} = \frac{a_0}{2 \cdot 4} \quad a_5 = 0 \quad a_6 = \frac{a_4}{6} = \frac{a_0}{2 \cdot 4 \cdot 6} = \frac{a_0}{2^3 \cdot 3!}$$

From the condition that $y(0) = 2$, it can be seen that $a_0 = 2$, and that a_1 must be 0 because there is no term with a corresponding power of x in Eq. (6.84). Therefore, $a^3 = a^5 = 0$ and substituting the preceding values for the coefficients gives

$$y = 2 + 0x + \frac{a_0}{2} x^2 + 0x^3 + \frac{a_0}{2 \cdot 4} x^4 + 0x^5 + \frac{a_0}{2 \cdot 4 \cdot 6} x^6 + \cdots \tag{6.85}$$

Because $a_0 = 2$, substituting for a_0 and factoring out 2 gives

$$y = 2\left(1 + \frac{x^2}{2!} + \frac{x^4}{2^3} + \frac{x^6}{2^3 \cdot 3!} + \cdots\right) = 2\left[\exp\left(\frac{x^2}{2}\right)\right] \tag{6.86}$$

The preceding discussion is intended to show how series solutions are obtained for relatively simple differential equations. The equations that arise in the quantum mechanical treatment of problems are often more complex than those used in the illustrations, but the brief introduction is sufficient to remove some of the mystery of using series in this way. It is not expected that the reader will know how to solve many of the complex equations of mathematical physics, but the approach will now be familiar, even if the details are not.

6.6 BACK TO THE HARMONIC OSCILLATOR

In the previous sections, illustrations have been given to show how to solve some of the simple problems dealing with vibrations and the differential equations that describe them. As will now be shown, the treatment of the harmonic oscillator in quantum mechanics is a quite different problem. As was shown earlier, the harmonic oscillator can be described in a relatively simple fashion (see Fig. 6.4).

Using Hooke's law, the restoring force is written as

$$F = -kx \tag{6.87}$$

and the potential energy V is given by the relationship

$$V = \frac{1}{2}kx^2 \tag{6.88}$$

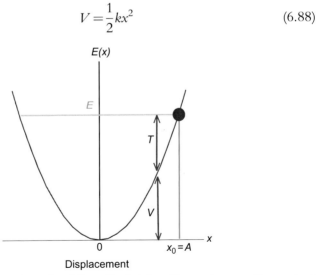

Fig. 6.4 A harmonic oscillator that follows Hooke's law. The total energy E is constant, but the potential and kinetic energies vary with displacement of the object from its equilibrium position so that $E = T + V$.

The angular frequency of vibration is $\omega = (k/m)^{1/2}$, so $k = m\omega^2$. By substitution, it is found that

$$V = \frac{1}{2}m\omega^2 x^2 \tag{6.89}$$

For this model, it is apparent that the total energy E is the sum of the kinetic and potential energies:

$$E = T + V \tag{6.90}$$

At the equilibrium position of the vibration the spring is not stretched, so $V = 0$ and T is a maximum, with the result being that the total energy is $E = T$. At the extremes of the vibration, the oscillator comes to rest for an instant and the kinetic energy is 0 ($T = 0$) before it changes direction. Therefore, at those positions the total energy is the potential energy, ($E = V$). It is near the extremes of the vibration where the velocity is low that the oscillator spends the majority of its time. As a result, near the equilibrium position where the velocity is the highest, the probability of finding the oscillator is lowest. As will be discussed later, the probability of finding the oscillator at any point along its motion is inversely proportional to the velocity at that point.

The *total* energy of the oscillator can be represented in terms of the amplitude A by the relationship

$$E = \frac{1}{2}kA^2 = \frac{1}{2}m\omega^2 A^2 \tag{6.91}$$

The *potential* energy can be expressed as

$$V = \frac{1}{2}m\omega^2 x^2 \tag{6.92}$$

As in the formulation of other problems in quantum mechanics, the Schrödinger equation can be written in general form as

$$\hat{H}\psi = E\psi \tag{6.93}$$

In this case, the Hamiltonian operator can be written as

$$\hat{H} = -\frac{\hbar^2}{2m}\frac{\partial^2}{\partial x^2} + V = -\frac{\hbar^2}{2m}\frac{\partial^2}{\partial x^2} + \frac{1}{2}m\omega^2 x^2 \tag{6.94}$$

Therefore, substituting this result into Eq. (6.93) gives

$$\left(-\frac{\hbar^2}{2m}\frac{\partial^2}{\partial x^2} + \frac{1}{2}m\omega^2 x^2\right)\psi = E\psi \tag{6.95}$$

Simplifying this equation by multiplying by $-2m$ and then dividing by $\hbar^2$ gives

$$\left(\frac{\partial^2}{\partial x^2} - \frac{m^2\omega^2 x^2}{\hbar^2}\right)\psi = -\frac{2mE}{\hbar^2}\psi \tag{6.96}$$

A rearrangement of this equation leads to

$$\frac{\partial^2\psi}{\partial x^2} = -\left(\frac{2mE}{\hbar^2} - \frac{m^2\omega^2 x^2}{\hbar^2}\right)\psi \tag{6.97}$$

This is *not* a linear differential equation, and there will be greater difficulty obtaining a solution. If it is *assumed* for the moment that the solution has the form

$$\psi = c\left[\exp\left(-bx^2\right)\right] \tag{6.98}$$

where b and c are constants, this solution can be verified by substituting for ψ in Eq. (6.97). This is accomplished by taking the required first and second derivatives,

$$\frac{\partial\psi}{\partial x} = -2bxc\left[\exp\left(-bx^2\right)\right] \tag{6.99}$$

$$\frac{\partial^2\psi}{\partial x^2} = -2bc\left[\exp\left(-bx^2\right)\right] + 4b^2cx^2\left[\exp\left(-bx^2\right)\right] \tag{6.100}$$

By working with the right-hand side of Eq. (6.97), it is found that

$$-\left(\frac{2mE}{\hbar^2} - \frac{m^2\omega^2}{\hbar^2}x^2\right)\psi = -\frac{2mE}{\hbar^2}c\left[\exp\left(-bx^2\right)\right] + \frac{m^2\omega^2}{\hbar^2}x^2c\left[\exp\left(-bx^2\right)\right] \tag{6.101}$$

It should be noted that both Eqs. (6.100) and (6.101) contain terms in x^2 and terms that do not contain x, except in the exponential. Therefore, the terms that contain x^2 can be set equal to give

$$\frac{m^2\omega^2}{\hbar^2}x^2c\left[\exp\left(-bx^2\right)\right] = 4b^2cx^2\left[\exp\left(-bx^2\right)\right] \tag{6.102}$$

By canceling factors that are common to both sides of this equation, the result is

$$4b^2 = \frac{m^2\omega^2}{\hbar^2} \tag{6.103}$$

Solving this equation for b gives

$$b = \frac{m\omega}{2\hbar} \tag{6.104}$$

By working with the terms that do not contain x as a factor, it is found that

$$E = b\frac{\hbar^2}{m} \tag{6.105}$$

When the value of b shown in Eq. (6.104) is substituted in this result, the result is

$$E = \frac{1}{2}\omega\hbar \tag{6.106}$$

Therefore, when $b = m\omega/2\hbar$ and $E = \omega\hbar/2$, the function

$$\psi = c\left[\exp\left(-bx^2\right)\right] \tag{6.107}$$

satisfies the Schrödinger equation for a harmonic oscillator. Using the value obtained for b, the solution can be written as

$$\psi = c\left[\exp\left(\frac{-m\omega x^2}{2\hbar}\right)\right] \tag{6.108}$$

This is, in fact, the solution for the harmonic oscillator in its lowest energy state. Although it has been *assumed* that the solution has this form, it is now necessary to show how the problem is solved.

The solution of the harmonic oscillator problem will now be addressed, starting with the wave equation, which can be written as

$$\frac{d^2\psi}{dx^2} + \frac{2m}{\hbar^2}\left(E - \frac{1}{2}kx^2\right)\psi = 0 \tag{6.109}$$

Substitutions are now made such that

$$\alpha = \frac{2mE}{\hbar^2} \tag{6.110}$$

and

$$\beta = \frac{2\pi(mk)^{1/2}}{h} \tag{6.111}$$

Therefore, Eq. (6.109) can be written in the form

$$\frac{d^2\psi}{dx^2} + \left(\alpha - \beta^2 x^2\right)\psi = 0 \tag{6.112}$$

The usual procedure employed at this point is to introduce a change in variable such that

$$z = \sqrt{\beta}x \tag{6.113}$$

After this is done, the second derivatives are related by the equation

$$\frac{d^2}{dx^2} = \beta \frac{d^2}{dz^2} \tag{6.114}$$

The wave equation can now be written as

$$\beta \frac{d^2\psi}{dz^2} + \left(\alpha - \beta^2 x^2\right)\psi = 0 \tag{6.115}$$

and rearrangement leads to

$$\frac{d^2\psi}{dz^2} + \left(\frac{\alpha}{\beta} - \beta x^2\right)\psi = 0 \tag{6.116}$$

Therefore, given that $z^2 = \beta x^2$, this leads to the equation

$$\frac{d^2\psi}{dz^2} + \left(\frac{\alpha}{\beta} - z^2\right)\psi = 0 \tag{6.117}$$

If the solution is now expressed as a function of z, the result is

$$\psi(z) = u(z)\exp\left(-\frac{z^2}{2}\right) \tag{6.118}$$

Using the simplified notation that $\psi = \psi(z)$ and $u = u(z)$, the necessary derivatives are represented by the equations

$$\psi' = u'\exp\left(-\frac{z^2}{2}\right) - uz\exp\left(-\frac{z^2}{2}\right)$$

and

$$\psi'' = u''\exp\left(-\frac{z^2}{2}\right) - u'z\exp\left(-\frac{z^2}{2}\right) - u'z\exp\left(-\frac{z^2}{2}\right)$$

$$- u\exp\left(-\frac{z^2}{2}\right) + uz^2\exp\left(-\frac{z^2}{2}\right) \tag{6.119}$$

Simplifying Eq. (6.119) gives

$$\psi'' = u'' \exp\left(-\frac{z^2}{2}\right) - 2u'z \exp\left(-\frac{z^2}{2}\right) - u \exp\left(-\frac{z^2}{2}\right) + uz^2 \exp\left(-\frac{z^2}{2}\right)$$

(6.120)

By substituting Eqs. (6.118) and (6.120) in Eq. (6.117), the result is

$$\frac{d^2u}{dz^2} - 2z\frac{du}{dz} + \left(\frac{\alpha}{\beta} - 1\right)u = 0$$

(6.121)

If the factor $[(\alpha/\beta) - 1]$ is represented as $2n$, Eq. (6.121) becomes

$$\frac{d^2u}{dz^2} - 2z\frac{du}{dz} + 2nu = 0$$

(6.122)

This equation has exactly the form of *Hermite's equation*, as shown in Table 6.1.

Before considering the solution of Eq. (6.122) by means of a series, the energy levels for the harmonic oscillator will be illustrated. By making use of the relationship

$$\frac{\alpha}{\beta} - 1 = 2n$$

(6.123)

the expression for α/β can be shown as

$$\frac{\alpha}{\beta} = 2n + 1$$

(6.124)

However, by making use of Eqs. (6.110) and (6.111) it can be shown that

$$\frac{\alpha}{\beta} = \frac{8\pi mE}{2h\sqrt{mk}} = \frac{4\pi E\sqrt{m}}{h\sqrt{k}}$$

(6.125)

Therefore, by equating the right hand sides of Eqs. (6.124) and (6.125), it can be seen that

$$2n + 1 = \frac{4\pi E\sqrt{m}}{h\sqrt{k}}$$

(6.126)

Solving this equation for the energy, the result is

$$E = \frac{h}{2\pi}\sqrt{\frac{k}{m}}\left(n + \frac{1}{2}\right) = \hbar\sqrt{\frac{k}{m}}\left(n + \frac{1}{2}\right)$$

(6.127)

Because the frequency of vibration is $\nu = (1/2\pi)(k/m)^{1/2}$ and $\omega = 2\pi\nu$, it is possible to express the energy as

$$E = \hbar\omega\left(n + \frac{1}{2}\right) \tag{6.128}$$

This equation is analogous to Eq. (6.106), which applied only to the ground state. The fact that the energy levels of the harmonic oscillator are *quantized* arises from the restrictions on the nature of Hermite's equation.

Although the quantum number has been represented by the integer n, the *vibrational* quantum number is usually designated as V. Therefore, the vibrational energy levels of the harmonic oscillator can be expressed in terms of the quantum number V by the equation

$$E = \left(V + \frac{1}{2}\right)\hbar\omega \tag{6.129}$$

Application of this equation results in a series of energy levels, as shown in Fig. 6.5. The spacing between the energy levels is $\hbar\omega$, and there is a zero-point energy at $(1/2)\,\hbar\omega$. In 1900, Planck's treatment of blackbody radiation (see Chapter 1) predicted the same arrangement of energy levels.

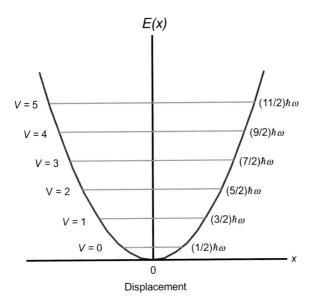

Fig. 6.5 The quantized energy levels of the harmonic oscillator. The energy when $V = 0$ is known as the *zero-point energy*.

Almost 30 years later the quantum mechanical solution of the harmonic oscillator problem gave rise to the same result.

The solution of Hermite's equation by a polynomial series will now be briefly addressed. Because of the nature of this equation and the complexity of its solution, an outline of the methods used will be provided. An advanced book on differential equations should be consulted for details of the solution of this type of equation.

The Hermite equation is written as

$$\frac{d^2u}{dz^2} - 2z\frac{du}{dz} - 2pz = 0 \qquad (6.130)$$

where p is an integer. It will be assumed that the solution can be expressed in a series that can be written as

$$H(z) = a_0 + a_1 z + a_2 z^2 + \cdots = \sum_{p=0}^{\infty} a_p z^p \qquad (6.131)$$

The required first and second derivatives are found to be

$$H'(z) = a_1 + 2a_2 z + 3a_3 z^2 \cdots = \sum_{p=0}^{\infty} p a_p z^{p-1} \qquad (6.132)$$

and

$$H''(z) = 2a_2 + 6a_3 z + 12a_4 z^2 + \cdots = \sum_{p=0}^{\infty} p(p-1) a_p z^{p-2} \qquad (6.133)$$

The terms involving a_0 and a_1 do not occur in the summation for H''. Therefore, the series can be written as

$$H''(z) = \sum_{p=0}^{\infty} (p+1)(p+2) a_{p+2} z^p \qquad (6.134)$$

Using these results, the Hermite equation can now be written as

$$\sum_{p=0}^{\infty} \left[(p+1)(p+2) a_{p+2} + (2n - 2p) a_p \right] z^p = 0 \qquad (6.135)$$

For this equation to be true for all values of z, the function in brackets must be zero:

$$\left[(p+1)(p+2) a_{p+2} + (2n - 2p) a_p \right] = 0 \qquad (6.136)$$

Therefore, solving for a_{p+2} yields

$$a_{p+2} = \frac{-(2n-2p)}{(p+1)(p+2)} a_p \qquad (6.137)$$

This is the *recursion formula* for the coefficients of the series to be calculated. Using this formula, the following relationships are obtained.

$$\text{for } p=0, \quad a_{p+2}=a_2 = -\frac{2(n-0)}{(1)(2)}a_0 = -na_0$$

$$\text{for } p=1, \quad a_{p+2}=a_3 = -\frac{n-1}{3}a_1$$

$$\text{for } p=2, \quad a_{p+2}=a_4 = -\frac{[2n-2(2)]}{(3)(4)}a_2 = -\frac{n-2}{6}a_2 = \frac{n(n-2)}{6}a_0$$

$$\text{for } p=3, \quad a_{p+2}=a_5 = -\frac{n-3}{10}a_0 = \frac{(n-1)(n-3)}{30}a_1$$

Two constants, a_0 and a_1, are not given by the recursion relation. As seen earlier, these are the two arbitrary constants that result from the solution of a second-order differential equation.

The next step in the solution is to show that the series can be written in terms of $\exp(-z^2)$ and that appropriate values can be assigned to the constants a_0 and a_1, resulting in a well-behaved wave function. The details of that rather tedious process will not be shown because it is more appropriately covered in advanced texts. However, the Hermite polynomials can be written in general form as

$$H_n(z) = (-1)^n \exp(z^2) \frac{d^n}{dz^n} \exp(-z^2) \qquad (6.138)$$

As a result, the first few Hermite polynomials can be written as

$$\begin{aligned}
H_0(z) &= 1 \\
H_1(z) &= 2z \\
H_2(z) &= 4z^2 - 2 \\
H_3(z) &= 8z^3 - 12z \\
H_4(z) &= 16z^4 - 48z^2 + 12 \\
H_5(z) &= 32z^5 - 160z^3 + 120z
\end{aligned} \qquad (6.139)$$

The wave functions for the harmonic oscillator, ψ_i, are written as a normalization constant, N_i, times $H_i(z)$ to give

$$\psi_0 = N_0 \exp\left(-z^2\right)$$

$$\psi_1 = N_1 (2z) \exp\left(-z^2\right)$$

$$\psi_2 = N_2 \left(4z^2 - 2\right) \exp\left(-z^2\right)$$

$$\psi_3 = N_3 \left(8z^3 - 12z\right) \exp\left(-z^2\right)$$

(6.140)

In this section, the mathematical procedures have been outlined that are necessary in order to obtain the full solution of the harmonic oscillator model using quantum mechanical methods. The details are not really required for a discussion at this level, but the general approach should be appreciated from the brief introduction to the solution of differential equations by series. Graphs of the first three wave functions and their squares are shown in Fig. 6.6. The squares of the wave functions, which are proportional to the probability density, show that the oscillator is not restricted to the classical limits of the vibration. For example, the plot of ψ^2 for the $V=0$ state shows that there is a slight but finite probability that the oscillator can be found beyond the classical limit (represented on the graphs as ± 2), the range of the vibration. For states with $V>0$, the probability of the oscillator tunneling through the classical limit is even greater (see Chapter 13).

It is interesting to note that as the position of a classical oscillator changes, its velocity continuously changes and reaches zero at the extremes of the vibration. The velocity has a maximum value at the center of the vibration (equilibrium position). Therefore, the time spent by the oscillator (and the probability of finding it) varies with position and has a minimum at the equilibrium distance.

The relationship between probability and displacement can be analyzed for a classical oscillator in the following way. Reference to Fig. 6.4 shows that the potential and kinetic energies vary during the oscillation whereas the total energy is constant. The total energy is

$$E = \frac{1}{2} kx_0^2$$

(6.141)

whereas the potential energy is

$$V = \frac{1}{2} kx^2$$

(6.142)

The kinetic energy, T, can be expressed as

$$T = E - V = \frac{1}{2} kx_0^2 - \frac{1}{2} kx^2 = \frac{1}{2} k\left(x_0^2 - x^2\right) = \frac{1}{2} mv^2$$

(6.143)

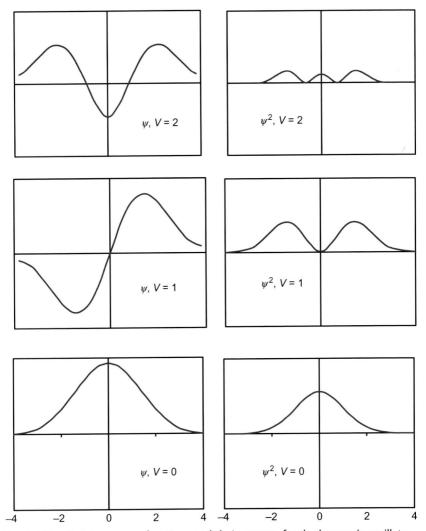

Fig. 6.6 The first three wave functions and their squares for the harmonic oscillator.

Solving Eq. (6.143) for the velocity v gives

$$v = \sqrt{\frac{k}{m}} (x_0^2 - x^2)^{1/2} \tag{6.144}$$

The probability of finding the oscillator at a given point is inversely proportional to its velocity, which can be expressed by the relationship

$$P \propto \frac{1}{\nu} \propto \frac{1}{\left(x_0^2 - x^2\right)^{1/2}} \qquad (6.145)$$

This equation as can also be written as

$$P \propto \frac{1}{\left(1 - q^2\right)^{1/2}} \qquad (6.146)$$

where $q = x/x_0$. Using this function, a plot of the probability of finding the oscillator as a function of q can be generated. Figure 6.7 shows this classical probability of finding the oscillator. For the lowest energy state shown in Fig. 6.6, the probability of finding the oscillator is given by the plot of ψ^2. This function is also shown in Fig. 6.7.

It is readily apparent that the probabilities obtained classically and by means of quantum mechanics are greatly different. For the states of higher energy, the probabilities become more similar, as they are in agreement with the principle that the quantum behavior approaches classical behavior under these conditions (referred to as the *correspondence principle*). According to the classical view of the harmonic oscillator, the probability of finding the oscillator is greatest near the limits of the vibration (represented as the vertical lines), and there is no probability that the oscillator is beyond those limits. According to quantum mechanics, the maximum probability density occurs at the equilibrium position, and there is a small but finite probability of finding the oscillator beyond the limits of the vibration.

In Fig. 6.5, the potential well for a harmonic oscillator is shown as a parabola, which is satisfactory for some purposes. However, such a relationship is not strictly correct because the potential is not symmetrical. More correctly,

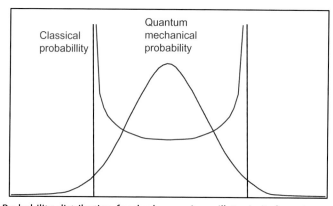

Fig. 6.7 Probability distribution for the harmonic oscillator in its lowest energy state.

a diatomic molecule has some degree of anharmonicity to the vibration. A better representation is provided by a relationship known as the Morse potential V, which is expressed as

$$V = D_e \left(1 - e^{-\beta(r-r_e)}\right)^2 \tag{6.147}$$

In this equation, D_e is the bond dissociation energy and $(r-r_e)$ is the distance of displacement from the equilibrium position (degree of bond stretching). β is a function of the reduced mass and D_e and represents the curvature near the bottom of the potential, and it can be expressed as

$$\beta = 1.2177 \times 10^{-7} \omega_e \sqrt{\frac{\mu}{D_e}} \tag{6.148}$$

In this relationship, D_e and ω_e are given cm^{-1} and μ is generally expressed in atomic mass units. As a result, D_e is frequently on the order of 50–1000 kJ mol^{-1}, 1–10 eV molecule^{-1} or 5000–$50,000$ cm^{-1}, whereas ω_e is much smaller, usually 1000–5000 cm^{-1} for most molecules. The energy of an anharmonic oscillator is given by

$$E = \left(V + \frac{1}{2}\right)\omega_e - \left(V + \frac{1}{2}\right)^2 \omega_e x_e \tag{6.149}$$

in which x_e represents a collection of constants that is given by $\omega_e/4D_0$. For example, for the H_2 molecule, the values for ω_e and $\omega_e x_e$ are 4395.2 and 118.0 cm^{-1}, respectively. The value of D_0 is the bond dissociation energy D_e minus the zero-point vibrational energy:

$$D_0 = D_e - \frac{1}{2}\hbar\omega = D_0 - \frac{1}{2}\bar{\nu}\left(in\, cm^{-1}\right) \tag{6.150}$$

Values for all of these spectroscopic constants are determined from a detailed analysis of spectra, and the harmonic oscillator and Morse potential are useful models. More sophisticated approaches can be found in the references listed at the end of this book.

6.7 POPULATION OF STATES

States of unequal energy are unequally populated. This principle is one of the most important ones in dealing with systems consisting of atoms and molecules. Experience indicates that this conclusion is true in application to a liquid and its vapor, to reactants and a transition state they form, or electrons populating orbitals in an atom. In its simplest form, the principle, known as the *Boltzmann distribution law*, gives the relative populations of two states as

$$\frac{n_1}{n_0} = e^{-\Delta E/kT} \tag{6.151}$$

where n is a population, ΔE is the difference in energy between the two states, k is Boltzmann's constant, and T is the temperature (K). When the energy is given on a molar basis, the equation becomes

$$\frac{n_1}{n_0} = e^{-\Delta E/RT} \tag{6.152}$$

Strictly speaking, this equation applies to situations where the states are single. If either of the states consists of a set of energy levels, the degeneracies of the states, g, are included to give an equation that can be written as

$$\frac{n_1}{n_0} = \frac{g_1}{g_0} e^{-\Delta E/kT} \tag{6.153}$$

For a harmonic oscillator with energy states separated by $2000\ \text{cm}^{-1}$ $(4.0 \times 10^{-13}\ \text{erg})$, the population of the first excited state (n_1) relative to the ground state (n_0) at 300 K is

$$\frac{n_1}{n_0} = \exp\left[\frac{-4.0 \times 10^{-13}\,\text{erg}}{1.38 \times 10^{-16}\,\text{erg/molecule K} \times 300\,\text{K}}\right] = e^{-9.66} = 6.4 \times 10^{-5}$$

At 600 K the relative population is only 7.0×10^{-3}. It is clear that unless the temperature is very high, a collection of a large number of such oscillators will be found almost totally in the ground state if the energy difference is rather large. If the oscillators represent vibrating molecules, then this has some significant implications for spectroscopic studies on the molecules, which will be discussed in Chapters 7, 11, and 12.

6.8 HEAT CAPACITY OF METALS

In Chapter 7, it will be shown that the heat capacity of a monatomic gas has a value of $\frac{3}{2}\,R$ as a result of three degrees of freedom, each of which contributes $\frac{1}{2}\,R$ to that value. It will also be shown that gases consisting of diatomic molecules do not follow that pattern as a result of the absorption of energy by changes in rotational and vibrational states. For a solid, there is some vibration of the lattice members about their equilibrium positions, and the amplitude of the vibrations increases with increasing temperature. The lattice structure of a metal consists of atoms in fixed positions, but they can increase in vibrational energy as the temperature increases. It follows that with there being three directions of motion, each of those degrees of freedom should contribute to the heat capacity. When a mole of particles is considered, $3N-6 \approx 3N$, so it would be expected that the heat capacity should have a value close to $3R$. With R having a value of 8.314 J mol^{-1} K^{-1}, the heat capacity would be approximately 25 J mol^{-1} K^{-1}. The heat capacities of several common metals are shown in Table 6.2. It is apparent that the crude approximation gives values for heat capacities that are about correct. However, this is true only at temperatures that are high enough so that lattice vibrations are sufficiently activated.

The heat capacity on a molar basis is the specific heat (per gram) multiplied by the number of grams in a mole (the atomic weight for a metal). Thus it follows that

$$\text{Specific heat } \left(\text{cal g}^{-1} \text{ K}^{-1}\right) \times \text{Atomic weight } \left(\text{g mol}^{-1}\right) = C_p \left(\text{cal mol}^{-1} \text{ K}^{-1}\right)$$
$$\text{Specific heat} \times \text{Atomic weight} \approx 6 \text{ cal mol}^{-1} \text{ deg}^{-1} \approx 25 \text{ Jmol}^{-1} \text{ K}^{-1}$$

This two century old "rule" is known as the Law of Dulong and Petit, and it shows that if the specific heat of a metal is determined experimentally that the atomic weight can be approximated. However, for binary salts such as LiF, NaCl, etc., one mole of solid consists of *two* moles of "particles" so that on a molar basis, the heat capacity should be approximately *twice* that of a solid consisting of atoms.

Vibrational energy levels are populated in a manner that depends on the temperature. Therefore, the heat capacity of a solid, be it a metal or an ionic solid, is a function of temperature. It is appropriate to provide some explanation of this phenomenon on the basis of the harmonic oscillator model. Approaches to this problem based on considering a solid as a collection of harmonic oscillators were developed by Einstein and Debye. The following discussion, attributable to Einstein in 1906, presents some of the salient features of the method and results.

Table 6.2 Heat capacity of selected metals at room temperature

Metal	C_p (J mol^{-1} K^{-1})	Metal	C_p (J mol^{-1} K^{-1})
Cu	24.5	Sb	25.1
Ni	25.8	Bi	25.6
Ag	25.8	Cd	25.8
Au	25.7	Sn	25.6
Pt	26.5	Pd	26.5

Although the simple approach outlined above indicates that a metal would have the heat capacity $3R$, this is not the case at low temperature. The heat capacity has a value of 0 at absolute zero, but it rises rapidly as the temperature is increased. Figure 6.8 shows the heat capacity of copper and aluminum as a function of temperature. A successful interpretation of the variation in heat capacity with temperature must explain the general features of such a graph.

The mean energy of a harmonic oscillator of frequency ν can be expressed as

$$E = \frac{1}{2}h\nu + \frac{\sum_{n} nh\nu \, e^{-nh\nu/kT}}{\sum_{n} h\nu \, e^{-nh\nu/kT}} \tag{6.154}$$

in which h and k are Planck's and Boltzmann's constants, respectively. This expression can be simplified to yield

$$E = \frac{1}{2}h\nu + \frac{h\nu}{e^{h\nu/kT} - 1} \tag{6.155}$$

The value of E in this equation represents (a) the average vibrational energy of a specific atom or (b) the average vibrational energy of all atoms at a specific instant. The $\frac{1}{2}\,h\nu$ represents the zero-point energy of the oscillator.

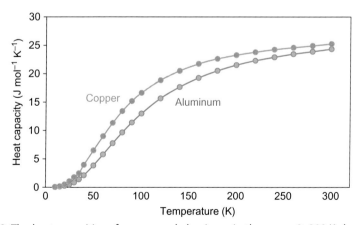

Fig. 6.8 The heat capacities of copper and aluminum in the range 0–300 K show that the value of approximately 25 J mol^{-1} K^{-1} is approached near ambient temperature. The rapid increase in heat capacity at low temperature is approximately a function of T^3.

If the temperature is very low, $h\nu \gg kT$ with the result that the average energy is just the zero-point energy, $\frac{1}{2} h\nu$. On the other hand, if the temperature is high, then $h\nu \ll kT$. In this case, the denominator in the second term of Eq. (6.155) can be simplified by making use of the series representation of an exponential

$$e^x = 1 + x + \tfrac{1}{2}x^2 + \cdots$$

Therefore,

$$E = \frac{1}{2}h\nu + \frac{h\nu}{e^{h\nu/kT} - 1} \approx \frac{1}{2}h\nu + \frac{h\nu}{1 + \dfrac{h\nu}{kT} - 1} \approx \frac{1}{2}h\nu + kT \qquad (6.156)$$

At a sufficiently high temperature, the zero-point energy is negligible compared to kT, so this reduces to the classical value.

$$E = \frac{1}{2}h\nu + kT \approx kT \qquad (6.157)$$

This is the classical limit because the energy levels expressed in terms of $h\nu$ are much smaller than the average energy of the oscillator kT (or RT on a molar basis).

It is a fact of life that limiting cases have their limits and do not represent the *intermediate* cases. At intermediate temperatures, $h\nu \approx kT$, and lattice vibrations have a frequency that is typically as high as 10^{13} Hz. Vibrational levels are considered to have frequencies that are multiples of a fundamental frequency, giving rise to energies that can be expressed as 0, $h\nu$, $2h\nu$, $3h\nu$, etc., with populations of the states being n_0, n_1, n_2, etc. The populations will be determined by exponential distributions and will be in the ratios $1: e^{-h\nu/kT}: e^{-2h\nu/kT}: e^{-3h\nu/kT}$, etc. For N atoms, there will be $3N$ vibrational states. Based on the Boltzmann distribution law, the relationship between the populations can be written as

$$n_1 = n_o\, e^{-h\nu/kT};\quad n_1 = n_o e^{-2h\nu/kT};\quad n_1 = n_o e^{-3h\nu/kT};\ \text{etc} \qquad (6.158)$$

It is now assumed that the total heat content of the crystal Q is the sum of the energies of the oscillators. This is obtained by multiplying the number of vibrators in each level by the energy of the level. The result is

$$Q = n_o\left[h\nu e^{-h\nu/kT} + 2h\nu e^{-2h\nu/kT} + 3h\nu e^{-3h\nu/kT} + \cdots \right] \qquad (6.159)$$

The total number of vibrators N can be expressed as the sum of the populations of occupied states

$$N = n_0 + n_1 + n_2 + n_3 + \ldots \tag{6.160}$$

Because there are three degrees of freedom, the simplified expression (in which we now let N equal Avogadro's number) giving the total internal energy per mole can be written as

$$Q = \frac{3Nh\nu}{e^{h\nu/kT} - 1} \tag{6.161}$$

The heat capacity at constant volume is the derivative of the internal energy with respect to temperature

$$C_v = \left(\frac{\partial Q}{\partial T}\right)_V = 3Nk\left(\frac{h\nu}{kT}\right)^2 \frac{e^{h\nu/kT}}{(e^{h\nu/kT} - 1)} \tag{6.162}$$

In the approach taken by Peter Debye (in whose honor the unit of dipole moment is named), it was assumed that a distribution of vibrational frequencies exists. When $x = h\nu/kT$ and θ (known as the Debye characteristic temperature) is defined so that $k\theta = h\nu_{max}$, where ν_{max} is the frequency with the highest population, it is possible to obtain the following expression for the total energy:

$$Q = \frac{9NkT^4}{\theta^3} \int_0^{x_{max}} \frac{x^3}{e^x - 1} dx \tag{6.163}$$

Differentiating this expression with respect to temperature at constant volume gives the following expression for C_v

$$C_v = 9R\left(\frac{T}{\theta}\right)^3 \int_0^{x_{max}} \frac{e^x x^4}{(e^x - 1)^2} dx \tag{6.164}$$

At temperatures where $T \gg \theta$ and $x \ll 1$, the integral is simplified in the following way:

$$\frac{x^4 e^x}{(e^x - 1)^2} \frac{x^4}{(e^x - 1)(1 - e^x)} = \frac{x^4}{2(x^2/2! + x4/4! + \cdots)} \tag{6.165}$$

If only the term in x^2 in the denominator is used, the result is

$$C_v = 9R\left(\frac{T}{\theta}\right) \int_0^\infty x^2 dx \tag{6.166}$$

Table 6.3 Debye characteristic temperatures for selected metals

Metal	θ (K)	Metal	θ (K)
Li	430	Be	980
Na	160	Mg	330
K	99	Ca	230
Au	170	Cr	405
Pb	86	Zn	240
Cd	165	Pt	225

This expression evaluates to $3R$ at high temperature in accord with the Dulong and Petit rule. At low temperature, $h\nu/kT=x$ and its value is replaced by ∞, as shown in Eq. (6.166). In that case, the integral evaluates to $(12/5)\,\pi^4\,R\,(T/\theta)^3$. This shows that at low temperature the heat capacity increases as T^3, which agrees well with the experimental values for most metals, as shown in Fig. 6.8. The Debye characteristic temperature varies significantly for different metals. Table 6.3 shows the values for several metals.

The harmonic oscillator is one of the important models that can be treated by the methods of quantum mechanics. It is the most useful model for describing vibrations in molecules, and as shown above, it is the starting point for dealing with heat capacities of solids. Because of the relationship between molecular vibrations and rotations, this application of the harmonic oscillator will be deferred until the model for rotating molecules (known as the rigid rotor) has been described in the next chapter. Both models will be proven to be essential in the discussion of spectroscopy.

LITERATURE CITED

Badger, R. M. A Relation Between Internuclear Distances and Bond Force Constants. *J. Chem. Phys.* **1934**, *2*, 128–131.

PROBLEMS

1. Calculate the zero-point vibrational energies for —O—H and —O—D bonds. If a reaction of these bonds involves breaking them, what does this suggest about the relative rates of the reactions of —O—H and —O—D bonds? What should be the ratio of the reaction rates?

2. The OH stretching vibration in gaseous CH_3OH is at 3687 cm^{-1}. Estimate the position of the O—D vibration in CH_3OD.

3. If $y = x^2 + 5x + 2e^x$, evaluate the following, where $D = d/dx$:
   (a) $(D^2 + 4D + 2)y$
   (b) $(D + 4)y$
   (c) $(2D^3 + 4D)y$

4. If $y = \sin 3x + 4 \cos 2x$, evaluate the following:
   (a) $(D^2 + 3D + 3)y$
   (b) $(D + 3)y$
   (c) $(2D^2 + 3D)y$

5. Use the auxiliary equation method to solve the following:
   (a) $\dfrac{d^2y}{dx^2} + 4\dfrac{dy}{dx} - 5y = 0$
   (b) $(4D^2 - 36)y = 0$
   (c) $\dfrac{d^2y}{dx^2} - y = 0, \quad y(0) = -1, \quad \text{and} \quad y'(0) = -3$
   (d) $(D^2 - 3D + 2)y = 0, \quad y(0) = -1, \quad \text{and} \quad y(0) = 0$

6. Use the series approach to solve the following:
   (a) $y + y = 0, \quad \text{with} \quad y(0) = 1$
   (b) $\dfrac{dy}{dx} - xy = 0, \quad \text{with} \quad y(0) = 2$
   (c) $\dfrac{d^2y}{dx^2} + y = 0$

7. Wave functions for which $\psi(x) = \psi(-x)$ are symmetric, whereas those for which $\psi(x) = -\psi(-x)$ are antisymmetric. Determine whether the first four normalized wave functions for the harmonic oscillator are symmetric or antisymmetric.

8. Find the normalization constant for the wave function corresponding to the lowest energy state of a harmonic oscillator $\psi = N_0 \exp(-bx^2)$, where b is a constant.

9. If a molecule has the lowest two vibrational states separated by 3000 cm^{-1}, calculate the fraction of the molecules that will be in the higher state at 300 K. If the total number of molecules is one mole, how many molecules will populate the higher state? Now repeat the calculations assuming that the temperature is 800 K. What do the results indicate with respect to reaction rates?

CHAPTER 7

Molecular Rotation and Spectroscopy

Spectroscopic studies of molecules usually involve the electromagnetic radiation that is absorbed or emitted as a result of changes in the energy levels associated with vibration and rotation. In fact, much of what has been determined about the structure of atoms and molecules has been obtained by studying the interaction of electromagnetic radiation with matter. In this chapter, molecular spectra will be considered after a discussion of the quantum mechanical problem of rotation and its combination with vibration. In this chapter the solution of the rigid rotor quantum mechanical model and its application to rotational states of diatomic molecules will be illustrated. As a preview of coming topics, the relationship of rotational energy states to vibrational and electronic states will be depicted. After discussions of bonding theories, the topic of molecular spectroscopy will be presented in Chapters 11 and 12.

7.1 ROTATIONAL ENERGIES

To introduce the principles associated with molecular rotation, the case of an object of mass m rotating around a fixed center will be considered, as shown in Fig. 7.1. In this case, it is simpler to consider the center of the rotation as being stationary, but this is only for convenience. The moment of inertia, I, in this case can be represented as

$$I = mr^2 \tag{7.1}$$

in which r is the radius of rotation and m is the mass of the object. The angular velocity, ω, is given as the change in angle φ, with time:

$$\omega = \frac{d\varphi}{dt} \tag{7.2}$$

The kinetic energy of rotation, T, can be expressed as

$$T = \frac{1}{2} I \frac{d^2\varphi}{dt^2} \tag{7.3}$$

Fundamentals of Quantum Mechanics
http://dx.doi.org/10.1016/B978-0-12-809242-2.00007-3

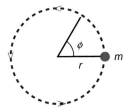

Fig. 7.1 A rotating object of mass m.

Analogous to linear momentum, the *angular* momentum P_φ can be represented as

$$P_\varphi = I\frac{d\varphi}{dt} = I\omega \tag{7.4}$$

and the kinetic energy T is given as

$$T = \frac{P_\varphi^2}{2I} \tag{7.5}$$

which is analogous to the expression $T = p^2/2\,m$ for linear momentum. In polar coordinates, the operator for angular momentum can be written as

$$\hat{P}_\varphi = \frac{\hbar}{I}\frac{\partial}{\partial\varphi} \tag{7.6}$$

and the operator for rotational kinetic energy can be expressed as

$$\hat{T} = -\frac{\hbar}{2I}\frac{\partial^2}{\partial\varphi^2} \tag{7.7}$$

Assuming that the potential energy of the object is 0, $V = 0$, the Hamiltonian operator is

$$\hat{H} = \hat{T} + \hat{V} = \hat{T} + 0 = \hat{T}$$

and the Schrödinger equation $\hat{H}\psi = E\psi$ can be written as

$$-\frac{\hbar}{2I}\frac{\partial^2\psi}{\partial\varphi^2} = E\psi \tag{7.8}$$

Rearranging Eq. (7.8) leads to

$$\frac{\partial^2\psi}{\partial\varphi^2} + \frac{2I}{\hbar}E\psi = 0 \tag{7.9}$$

An equation of this form has been solved before in relation to the particle in a one-dimensional box model. Therefore, the characteristic equation, where $k^2 = 2\,IE$, can be written as

$$m^2 + k^2 = 0 \tag{7.10}$$

from which it is found that

$$m = \pm ik \tag{7.11}$$

Therefore, the general form of the solution can be written as

$$\psi = C_1 e^{ik\phi} + C_2 e^{-ik\phi} \tag{7.12}$$

Because φ is an angular measure, $\psi\,(\varphi) = \psi\,(\varphi + 2\pi)$, where φ is in radians. As a result of $k^2 = 2\,IE$. It follows that

$$(2IE)^{1/2} = J(J = 0,\ 1,\ 2, \ldots) \tag{7.13}$$

Solving for E leads to

$$E = \frac{J^2}{2I} \tag{7.14}$$

If the assumption made by Bohr in regard to angular momentum being quantized is followed,

$$mvr = nh/2\pi$$

and so the angular momentum $I\omega$ can be expressed as

$$I\omega = mvr = nh/2\pi$$

in which n is an integer. Using J as the quantum number, the result obtained is

$$I\omega = J\frac{h}{2\pi} \tag{7.15}$$

Consequently,

$$\omega = \frac{hJ}{2\pi I} \tag{7.16}$$

With Eqs. (7.3) and (7.4), it is found that the energy of rotation E_{rot} is

$$E_{\text{rot}} = \frac{1}{2}I\omega^2 = \frac{1}{2I}\left(I^2\omega^2\right) \tag{7.17}$$

Therefore, by combining Eqs. (7.16) and (7.17), it can be seen that

$$E_{rot} = \frac{1}{2I}\left(\frac{hJ}{2\pi}\right)^2 = \frac{h^2}{8\pi^2 I}J^2 \tag{7.18}$$

This model is oversimplified in that it does not represent the rotation of a diatomic molecule around a center of mass, as it is considered from the quantum mechanical point of view, so angular momentum has been *assumed* to be quantized. The rotation of a diatomic molecule will be described in the next section by using the principles of quantum mechanics.

7.2 QUANTUM MECHANICS OF ROTATION

In order to show the applicability of quantum mechanical methods to the rigid rotor problem, a diatomic molecule, as shown in Fig. 7.2, will be considered. The bond in the molecule will be considered rigid so that molecular dimensions remain unchanged during rotation. Because of the moments around the center of gravity,

$$m_1 r_1 = m_2 r_2 \tag{7.19}$$

where m is mass and r is the distance from the center of gravity. It should be obvious that

$$R = r_1 + r_2 \tag{7.20}$$

By substitution and solving for r_1 and r_2, it is found that

$$r_1 = \frac{m_2 R}{m_1 + m_2} \tag{7.21}$$

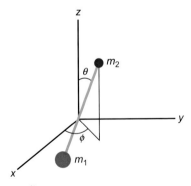

Fig. 7.2 The rigid rotor coordinate system.

$$r_2 = \frac{m_1 R}{m_1 + m_2} \qquad (7.22)$$

The moment of inertia, I, for this model can be described by the equation

$$I = m_1 r_1^2 + m_2 r_2^2 \qquad (7.23)$$

When the results obtained above for r_1 and r_2 are used, I can be written as

$$I = \mu R^2 \qquad (7.24)$$

in which μ is the reduced mass,

$$\mu = \frac{m_1 m_2}{m_1 + m_2}$$

For convenience, the center of gravity will be placed at the origin of the coordinate system, as shown in Fig. 7.2. However, the solution of this problem usually comes from transforming the representation of the model into polar coordinates (see Chapter 4). For each atom, the kinetic energy can be written as

$$T = \frac{1}{2} m \left[\left(\frac{\partial x}{\partial t} \right)^2 + \left(\frac{\partial y}{\partial t} \right)^2 + \left(\frac{\partial z}{\partial t} \right)^2 \right] \qquad (7.25)$$

In terms of polar coordinates, this becomes for m_1

$$T_1 = \frac{1}{2} m_1 r_1^2 \left[\left(\frac{\partial \theta}{\partial t} \right)^2 + \sin^2 \theta \left(\frac{\partial \varphi}{\partial t} \right)^2 \right] \qquad (7.26)$$

When both atoms are included, the kinetic energy is given by

$$T = \frac{1}{2} \left(m_1 r_1^2 + m_2 r_2^2 \right) \left[\left(\frac{\partial \theta}{\partial t} \right)^2 + \sin^2 \theta \left(\frac{\partial \varphi}{\partial t} \right)^2 \right] \qquad (7.27)$$

When this equation is modified to incorporate the moment of inertia, the result is

$$T = \frac{1}{2} I \left[\left(\frac{\partial \theta}{\partial t} \right)^2 + \sin^2 \theta \left(\frac{\partial \varphi}{\partial t} \right)^2 \right] \qquad (7.28)$$

The derivatives are found from the conversions from Cartesian to polar coordinates. For example, the relationship for the x direction is

$$x = r \sin \theta \cos \varphi \qquad (7.29)$$

so that

$$\frac{\partial x}{\partial t} = (r \cos\theta \cos\varphi)\frac{\partial \theta}{\partial t} - (r \sin\theta \sin\varphi)\frac{\partial \varphi}{\partial t} \tag{7.30}$$

The derivatives for the other variables are found in a similar manner. Using these results, it is possible to write the operator for the kinetic energy as

$$\hat{T} = -\frac{\hbar^2}{2m}\left[\frac{1}{r^2}\frac{\partial}{\partial r}r^2\frac{\partial}{\partial r} + \frac{1}{r^2\sin\theta}\frac{\partial}{\partial\theta}\sin\theta\frac{\partial}{\partial\theta} + \frac{1}{r^2\sin^2\theta}\frac{\partial}{\partial\varphi^2}\right] \tag{7.31}$$

For a constant internuclear distance, the first term inside the brackets is 0, because r does not change and its derivative is 0. Furthermore, the kinetic energy must be described in terms of angular momenta in order to write the Hamiltonian operator. When this is done, it is found that

$$T = \frac{1}{2I}\left[p_\theta^2 + \frac{p_\varphi^2}{\sin^2\theta}\right] \tag{7.32}$$

and the operators for the angular momenta are

$$\hat{p}_\theta = \frac{\hbar}{i}\frac{\partial}{\partial\theta} \quad \text{and} \quad \hat{p}_\varphi = \frac{\hbar}{i}\frac{\partial}{\partial\varphi} \tag{7.33}$$

However, to describe rotation for which the moment of inertia is used, it is assumed that no forces are acting on the rotor so that the potential energy is given by $V=0$. Under these conditions, $H=T+V=T$ and the Hamiltonian operator is

$$\hat{H} = -\frac{\hbar^2}{2I}\left[\frac{1}{\sin\theta}\frac{\partial}{\partial\theta}\sin\theta\frac{\partial}{\partial\theta} + \frac{1}{\sin^2\theta}\frac{\partial}{\partial\varphi^2}\right] \tag{7.34}$$

The careful reader will observe that the form of the operator exactly replicates the angular portion of the Hamiltonian shown for the hydrogen atom in Eq. (4.11). Using this operator, the Schrödinger equation $\hat{H}\psi = E\psi$ becomes

$$-\frac{\hbar^2}{2I}\left[\frac{1}{\sin\theta}\frac{\partial}{\partial\theta}\sin\theta\frac{\partial}{\partial\theta} + \frac{1}{\sin^2\theta}\frac{\partial}{\partial\varphi^2}\right]\psi = E\psi \tag{7.35}$$

It should also come as no surprise that the technique used in solving the equation is the separation of variables. Therefore, it is assumed that a solution can be written as

$$\psi(\theta, \varphi) = Y(\theta, \varphi) = \Theta(\theta)\,\Phi(\varphi) \tag{7.36}$$

This product of two functions can be substituted into Eq. (7.35), and by dividing by the product, separating terms, and rearranging, it is found that both factors of the assumed solution can be set equal to a constant, which will be represented by $-m^2$. The two equations that are obtained by this separation are

$$\frac{d^2\Phi}{d\varphi^2} = -m^2\Phi \tag{7.37}$$

$$\frac{1}{\sin\theta}\frac{\partial}{\partial\theta}\sin\theta\frac{\partial\Theta}{\partial\theta} - \frac{m^2}{\sin^2\theta}\Theta + \frac{2IE}{\hbar^2}\Theta = 0 \tag{7.38}$$

The equation in φ is of a form that has already been solved several times in earlier chapters. The equation can be written as

$$\frac{d^2\Phi}{d\varphi^2} + m^2\Phi = 0 \tag{7.39}$$

and the auxiliary equation can be written as

$$x^2 + m^2 = 0 \tag{7.40}$$

so that $x = (-m^2)^{1/2}$ and $x = im$. Therefore,

$$\Phi = e^{im\varphi} \tag{7.41}$$

but after a complete rotation through 2π rad, the molecule has the same orientation. Therefore, it can be seen that

$$e^{im\varphi} = e^{im(\phi + 2\pi)} \tag{7.42}$$

This equation signifies that

$$e^{im2\pi} = 1 \tag{7.43}$$

Using Euler's formula,

$$e^{ix} = \cos x + i\sin x \tag{7.44}$$

it is found that

$$e^{im2\pi} = \cos 2\pi m + i \sin 2\pi m \tag{7.45}$$

The right-hand side of this equation must also be equal to 1, but this is true only when m is an integer, which means that $m = 0, \pm 1, \pm 2, \ldots$

The second equation obtained by the separation of variables is simplified by letting $2 IE/\hbar^2$ be equal to $l(l+1)$. When this is done, the equation can be written in terms of ordinary derivatives because only one variable is present, and the result is

$$\frac{1}{\sin\theta}\frac{d}{d\theta}\sin\theta\frac{d}{d\theta}\Theta - \frac{m^2}{\sin^2\theta}\Theta + l(l+1)\,\Theta = 0 \tag{7.46}$$

A rearrangement of this equation, which by taking derivatives and collecting terms, gives the equation

$$\frac{d^2\Theta}{d\theta^2} + \frac{\cos\theta}{\sin\theta}\frac{d\Theta}{d\theta} + \left[l(l+1) - \frac{m^2}{\sin^2\theta}\right]\Theta = 0 \tag{7.47}$$

A transformation of the variable from θ to x is accomplished by the following changes:

$$x = \cos\theta \tag{7.48}$$

$$\sin^2\theta = 1 - x^2 \tag{7.49}$$

$$\frac{dx}{d\theta} = -\sin\theta \tag{7.50}$$

$$\frac{d}{d\theta} = \frac{d\theta}{dx}\frac{d}{d\theta} = -\sin\theta\frac{d}{dx} \tag{7.51}$$

$$\frac{d^2}{d\theta^2} = \frac{d}{d\theta}\left[-\sin\theta\frac{d}{dx}\right] = \cos\theta\frac{d}{dx} - \sin\theta\frac{d}{d\theta}\frac{d}{dx} \tag{7.52}$$

Substituting for $d/d\theta$, the last relationship gives

$$\frac{d^2}{d\theta^2} = \sin^2\theta\frac{d^2}{dx^2} = -\cos\theta\frac{d}{dx} \tag{7.53}$$

Substituting these quantities into Eq. (7.47), it is found that

$$\frac{d^2\Theta(\theta)}{d\theta^2} + \frac{\cos\theta}{\sin\theta}\frac{d\Theta(\theta)}{d\theta} + \left[l(l+1) - \frac{m^2}{1-x^2}\right]\Theta(\theta) = 0 \tag{7.54}$$

This equation can also be written as

$$\sin^2\theta\frac{d^2\Theta(x)}{dx^2} - \cos x\frac{d\Theta(x)}{dx} + \frac{\cos\theta}{\sin\theta}\frac{d\Theta(\theta)}{d\theta}\left[l(l+1) - \frac{m^2}{1-x^2}\right]\Theta(x) = 0 \tag{7.55}$$

Now, by replacing $d\Theta(\theta)/d\theta$ with $-\sin\theta\,d\Theta(\theta)/dx$ and $\sin^2\theta$ with $1-x^2$, Eq. (7.55) becomes

$$(1-x^2)\frac{d^2\Theta(x)}{dx^2} - 2x\frac{d\Theta(x)}{dx} + \left[l(l+1) - \frac{m^2}{1-x^2}\right]\Theta(x) = 0 \qquad (7.56)$$

This equation has the form shown in Table 6.1 for Legendre's equation. Once again, it has been possible to reduce a problem in quantum mechanics to one of the famous differential equations shown in Table 6.1. For example, the solution of the equation for the radial portion of the hydrogen atom problem was equivalent to solving Laguerre's equation. The angular portion of the hydrogen atom problem involved Legendre's equation, and the solution of the harmonic oscillator problem required the solution of Hermite's equation.

A series solution for Legendre's equation is well known and requires the series of polynomials known as the *Legendre polynomials*, which can be written as

$$P_l^{|m|}(\cos\theta) \qquad (7.57)$$

Therefore, the wave functions obtained by solving the wave equation for the rigid rotor are written as

$$\psi_{l,m}(\theta, \varphi) = N\, P_l^{|m|}(\cos\theta)\, e^{im\varphi} \qquad (7.58)$$

where N is a normalization constant. The solutions $\psi_{l,m}(\theta, \varphi)$ are known as the *spherical harmonics* that were first encountered in this book in the solution of the wave equation for the hydrogen atom. It can be shown that the normalization constant can be written as functions of l and m that are described by the equation

$$N = \left[\frac{(2l+1)(l-|m|)!}{4\pi\,(l+|m|)!}\right]^{1/2} \qquad (7.59)$$

Therefore, the complete solutions can be written as

$$\psi_{l,m} = \left[\frac{(2l+1)(l-|m|)!}{4\pi\,(l+|m|)!}\right]^{1/2} P_l^{|m|}(\cos\theta)\, e^{im\varphi} \qquad (7.60)$$

As illustrated, the solution of several quantum mechanical problems involves a rather heavy investment in mathematics, especially the solution of several famous differential equations. Although the complete details have not been presented in this book, the procedures have been outlined in sufficient detail so that the reader has an appreciation of the methods adequate for quantum mechanics at this level. The references at the end of this book should be consulted for more detailed treatment of the problems.

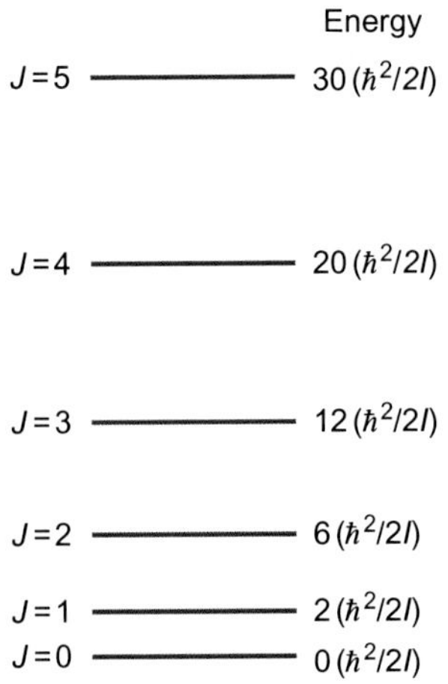

Energy

| J=5 | ——————— | $30\,(\hbar^2/2I)$ |

| J=4 | ——————— | $20\,(\hbar^2/2I)$ |

| J=3 | ——————— | $12\,(\hbar^2/2I)$ |

| J=2 | ——————— | $6\,(\hbar^2/2I)$ |

| J=1 | ——————— | $2\,(\hbar^2/2I)$ |
| J=0 | ——————— | $0\,(\hbar^2/2I)$ |

Fig. 7.3 Rotational energies for a diatomic molecule (drawn to scale).

By making a comparison of Eqs. (7.38) and (7.46), it can be seen that

$$\frac{2IE}{\hbar^2} = l(l+1) \tag{7.61}$$

Although it will not be proven here, the restrictions on the values of m and l imposed by the Legendre polynomials require that l be a nonnegative integer, and in this case, $l = 0, 1, 2, \ldots$. For a rotating diatomic molecule, the rotational quantum number is usually expressed as J, so the energy levels are given in terms of that quantum number as

$$E = \frac{\hbar^2}{2I}J(J+1) \tag{7.62}$$

Therefore, the allowed rotational energies are

$$E_0 = 0; \quad E_1 = 2\frac{\hbar^2}{2I}; \quad E_2 = 6\frac{\hbar^2}{2I}; \quad E_3 = 12\frac{\hbar^2}{2I}; \quad etc.$$

Figure 7.3 shows the first few rotational energy levels.

7.3 HEAT CAPACITIES OF GASES

When gaseous molecules absorb heat, they undergo changes in rotational energies. Therefore, studying the thermal behavior of gases provides information on rotational states of molecules. It can be shown that for an ideal gas,

$$PV = nRT = \left(\frac{2}{3}\right)E \qquad (7.63)$$

where E is the total kinetic energy and P, V, and R have their usual meanings. Therefore, this equation can be put in the form

$$\frac{E}{T} = \frac{3}{2}nR \qquad (7.64)$$

If n is 1 mol and the temperature of the gas is changed by 1 K, there will be a corresponding change in E, which can be represented as ΔE. Therefore,

$$\frac{\Delta E}{\Delta T} = \frac{3}{2}R \qquad (7.65)$$

The amount of heat needed to raise the temperature of 1 g of some material by one degree is the *specific heat* of the material. The *molar* quantity is called the *heat capacity* and is measured in $J\ mol^{-1}\ K^{-1}$ or $cal\ mol^{-1}\ K^{-1}$. Equation (7.65) shows that the heat capacity of an ideal gas should be $(3/2)R$, which is $12.47\ J\ mol^{-1}\ K^{-1}$ or $2.98\ cal\ mol^{-1}\ K^{-1}$. Table 7.1 shows the heat capacities of several gases at constant volume. There are two different heat capacities in use: C_v is the heat capacity at constant *volume*, and C_p is the heat capacity at constant *pressure*. It can be concluded that $C_p = C_v + R$. If the gas is at constant pressure, heating the gas to change the temperature by 1 K causes an expansion of the gas, which requires work to push back the surroundings of the gas. Therefore, the heat capacity at constant pressure is greater than the heat capacity at constant volume, where the absorbed heat changes only the kinetic energy of the gas.

The experimental heat capacities for helium and argon are identical to those predicted by the ideal gas equation ($12.5\ J\ mol^{-1}\ K^{-1}$). However, for all of the other gases listed in the table, the values do not agree with the heat capacity for an ideal gas. At first glance, it appears that the

Table 7.1 Molar heat capacities of gases at 25°C.

Gas	C_v (cal mol^{-1} deg^{-1})	C_v (J mol^{-1} deg^{-1})
Helium	2.98	12.5
Argon	2.98	12.5
Hydrogen	4.91	20.5
Oxygen	5.05	21.1
Nitrogen	4.95	20.7
Chlorine	6.14	25.7
Ethane	10.65	44.6

monatomic gases have heat capacities that agree with the simple model based on increasing the kinetic energy of the molecules, whereas gases consisting of diatomic and polyatomic molecules do not.

The motion of gaseous molecules through space is described in terms of motion in three directions. Because the absorbed heat increases the kinetic energy in three directions, on average, the same amount of heat goes into increasing the energy in each direction. That is, $(3/2)R = 3(1/2)R$, with $(1/2)R$ going to increase kinetic energy in each direction. Each direction is called a *degree of freedom*, and the overall kinetic energy is the sum of the energy in each direction. However, this is not the only way in which the heat is absorbed by molecules if they consist of more than one atom. There are other degrees of freedom in addition to linear motion through space (*translation*). The other ways in which molecules absorb heat are by changing *rotational* and *vibrational* energies.

A principle known as the law of equipartition of energy can be stated quite simply as follows: *If a molecule can absorb energy in more than one way, it can absorb equal amounts in each way.* This is also true for translation, where $(1/2)R$ absorbed goes toward increasing the kinetic energy in each of the three directions. In order to use this principle, it is necessary to know the number of "ways" (i.e., degrees of freedom) a molecule can rotate and vibrate. As will be shown later, for a diatomic molecule (which is linear), there are only two degrees of rotational freedom, and it can be inferred that this would also be true for other *linear* molecules. In the case of linear molecules for which absorbed heat can change their rotational energy, there will be $2(1/2)R$ absorbed. The total heat absorbed for a mole of a gas composed of linear molecules will be

$$3\left(\frac{1}{2}\right)R = \left(\frac{3}{2}\right)R = 12.5 \text{ J mol}^{-1}\text{K}^{-1} \text{ (translation)}$$

$$2\left(\frac{1}{2}\right)R = R = 8.3 \text{ J mol}^{-1}\text{K}^{-1} \text{ (rotation)}$$

Thus, the total heat capacity is $20.8 \text{ J mol}^{-1}\text{ K}^{-1}$, which is equal to $(5/2)R$. Note that this value is very close to the actual heat capacities shown in Table 7.1 for H_2, O_2, and N_2. It is reasonable to conclude that for these diatomic molecules, absorbed heat is changing only their translational and rotational energies. It should be noted that for nonlinear molecules, there are three degrees of rotational freedom, each of which can absorb $(1/2)R$.

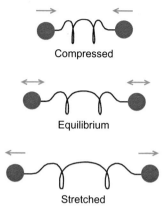

Fig. 7.4 Motion of a diatomic molecule during vibration.

Although the value of $(5/2)R$ for the heat capacity of diatomic molecules at 25°C has been rationalized, it should be noted that at very high temperatures (1500 K), the heat capacity of hydrogen is about 29.2 J mol^{-1} K^{-1}. This shows that at high temperature, the H_2 molecule can absorb energy in some way other than by changing only its translational and rotational energies. The additional way in which H_2 molecules can absorb energy is by changing vibrational energy. A chemical bond is not totally rigid, and in a diatomic molecule the bond can be represented as a spring (see Fig. 7.4). However, the vibrational energy (as well as the rotational energy) is quantized (see Sections 6.6 and 7.2). Because the observed heat capacity of H_2 at 25°C can be accounted for in terms of changes in only translational and rotational energies, it can be concluded that the molecules cannot change vibrational energy at this low temperature. Therefore, it should be apparent that the rotational energy states must be separated by an energy *smaller* than that which separates vibrational energy levels. The discussion will now be turned to a more detailed description of the nature of rotational and vibrational energy states for gaseous diatomic molecules.

7.4 ENERGY LEVELS IN GASEOUS ATOMS AND MOLECULES

Emission spectra for atoms appear as a series of lines, because electrons fall from higher energy states to lower ones and emit energy as electromagnetic radiation. The reader should recall the line spectrum of hydrogen (see

Chapter 1) and the fact that the Lyman series is in the ultraviolet (UV) region, whereas the Balmer series is in the visible region. Consequently, spectroscopy that is carried out to observe the transitions between the electronic energy levels often involves radiation in the visible and UV regions of the electromagnetic spectrum.

For a spectral line of 6000 Å (600 nm), which is in the visible light region of the electromagnetic spectrum, the corresponding energy is

$$E = h\nu = \frac{hc}{\lambda} = \frac{(6.63 \times 10^{-27}\,erg\,s) \times (3.00 \times 10^{10}\,cm/s)}{6.00 \times 10^{-5}\,cm} = 3.3 \times 10^{-12}\,erg$$

The conversion of this energy to a molar quantity by multiplying by Avogadro's number gives 2.0×10^{12} erg mol^{-1}. Converting this value to kilojoules per mole gives an energy of about 200 kJ mol^{-1}:

$$E = \frac{(3.3 \times 10^{-12}\,erg\,s/molecule) \times (6.02 \times 10^{23}\,molecules/mol)}{10^{10}\,erg/kJ}$$

$$= 200kJ/mol$$

This is within the typical range of energies separating electronic states, which is about 200–400 kJ mol^{-1}.

Although the discussion has concerned electronic energy levels in atoms, the electronic states in molecules are usually separated by similar energies. In general, the electromagnetic radiation *emitted* from atoms is usually studied, but it is the radiation *absorbed* by molecules that is usually examined by UV/visible spectroscopy. In addition to electronic energy levels, molecules also have vibrational and rotational energy states. As shown in Fig. 7.4, the atoms in a diatomic molecule can be viewed as though held together by bonds that have some stretching and bending (vibrational) capability, and the whole molecule can rotate as a unit. Figure 7.5 shows the relationship between the bond length and the potential energy for a vibrating molecule. The bottom of the potential well is rather closely approximated by a parabolic potential (see Chapter 6).

The difference in energy between adjacent vibrational levels ranges from about 10 to 40 kJ mol^{-1}. Consequently, the differences in energy between two vibrational levels correspond to radiation in the infrared region of the spectrum. Rotational energies of molecules are also quantized, but the difference between adjacent levels is typically only about 10–40 J mol^{-1}. These small energy differences correspond to electromagnetic radiation in the far infrared (or in some cases the microwave) region of the spectrum.

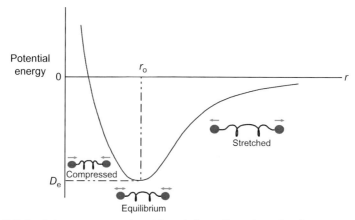

Fig. 7.5 Potential energy versus bond length for a diatomic molecule.

Therefore, an infrared spectrometer is needed to study changes in vibrational or rotational states in molecules.

The experimental technique known as infrared spectroscopy is concerned with changes in vibrational and rotational energy levels in molecules. Figure 7.6 shows the relationship between the electronic, vibrational, and rotational energy levels for molecules and the approximate range of energy for each type of level.

In the study of heat capacities of gases, it was shown how the existence of rotational states affects the heat capacity. For He the heat capacity is $(3/2)R$; the heat capacity for H_2 is $(5/2)R$ at room temperature, but it approaches $(7/2)R$ at high temperatures. The reason is that for H_2, the absorbed energy will not only change the kinetic energies (translation) of the molecules, but it will also change their rotational energies. At quite high temperatures, the vibrational energies of the molecules can also change. The law of equipartition of energy states that if a molecule can absorb energy in more than one way, it can absorb equal amounts in each way. As a result of there being three degrees of translational freedom (x, y, and z to velocity or kinetic energy), each degree of freedom can absorb $(1/2)R$ as the molecules change their kinetic energies. Because H_2 (and all other diatomic molecules) are linear, there are only two degrees of rotational freedom, as shown in Fig. 7.7. If the possibility of rotation around the z-axis (the internuclear axis) is considered, it is found that the moment of inertia is extremely small and results in no change in the positions of the atoms. Using a nuclear radius of 10^{-13} cm, the value for I is several orders of magnitude larger for rotation around the x- and y-axes because the bond length is generally on the order of 10^{-8} cm.

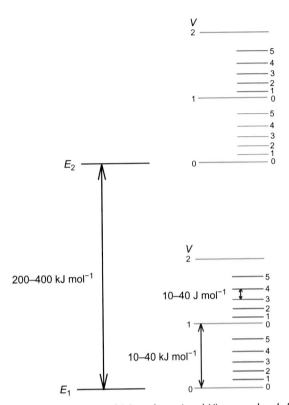

Fig. 7.6 The electronic (E), vibrational (V), and rotational (J) energy levels for a diatomic molecule with typical ranges of energies.

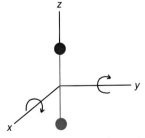

Fig. 7.7 The rotational degrees of freedom for a diatomic molecule.

The energy of rotation is given by

$$E = \frac{\hbar^2}{2I} \qquad (7.66)$$

so by making use of this equation, it is possible to calculate the difference in energy between two adjacent rotational levels ($J=0$ to $J=1$), which can be expressed as

$$E = \frac{h^2}{8\pi^2 I} \qquad (7.67)$$

As a result of the very small value for I around the z-axis, the energy is very large for rotation around that axis. Therefore, an increase in rotational energy around the internuclear axis for linear molecules (the z-axis because that is the axis of highest symmetry) is not observed and linear molecules have only two degrees of rotational freedom.

There is only one degree of vibrational freedom, but it counts double $[2(1/2)R]$ because an increase in both kinetic and potential energy is involved. Therefore, when H_2 is absorbing heat in *all* possible ways, the heat capacity is given as the sum

$$(3/2)R + (2/2)R + (2/2)R = (7/2)R$$

At a very high temperature, the heat capacity will approach this value. At room temperature (about 300 K), RT is the thermal energy available, which can be calculated as $(8.3144\,\text{J mol}^{-1}\,\text{K}^{-1}) \times (300\text{ K}) = 2500\,\text{J mol}^{-1}$. Therefore, the very large separation between electronic states (typically 100–300 kJ mol^{-1}) means that only the *lowest* electronic state will be populated. Likewise, for most molecules, the usual difference of 10–40 kJ mol^{-1} between vibrational states means that only the *lowest* vibrational state will be populated at low temperatures. The relatively small differences between rotational states make it possible for several rotational states to be populated (although unequally) even at room temperature.

7.5 ROTATIONAL SPECTRA OF DIATOMIC MOLECULES

At this point the types of energy levels for molecules have been described, and the differences between these states have been shown to correspond to different regions of the electromagnetic spectrum. Electronic energy levels are usually separated by sufficient energy to correspond to radiation in the visible and UV regions of the spectrum. Vibrations have energies of such magnitudes that the changes in energy levels are associated with the infrared region of the spectrum. Therefore, infrared spectroscopy deals primarily with the changes in vibrational energy levels in molecules. The energy level diagram shown in Fig. 7.6 reflects the fact that the rotational energy levels are much more closely spaced than are the vibrational energy levels. Accordingly, it is easy to produce a change in the rotational state of a small molecule as the vibrational energy changes. In fact, for some molecules, there is a restriction (known as a *selection rule*) that permits the vibrational state of the molecule to change *only* if the rotational state is changed as well, and such a molecule is HCl. Therefore, if one studies the infrared spectrum of gaseous HCl, a series of peaks is seen that corresponds to the absorption of the infrared radiation as the rotational energy level changes at the same time as the vibrational state changes. There is *not* one single absorption peak due to the change in vibrational state, but rather there is a series of peaks as the vibrational state and rotational states change. Figure 7.8 shows how these transitions are related. As shown in Fig. 7.8, the quantized vibrational states are characterized by the quantum number V, whereas the rotational states are identified by the quantum number J.

For a diatomic molecule, the rotational energies are determined by the masses of the atoms and the distance of their separation, which is the bond length. Therefore, from the experimentally determined energies separating the rotational states, it is possible to calculate the distance of separation of the

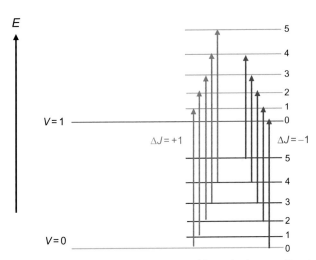

Fig. 7.8 Changes in energy as molecules are excited from the lowest vibrational state to the next higher one. All of the molecules are increasing in vibrational energy, but when $\Delta J=+1$, the molecules are increasing in rotational energy and when $\Delta J=-1$, the molecules are decreasing in rotational energy.

atoms if their masses are known. However, if a molecule rotates with a higher rotational energy, the bond length will be slightly longer because of the centrifugal force caused by the rotation. As a result, the spacing between rotational energy states for $J=1$ and $J=2$ is slightly different than it is for $J=4$ and $J=5$. Figure 7.8 shows this effect graphically in which the difference between adjacent rotational states increases slightly at higher J values.

As molecules change in vibrational energy states, the selection rule specifies that they must also change their rotational state. However, because several rotational states are populated, some of the molecules will increase in rotational energy, whereas some will decrease in rotational energy as *all* of the molecules *increase* in vibrational energy. The rotational states are designated by a quantum number J so that the levels are described by $J_0, J_1, J_2, \ldots,$ and the vibrational energy states are denoted by the vibrational quantum numbers, V_0, V_1, V_2, etc. For the transition in which the vibrational energy is increasing from $V=0$ to $V=1$, $(\Delta V=+1)$, it is possible for ΔJ to be $+1$ or -1, depending on whether the molecules are *increasing* in rotational energy $(\Delta J=+1)$ or *decreasing* in rotational energy $(\Delta J=-1)$. These types of transitions are illustrated in Fig. 7.8.

Figure 7.9 shows the vibration-rotation spectrum obtained by the author in 1969 for gaseous HCl under moderately high resolution using a

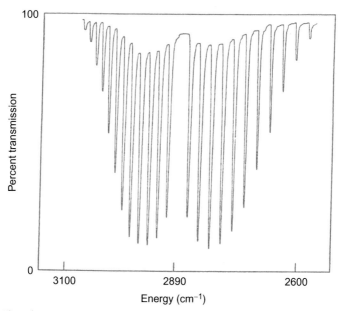

Fig. 7.9 The vibration-rotation spectrum of gaseous HCl. The spacing between adjacent peaks is about 20.7 cm^{-1}.

Beckman IR8 infrared spectrophotometer. The spectrum shows a series of sharp peaks appearing in two portions. All of the molecules are increasing in vibrational energy, but some are increasing in rotational energy and some are decreasing in rotational energy. Note that there appears to be a "gap" in the middle of the spectrum where a peak is missing. The missing peak would represent the transition between the first two vibrational states with no change in rotational state. That peak is missing as a result of this type of transition being prohibited for the HCl molecule.

If a photograph is taken using a camera with a poor lens, details of the subject are not visible. Closely spaced lines appear as a blur. The same subject photographed with a camera having a lens of high quality will show much better resolution so that small details are visible. A similar situation exists with spectra. If a spectrometer having poor resolution is used to record the spectrum of gaseous HCl, the individual sharp peaks are not resolved, and only two large peaks are observed. With a better spectrometer, all of the peaks are resolved (as shown in Fig. 7.9), and the two series of sharp peaks are observed. A spectrum such as that in Fig. 7.9 showing the absorption of infrared radiation as the molecules change vibrational and rotational state is said to show *rotational fine structure*. If the instrument is capable of ultrahigh resolution, the individual sharp peaks are seen to split into two smaller peaks. The reason for this is that chlorine exists as a mixture of ^{35}Cl and ^{37}Cl, and the rotational energy of HCl depends on the masses of the atoms. Therefore, $H^{35}Cl$ and $H^{37}Cl$ have slightly different rotational energies, which causes each peak to be split into two closely spaced peaks. Figure 7.10 shows how the peaks split for gaseous DBr. The reason for the peaks splitting in this case is that bromine occurs naturally as two isotopes, ^{79}Br (mass 78.92) and ^{81}Br (mass 80.92). Therefore, the moments of inertia are slightly different for $D^{79}Br$ and $D^{81}Br$, resulting in rotational states that are also slightly different. In this case, the splitting of the peaks due to isotopes of bromine is sufficiently large for even an instrument like the Beckman IR 8 to show some resolution of peak splitting that results from different isotopic masses.

In this chapter the solution of the problem known as the quantum mechanical rigid rotor was carried out. Along with the model systems discussed in earlier chapters, this model and the harmonic oscillator are basic tools for understanding rotations and vibrations of molecules. Additional aspects of spectroscopy will treated in greater detail in Chapters 11 and 12. It is interesting to note, however, that the existence of rotational and vibrational levels in molecules is indicated by the study of a topic as basic as the heat capacities of gases.

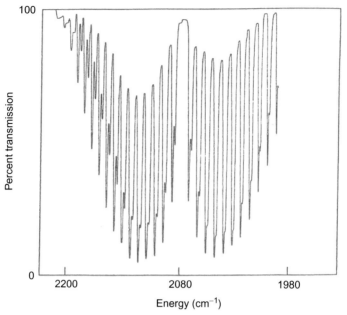

Fig. 7.10 The vibration-rotation spectrum of gaseous DBr.

PROBLEMS

1. The force constant for the C—H radical is 4.09×10^5 dyn cm^{-1}. What would be the wave number for the fundamental stretching vibration?

2. For HI, the bond length is about 1.60 Å or 160 pm. What would be the spacing between the consecutive rotational bands in the IR spectrum of HI?

3. For CO, the change in rotational state from $J=0$ to $J=1$ gives rise to an absorption band at 0.261 cm, and that for $J=1$ to $J=2$ is associated with an absorption band at 0.522 cm. Use this information to determine the bond length of the CO molecule.

4. The vibrational-rotational spectrum shown in Fig. 7.9 was obtained using an infrared spectrometer of limited resolution. Therefore, the bands do not show the separation that actually exists due to H^{35}Cl and H^{37}Cl. What degree of resolution would the spectrometer need in order to show that separation? Assume that the bond lengths of the molecules are the same: 127.5 pm.

5. Figure 7.10 shows the vibration-rotation spectrum of DBr with the splitting caused by ^{79}Br and ^{81}Br. What would be the splitting of the

peaks caused by the difference in isotopic masses in $D^{35}C1$ and $D^{37}Cl$ when measured in cm^{-1}?

6. The spacing between vibrational levels for HCl is about 2890 cm^{-1}, which is where the missing peak would be in the spectrum shown in Fig. 7.9. Calculate the force constant for the H—Cl bond in
   (a) dyn cm^{-1}
   (b) mdyn $Å^{-1}$
   (c) N m^{-1}

7. The spacing between rotational levels for HCl is about 20.7 cm^{-1}. Use this value to calculate the bond length in HCl.

8. For a rigid rotor, the rotational energy can be written as $L^2/2I$, where I is the moment of inertia and L is the angular momentum whose operator is

$$L = \frac{\hbar}{i}\frac{\partial}{\partial \varphi}$$

Obtain the operator for rotational energy. Write the Schrödinger equation, and from the form of the equation, tell what functions could give acceptable solutions. Use the functions to evaluate the energy levels for rotation.

9. For CO, the bond length is 113 pm for both $^{12}C^{16}O$ and $^{14}C^{16}O$.
   (a) Determine the moments of inertia for the two molecules.
   (b) Determine the difference in the energy between the $J=1$ and $J=2$ states for $^{12}C^{16}O$ and $^{14}C^{16}O$.

10. Calculate the energy of the first three rotational states $H^{35}Cl$ and $H^{37}Cl$. Assume that the bond length is 129 pm.

11. Calculate the reduced mass for (a) LiH and (b) CO molecules.

12. By referring to Fig. 7.9, the vibrational-rotational spectrum for gaseous HCl, and Fig. 7.8, it is evident that numerous rotational states must be populated even at room temperature. How can you account for that? Why is the most highly populated state not the one with the lowest J value?

13. Explain why a change in vibrational energy for the hydrogen molecule is not observable in an infrared spectrometer.

CHAPTER 8

Bonding and Properties of Diatomic Molecules

In describing the characteristics of matter by means of quantum mechanics, the presentation has progressed from the behavior of particles in boxes to atoms. Subsequent chapters will deal with applications of quantum mechanics to molecules. As will be seen, there are two major approaches to the application of quantum mechanics to the description of molecules. These are the valence bond approach developed by Heitler, London, Slater, and Pauling, and the molecular orbital (MO) approach developed primarily by Robert Mulliken. Most of what will be covered in this chapter deals with the molecular orbital approach because it is much simpler to use from a computational standpoint.

8.1 AN ELEMENTARY LOOK AT COVALENT BONDS

To begin the task of describing how molecules exist, it is useful to look first at a very simple picture. Consider the formation of the H_2 molecule as two hydrogen atoms are brought together from a very large distance, as shown in Fig. 8.1. As the atoms get closer together, the interaction between them increases until a covalent bond forms. Each atom is considered complete in itself, but an additional interaction occurs as the atoms get closer together, until the *molecule* can be represented as shown in Fig. 8.1C. The question should now be asked, "What do we know about this system?"

First, the interaction energy between a hydrogen nucleus and its electron is -13.6 eV, the binding energy of an electron in a hydrogen atom. The hydrogen *molecule* has a bond energy (BE) of 4.51 eV, which is required to *break* the bond or -4.51 eV when the bond *forms*. If the H_2 molecule is represented as shown in Fig. 8.1C, it becomes clear that the *molecule* differs from two *atoms* by the lines connecting each electron to both nuclei as well as the repulsion of the two nuclei and the two electrons. The dashed lines represent the interactions of the electron (1) with the nucleus (2) and the electron (2) with the nucleus (1). Despite the repulsions between the nuclei and between the electrons, the bond energy for the hydrogen molecule is 4.51 eV bond^{-1} (432 kJ mol^{-1}). Therefore, the attractions between the

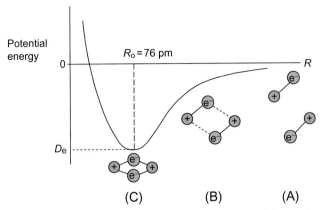

Fig. 8.1 The formation of the bond in H_2 as two H atoms approach from a long distance (A), at an intermediate distance (B), and at the equilibrium distance (C).

nuclei and the electrons more than offset the repulsions of like charged particles at the normal bond length of 0.76 Å (76 pm).

After the quantum mechanical treatment of the models shown in earlier chapters, it should not be expected that the quantum mechanical treatment of even diatomic molecules will be simple. The preceding discussion was intended to show that some of the basic parameters are intuitively known, so a great deal is already known about diatomic molecules. Some reference points are provided by the energies of electrons in atoms and the bond energies of the molecules.

The energies of chemical bonds are usually expressed as positive numbers that represent the energy required to break the bond:

$$A:B \rightarrow A + B, \quad \Delta H = \text{bond enthalpy(positive)} \qquad (8.1)$$

When the bond forms, the same quantity of energy is involved, but the energy is liberated (this is the negative of the bond enthalpy).

8.2 SOME SIMPLE RELATIONSHIPS FOR BONDS

It should be expected that some of the simplest diatomic molecules are those in which only one type of atomic orbital is utilized. Therefore, the discussion will begin by considering several molecules that are relatively simple from the standpoint of the orbitals used. Because s orbitals are singly degenerate and spherically symmetric, the simplest bonds between atoms are those in which only s orbitals are used. Fortunately, there is a rather wide range of molecules to consider. These include H_2, Li_2 ... Cs_2, LiH ... CsH, NaLi,

KNa, and RbNa, all bound with single bonds. Searching for relationships between properties of molecules has long been an honorable activity for persons seeking to understand chemical bonding, and the literature contains an enormous number of such relationships. Moreover, many of these relationships, whether empirical, semiempirical, or theoretical, are of great heuristic value. Before engaging the gears of the quantum machinery, a little chemical intuition will be applied.

Based on the simplest of assumptions, it would be expected that if everything else is equal, the *longer* a bond, the *weaker* it is likely to be. This is to be expected because even the most naive view of orbital interaction (overlap) would lead to the conclusion that the larger, more diffuse orbitals do not interact (overlap) as well as smaller, more compact ones. For the first attempt at verifying our intuition, a graph will be made showing bond energy versus bond length for the molecules previously listed. Table 8.1 shows the pertinent data for the atoms considered, and the graph that results is shown in Fig. 8.2. Although the study has been simplified by considering only those molecules using *s* orbitals in bonding, it is obvious that there is, as expected, a reasonably good relationship between bond energy and bond length for such cases. A relationship like that shown in Fig. 8.2 confirms that some of the elementary ideas about chemical bonds are correct. It would now be possible to find the equation for the line to obtain an empirical algebraic relationship.

Table 8.1 Properties of molecules bonded by *s* orbitals

Molecule	Bond length (pm)	Bond energy (kJ mol^{-1})	Bond energy (eV bond^{-1})	Average IP[a] (kJ mol^{-1})
H_2	74	432	4.48	1312
Li_2	267	105	1.09	520
Na_2	308	72.4	0.750	496
K_2	392	49.4	0.512	419
Rb_2	422	45.2	0.468	403
Cs_2	450	43.5	0.451	376
LiH	160	234	2.43	826
NaH	189	202	2.09	807
KH	224	180	1.87	741
RbH	237	163	1.69	727
CsH	249	176	1.82	702
NaLi	281	88.1	0.913	508
NaK	347	63.6	0.659	456
NaRb	359	60.2	0.624	447

[a]The ionization potentials (IPs) for H, Li, Na, K, Rb, and Cs atoms are the same as the average values for the diatomic molecules.

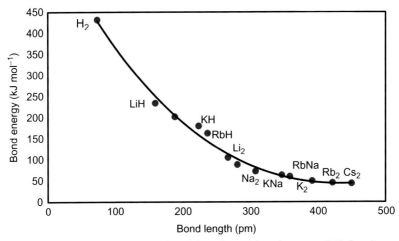

Fig. 8.2 The relationship between bond length and bond energy (BE) for diatomic molecules bonded by overlap of s orbitals.

Having "discovered" one relationship that fits bonds between atoms using s orbitals, a different approach will now be taken. It is known that metals, particularly those in Group IA, have low ionization potentials (IP), and therefore, do not have great attraction for electrons. If two alkali metals form a bond, it would be expected that the bond should be weak. However, the discussion will not be limited to alkali metals themselves, but rather include all the molecules for which data are shown in Table 8.1.

As a first and very crude approximation, it can be considered that an electron pair bond between atoms A and B results from a mutual attraction of these atoms for the pair of electrons. It is logical to assume that the attraction of an atom A for its outer electron can be approximated from its ionization potential. The same situation exists for atom B. Because the strength of the bond between atoms A and B reflects their mutual attraction for the electron pair, one might guess that the electron pair would be attracted by the two atoms with an energy that is related to the average of the ionization potentials of atoms A and B. Like most things in life, the behavior of atoms is not usually this simple, so it is to be hoped that the bond energy will be *related* to the average ionization in some simple way.

The first problem to be confronted is how to calculate the "average" ionization potential of two atoms. The two approaches to getting a mean value for the two parameters x_1 and x_2 are the *arithmetic mean*, which is $(x_1 + x_2)/2$, and the *geometric mean*, which is $(x_1 x_2)^{1/2}$. When the two quantities whose average value is to be determined are of similar magnitudes, the ways of

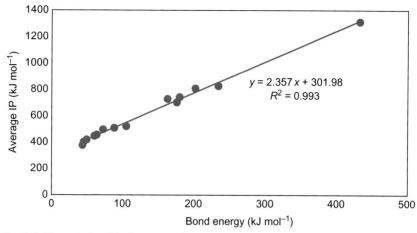

Fig. 8.3 The relationship between the average ionization potential (IP) (geometric mean) and the bond energy for molecules having overlap of s orbitals.

getting an average value give about the same result. In fact, if $x_1 = x_2$, the averages are identical. On the other hand, if one of the quantities is zero ($x_2 = 0$), the arithmetic mean is $x_1^{1/2}$, but the geometric mean is zero. Because bonds between atoms having greatly differing ionization potentials are to be considered, the geometric mean of the ionization potentials will be used. Figure 8.3 shows a graph of the average ionization potential versus bond energy for a series of diatomic molecules in which only s orbitals are used in bonding.

In this case, the relationship is indeed very good. Linear regression applied to the data gives the equation

$$IP_{av}\left(kJ\ mol^{-1}\right) = 2.357E\left(kJ\ mol^{-1}\right) + 301.98\left(r = 0.993\right) \qquad (8.2)$$

Not only is the intuitive approach correct, but also a relationship has been obtained that can be used for predictive purposes. For example, francium is a radioactive element for which a relatively small amount of data exists. The ionization potential for francium is about 3.83 eV or 369 kJ mol^{-1}. Using the relationship shown in Eq. (8.2), a bond energy of only 28.5 kJ mol^{-1} (6.81 kcal mol^{-1} or 0.295 eV bond^{-1}) would be predicted for the Fr$_2$ molecule. In a similar manner, the equation could estimate the bond energies of CsK, CsLi, etc.

It is interesting to speculate on what the intercept of 302.2 kJ mol^{-1} means. One interpretation is that the intercept, which corresponds to a hypothetical bond energy of 0, occurs when two atoms have such a low average ionization potential (302.2 kJ mol^{-1}) that they have no residual attraction for other electrons. The atom having the lowest ionization potential is F_I (369 kJ mol^{-1}), and the Fr_2 bond is very weak. If there were two atoms having an average ionization energy of 302.2 kJ mol^{-1}, their attraction for a bonding pair of electrons would be so slight that they should form no bond between them. Of course, this assumes that the relationship is valid outside the range for which data are available to test it.

In this instance, our intuition that the ionization potentials of the atoms forming the bonds ought to be related to the bond energy is completely justified. Of course, the analysis was restricted to molecules using only s orbitals, but the results are still gratifying. Any relationship between bond energies and atomic properties that gives a correlation coefficient of 0.993 is interesting. Throughout the study of the chemical bond, many workers have sought to correlate bond energy with such properties as the difference in electronegativity between the atoms and force constants for bond stretching. In most cases, a rather restricted list of molecules must be considered if a reasonably good relationship is expected. The discussion will now turn to a quantum mechanical description of diatomic molecules.

8.3 THE LCAO-MO METHOD

The linear combination of the atomic orbitals–molecular orbital (LCAO–MO) method is based on the idea that a wave function for a molecule (ψ) can be written as a *linear combination* of atomic wave functions (φ_i). This can be expressed by the equation

$$\psi = \Sigma a_i \varphi_i \tag{8.3}$$

For a diatomic molecule, this relationship reduces to

$$\psi = a_1 \varphi_1 + a_2 \varphi_2 \tag{8.4}$$

where φ is a wave function for an atomic orbital and a is a weighting or mixing coefficient. As will also be shown, $a_1 \varphi_1 - a_2 \varphi_2$ is also a possible linear combination. The values of the coefficients must be determined, and they are treated as parameters to be obtained by the optimization of the molecular wave function using the variation method (see Section 4.5). In applying the variation method, the first step is representing the value of the energy using

$$E = \frac{\int \psi^* \hat{H} \psi \, d\tau}{\int \psi^* \psi \, d\tau} \tag{8.5}$$

Substituting the trial wave function for ψ gives

$$E = \frac{\int (a_1\varphi_1{}^* + a_2\varphi_2{}^*)\hat{H}(a_1\varphi_1 + a_2\varphi_2)d\tau}{\int (a_1\varphi_1{}^* + a_2\varphi_2{}^*)(a_1\varphi_1 + a_2\varphi_2)d\tau} \tag{8.6}$$

Expansion of the binomials leads to

$$E = \frac{a_1{}^2\int \varphi_1{}^*\hat{H}\varphi_1 d\tau + 2a_1 a_2\int \varphi_1{}^*\hat{H}\varphi_2 d\tau + a_2{}^2\int \varphi_2{}^*\hat{H}\varphi_2 \, d\tau}{a_1{}^2\int \varphi_1{}^*\varphi_1 \, d\tau + 2a_1 a_2\int \varphi_1{}^*\varphi_2 d\tau + a_2{}^2\int \varphi_2{}^*\varphi_2 d\tau} \tag{8.7}$$

In writing this equation in this form, it has already been assumed that

$$\int \varphi_1{}^*\hat{H}\varphi_2 d\tau = \int \varphi_2{}^*\hat{H}\varphi_1 d\tau \tag{8.8}$$

and that

$$\int \varphi_1{}^*\varphi_2 d\tau = \int \varphi_2{}^*\varphi_1 d\tau \tag{8.9}$$

In other words, the discussion has been restricted to a *homonuclear* diatomic molecule. By an inspection of Eq. (8.8), it is apparent that it will be necessary to deal with integrals that have the forms

$$H_{11} = \int \varphi_1{}^*\hat{H}\varphi_1 d\tau \tag{8.10}$$

and

$$H_{12} = \int \varphi_1{}^*\hat{H}\varphi_2 d\tau \tag{8.11}$$

These integrals are frequently represented in a kind of shorthand notation as

$$\langle \varphi_1{}^*|\hat{H}|\varphi_1 \rangle \quad \text{and} \quad \langle \varphi_1{}^* |\hat{H}|\varphi_2 \rangle$$

Similarly, the expansion of the denominator of Eq. (8.7) leads to integrals of the type

$$S_{11} = \int \varphi_1{}^*\varphi_1 d\tau \tag{8.12}$$

and

$$S_{12} = \int \varphi_1{}^* \varphi_2 d\tau \tag{8.13}$$

These integrals are represented in shorthand notation as

$$\langle \varphi_1{}^* | \varphi_1 \rangle \quad \text{and} \quad \langle \varphi_1{}^* | \varphi_2 \rangle$$

respectively. An explanation of the meaning of integrals of these types will now be given.

The integrals represented as H_{11} and H_{22} represent the energies with which an electron is held in atoms 1 and 2, respectively. This is known because these integrals contain the wave function for an electron in each atom, and the Hamiltonian operator is the operator for total energy. Moreover, the binding energy of an electron in an atom is simply the reverse of its ionization potential, which is *positive* because it requires work to remove an electron from an atom. Therefore, these *binding* energies are *negative*, with their magnitudes being determined from the *valence state ionization potential* (VSIP), an application of *Koopmans' theorem* that states the ionization potential is equal in magnitude to the orbital energy. These integrals represent the Coulombic attraction for an electron in an atom and are accordingly called *Coulomb integrals.*

In a very loose way, the integrals represented as H_{12} and H_{21} are indicative of the attraction that nucleus 1 has for electron 2 and vice versa. These integrals are called the *exchange integrals* because they represent types of "exchange" attractions between the two atoms. It will be shown later that these integrals are of paramount importance in determining the bond energy for a molecule. Determining the magnitude of the exchange integrals represents a considerable part of the challenge of finding a quantum mechanical approach to bonding. It should be apparent that the values of H_{12} and H_{21} are related in *some* way to the bond length. It follows that if the two atoms are pulled completely apart, the nucleus of atom 1 would not attract the electron from atom 2 and vice versa.

The type of integral represented as S_{11} and S_{22} is called an *overlap integral.* These integrals give a view of how effectively the orbitals on the two atoms overlap. It should be clear that if the atomic wave functions are normalized, then

$$S_{11} = S_{22} = \int \varphi_1{}^* \varphi_1 d\tau = \int \varphi_2{}^* \varphi_2 d\tau = 1 \tag{8.14}$$

Therefore, these integrals do not present a difficulty, but those of the type

$$S_{21} = \int \varphi_2{}^* \varphi_1 \, d\tau \qquad (8.15)$$

$$S_{12} = \int \varphi_1{}^* \varphi_2 \, d\tau \qquad (8.16)$$

do pose a problem. It should be inferred that these integrals represent the overlap of the wave function of the orbital from atom 1 with that of an orbital from atom 2 and vice versa. Also, it is no surprise that the values of these integrals depend on the internuclear distance. If the two atoms were pushed closer and closer together until the two nuclei were at the same point, then the overlap would be complete, and the integrals S_{12} and S_{21} would be equal to 1. On the other hand, if the nuclei are pulled farther and farther apart to a distance of infinity, there would be no overlap of the atomic orbitals from atoms 1 and 2; therefore, S_{12} and S_{21} would be equal to 0. Thus, the overlap integral must vary from 0 to 1, and its value must be a function of the distance between the two atoms. Furthermore, because the values of integrals of the type H_{12}, etc., are also determined by bond length, it is apparent that it should be possible to find a relationship between H_{12} and S_{12}. As will be shown, this is precisely the case.

Making the substitutions for the integrals that appear in Eq. (8.7) leads to

$$E = \frac{a_1{}^2 H_{11} + 2a_1 a_2 H_{12} + a_2{}^2 H_{22}}{a_1{}^2 + 2a_1 a_2 S_{12} + a_2{}^2} \qquad (8.17)$$

Note that S_{11} and S_{22} have been omitted from the denominator in Eq. (8.17), their values being 1 if the atomic wave functions are normalized. It must be kept in mind that it is necessary to find the values for the weighting parameters a_1 and a_2 introduced in Eq. (8.4) that make the energy a minimum. This is done by taking the derivative of Eq. (8.17) with respect to a_1 and a_2 and setting them equal to 0:

$$\left(\frac{\partial E}{\partial a_1}\right)_{a_2} = 0 \quad \text{and} \quad \left(\frac{\partial E}{\partial a_2}\right)_{a_1} = 0 \qquad (8.18)$$

Differentiating Eq. (8.17) with respect to a_1 and simplifying gives Eq. (8.19). Repeating the differentiation with respect to a_2 gives Eq. (8.20):

$$a_1(H_{11} - E) + a_2(H_{12} - S_{12}E) = 0 \qquad (8.19)$$

$$a_1(H_{21} - S_{21}E) + a_2(H_{22} - E) = 0 \qquad (8.20)$$

Equations (8.19) and (8.20), known as the *secular equations*, constitute a pair of linear equations that can be represented by the general forms

$$ax + by = 0$$

$$cx + dy = 0 \qquad (8.21)$$

Of course, it is obvious that the equations are satisfied if all of the coefficients are 0 (the so-called *trivial solution*). A theorem of algebra, stated here without proof, requires that for a nontrivial solution, the determinant of the coefficients must be 0. Therefore,

$$\begin{vmatrix} a & b \\ c & d \end{vmatrix} = 0 \qquad (8.22)$$

In the secular equations, the values of a_1 and a_2 are the unknown quantities that need to be evaluated. Therefore, the determinant of the coefficients can be written as

$$\begin{vmatrix} H_{11} - E & H_{12} - S_{12}E \\ H_{21} - S_{21}E & H_{22} - E \end{vmatrix} = 0 \qquad (8.23)$$

This determinant is known as the *secular determinant*.

If the two atoms are identical (a homonuclear diatomic molecule), it is apparent that $H_{11} = H_{22}$ and $H_{12} = H_{21}$. Although it may not be quite as obvious, $S_{12} = S_{21}$ also, and both of these integrals will be represented simply as S at this time. Therefore, expanding the determinant yields

$$(H_{11} - E)^2 - (H_{12} - SE)^2 = 0 \qquad (8.24)$$

or

$$(H_{11} - E)^2 = (H_{12} - SE)^2 \qquad (8.25)$$

Taking the square root of both sides of Eq. (8.25) gives

$$H_{11} - E = \pm(H_{12} - SE) \qquad (8.26)$$

Solving this equation for E gives the values

$$E_b = \frac{H_{11} + H_{12}}{1 + S} \quad \text{and} \quad E_a = \frac{H_{11} - H_{12}}{1 - S} \qquad (8.27)$$

For these two energy states, E_b is referred to as the *symmetric* or *bonding* state, and E_a is called the *asymmetric* or *antibonding* state. The combinations of the wave functions are shown graphically in Fig. 8.4.

For the moment, it will be assumed that the molecule being considered is the simplest molecule, H_2^+. The integral H_{11} represents the binding energy

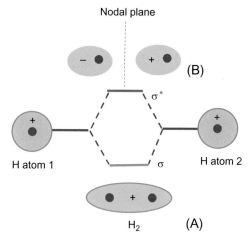

Fig. 8.4 Electron density contours for the bonding (A) and antibonding (B) σ orbitals formed from a linear combination of two s wave functions. *(Reproduced with permission from House, J. E. Inorganic Chemistry. 2nd ed.; Academic Press/Elsevier: Amsterdam, 2013.)*

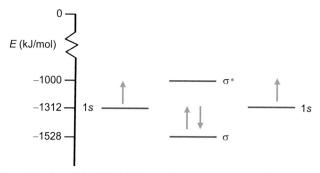

Fig. 8.5 The molecular orbital (MO) diagram for a hydrogen molecule.

of the electron in the $1s$ state to nucleus 1 and H_{12} the additional attraction of the second nucleus to this negative charge. The integral H_{12} is the expression for the energy that results from the fact that the electron can also reside in the $1s$ state of the second hydrogen atom. As such, it is often called an *exchange integral*, and it also represents a negative energy or a binding energy. Consequently, the energy denoted as E_b represents the lower energy, and the state is referred to as the *bonding* state (where electrons reside). The state of higher energy E_a is called the *antibonding* state. The orbitals used in H_2 are identical to those used in H_2^+, although the energies are obviously different. Figure 8.5 shows the molecular orbital diagram for the H_2 molecule.

Substituting the values for the energies shown in Eq. (8.27) into Eqs. (8.19) and (8.20) yields

$$a_1 = a_2 \text{(symmetric state)} \tag{8.28}$$

$$a_1 = -a_2 \text{(antisymmetric state)} \tag{8.29}$$

Consequently, the wave functions corresponding to the energy states E_b and E_a can be written as (in which S is used in place of both S_{12} and S_{21}):

$$\psi_b = a_1\varphi_1 + a_2\varphi_2 = \frac{1}{\sqrt{(2+2S)}}(\varphi_1 + \varphi_2) \tag{8.30}$$

$$\psi_a = a_1\varphi_1 - a_2\varphi_2 = \frac{1}{\sqrt{(2-2S)}}(\varphi_1 - \varphi_2) \tag{8.31}$$

The normalization constants represented as A are obtained as follows:

$$1 = \int A^2(\varphi_1 + \varphi_2)^2 d\tau = A^2\left[\int \varphi_1^2 d\tau + \int \varphi_2^2 d\tau + 2\int \varphi_1\varphi_2 d\tau\right] \tag{8.32}$$

If the atomic wave functions φ_1 and φ_2 are normalized, then

$$\int \varphi_1\varphi_1 d\tau = 1 \quad \text{and} \quad \int \varphi_2\varphi_2 d\tau = 1$$

so that

$$1 = A^2[1 + 1 + 2S]$$

Solving for A gives the normalization constant:

$$A = \frac{1}{\sqrt{(2+2S)}} \tag{8.33}$$

The combination of two $2s$ orbitals leads to the same type of molecular wave functions that we have already shown. Therefore, the molecular orbital diagram for Li_2, shown in Fig. 8.6, has the same appearance as that for H_2, except that the energies of the atomic orbitals and the molecular orbitals are quite different because the binding energy of the $2s$ electron in lithium is quite low. Information about the energies of molecular orbitals is obtained from photoelectron spectroscopy (House, 2013).

8.4 DIATOMIC MOLECULES OF THE SECOND PERIOD

In considering molecules composed of second-row atoms, it is necessary to include the combination of the p orbitals in forming MOs. In accord with the convention regarding symmetry (see Chapter 10), the direction lying

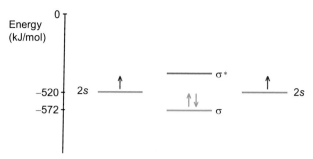

Fig. 8.6 The molecular orbital diagram for the lithium molecule, Li_2.

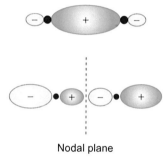

Nodal plane

Fig. 8.7 The bonding *(top)* and antibonding *(bottom)* orbitals that result from combining two $2p_z$ orbitals.

along the bond is taken to be the z-axis. Therefore, when the $2p_z$ orbitals overlap (i.e., the wave functions are combined), a σ bond is formed that is symmetric around the internuclear axis. The wave functions produced by the combination of two $2p_z$ wave functions can be written as

$$\psi(p_z) = \frac{1}{\sqrt{2+2S}}[\varphi(z_1) + \varphi(z_2)] \tag{8.34}$$

$$\psi^*(p_z) = \frac{1}{\sqrt{2-2S}}[\varphi(z_1) - \varphi(z_2)] \tag{8.35}$$

A pictorial representation of the bonding and antibonding orbitals is shown in Fig. 8.7

After the p_z orbitals have formed a σ bond, the p_x and p_y orbitals can form π bonds. The molecular orbital produced by the combination of two p_y orbitals produces a node in the xz plane, and the combination of the p_x orbitals produces a node in the yz plane. The combinations of atomic wave functions have the same form as those shown in Eqs. (8.34) and (8.35),

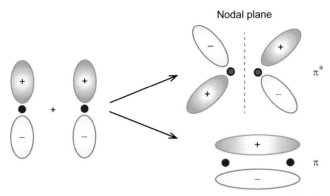

Fig. 8.8 Electron density contours for bonding and antibonding π orbitals.

except for the designation of the atomic orbitals. The electron density contours for these orbitals are symmetric on either side of the internuclear axis and represent π bonds. The electron density plots for the two π orbitals are shown in Fig. 8.8.

The combination of the $2p_y$ wave functions also produces π and π* orbitals, so the combination of two sets of three p orbitals form six molecular orbitals, showing the number of molecular orbitals produced is equal to the number of atomic orbitals combined. The antibonding states lie higher in energy, but there is still a problem in ordering the energies of the σ and two π bonding molecular orbitals. For some molecules, the σ orbital lies below the two degenerate π orbitals, whereas in other cases the two π orbitals lie lower in energy than the σ orbital. The reason for this is that for atoms early in the period (e.g., B and C), the 2s and 2p atomic orbitals are not greatly different in energy. This allows for partial "mixing" (or hybridization) of the atomic states *before* the molecular orbitals form. In this case, the π orbitals are stabilized whereas the σ orbital is destabilized with the result that the two π orbitals lie lower in energy. For atoms later in the period (e.g., O and F), the higher nuclear charge causes the 2s orbital to have an energy well below that of the 2p orbitals so that there is no effective mixing of the atomic states. In these cases, the σ orbital lies lower in energy than the two π orbitals.

For the nitrogen molecule, some uncertainty might be expected in regard to the relative energies of the σ and π orbitals. However, the experimental evidence from photoelectron spectroscopy indicates that when an electron is removed, it is from a molecular orbital having σ character rather than one of the π orbitals. Therefore, it is believed that the π orbitals in N_2 have lower energy than that of the σ orbital.

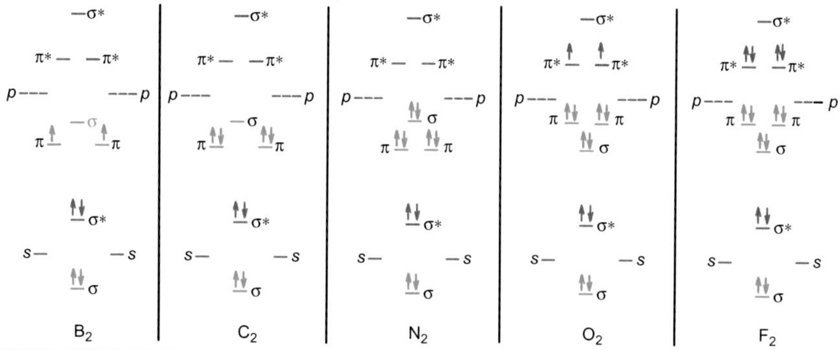

	B_2	C_2	N_2	O_2	F_2
BO	1	2	3	2	1
R, pm	159	131	109	121	142
BE, eV	3.0	5.9	9.8	5.1	1.6

Fig. 8.9 Molecular orbital diagrams for second-row homonuclear diatomic molecules. *(Reproduced with permission from House, J. E. Inorganic Chemistry. 2nd ed.; Academic Press/Elsevier: Amsterdam, 2013.)*

Figure 8.9 shows the molecular orbital diagrams for the diatomic molecules of the second period that involve $2s$ and $2p$ atomic orbitals to form molecular orbitals.

Now that the order of orbitals with respect to increasing energy has been determined, it can also be shown how the orbitals are occupied with the appropriate number of electrons according to Hund's rule. Figure 8.9 also includes data for the diatomic molecules that provide information regarding the population of the molecular orbitals. For example, the fact that the B_2 molecule has two unpaired electrons clearly indicates that the degenerate π orbitals lie below the σ orbital. If the σ orbital were lower in energy, it would contain a pair of electrons, and the molecule would be diamagnetic. The C_2 molecule is diamagnetic, but if the σ orbital were lower in energy than the two π orbitals, that orbital would be filled, and the two π orbitals would hold one electron each. Therefore, in the C_2 molecule, the π orbitals have lower energy than the σ orbital. Moreover, the O_2 molecule has two unpaired electrons in the two π^* orbitals and it is paramagnetic. Figure 8.9 also gives values for the *bond order (B)*, which is defined as

$$B = \frac{N_b - N_a}{2} \tag{8.36}$$

where N_b is the number of electrons in bonding orbitals and N_a is the number of electrons in antibonding orbitals. Also shown in Fig. 8.9 are the bond energies for the molecules, from which it is clear that the bond energy

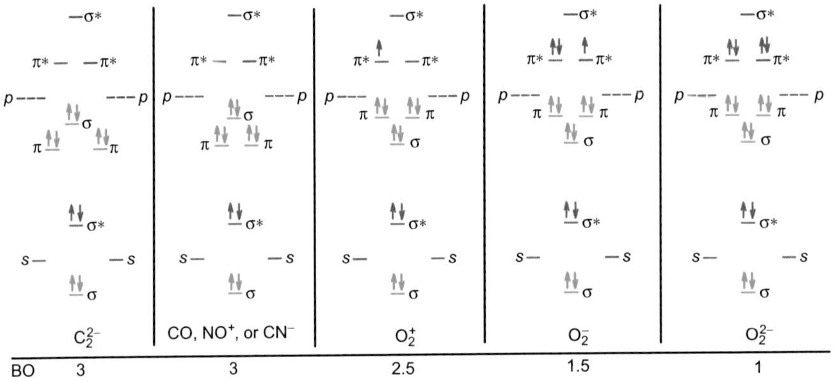

C_2^{2-}	CO, NO$^+$, or CN$^-$	O_2^+	O_2^-	O_2^{2-}

BO	3	3	2.5	1.5	1

Fig. 8.10 Molecular orbital diagrams for diatomic ions and heteronuclear molecules of second-row elements. *(Reproduced with permission from House, J. E. Inorganic Chemistry. 2nd ed.; Academic Press/Elsevier: Amsterdam, 2013.)*

increases roughly with bond order. This is an expected result because the greater the number of electrons involved in bond formation, the stronger the bond between the two atoms.

Several important heteronuclear diatomic species are derived from the atoms of the second period. These include heteronuclear molecules like CO and NO, for which the molecular orbital diagrams are shown in Fig. 8.10. Moreover, there are well-characterized chemical species that are derived from the O_2 molecule by either the loss or gain of electrons. Such species include O_2^+ (dioxygenyl ion) and O_2^- (superoxide ion), for which molecular orbital diagrams are also shown in Fig. 8.10. For example, the O_2^+ ion is generated by reaction of oxygen with PtF_6:

$$O_2 + PtF_6 \rightarrow O_2^+ + PtF_6^- \tag{8.37}$$

Potassium reacts with oxygen to form O_2^-:

$$K + O_2 \rightarrow KO_2 \tag{8.38}$$

The electron lost from O_2 to produce the O_2^+ ion comes from an antibonding orbital, which results in an increase in the bond order from 2 to 2.5. By the addition of two electrons to the O_2 molecule, the peroxide ion O_2^{2-} is produced. As a result of the peroxide ion having a bond order of 1, it would be expected that it would not be a very stable species. In agreement with that prediction, the bond strength is weak and some peroxides are not very stable chemical species. All of these oxygen-containing species are important in the chemistry of oxygen.

Another interesting diatomic species is the NO molecule, which has a single electron in a π^* orbital. In this case, the molecule can be ionized rather easily, and the resulting NO^+ is then isoelectronic with CN^- and CO. Therefore, in some metal complexes, NO behaves as a three-electron donor, with one electron donated by way of ionization of NO and an electron pair donated in the usual coordinate bond formation. The properties of many diatomic molecules are presented in Table 8.2.

Table 8.2 Properties for diatomic molecules and ions

Species	N_b	N_a	B^a	R (pm)	DE^b (eV)
H_2^+	1	0	0.5	106	2.65
H_2	2	0	1	74	4.75
He_2^+	2	1	0.5	108	3.1
Li_2	2	0	1	262	1.03
B_2	4	2	1	159	3.0
C_2	6	2	2	131	5.9
N_2	8	2	3	109	9.76
O_2	8	4	2	121	5.08
F_2	8	6	1	142	1.6
Na_2	2	0	1	308	0.75
Rb_2	2	0	1	–	0.49
S_2	8	4	2	189	4.37
Se_2	8	4	2	217	3.37
Te_2	8	4	2	256	2.70
N_2^+	7	2	2.5	112	8.67
O_2^+	8	3	2.5	112	6.46
BN	6	2	2	128	4.0
BO	7	2	2.5	120	8.0
CN	7	2	2.5	118	8.15
CO	8	2	3	113	11.1
NO	8	3	2.5	115	7.02
NO^+	8	2	3	106	–
SO	8	4	2	149	5.16
PN	8	2	3	149	5.98
SiO	8	2	3	151	8.02
LiH	2	0	1	160	2.5
NaH	2	0	1	189	2.0
PO	8	3	2.5	145	5.42

[a] B is the bond order, $(N_a - N_b)/2$.
[b] DE is the dissociation energy (1 eV = 96.48 kJ mol^{-1}).

8.5 OVERLAP AND EXCHANGE INTEGRALS

By now it should be apparent that the values of overlap integrals must be available if the energies of MOs are to be calculated. Over 60 years ago, Mulliken and colleagues evaluated the overlap integrals for a large number of cases using the Slater-type orbitals described in Chapter 5 (Mulliken et al., 1949). Because orbital overlap is crucial to describing covalent bonds, a convenient system of adjustable parameters was developed for determining overlap integrals, which depend on the type of orbitals and the nature of the atoms bonded. Two parameters, designated as p and t, are used in the Mulliken tables (Mulliken et al., 1949). These values are determined by the value of μ in the Slater wave functions written in the form

$$\phi = N_{nl} r^{n-1} e^{-\mu r/a_0} \tag{8.39}$$

as well as the internuclear distance R. For two atoms denoted by i and j, the parameters are defined as

$$p = 0.946R\left(\mu_i + \mu_j\right) \tag{8.40}$$

$$t = \frac{\left(\mu_i - \mu_j\right)}{\left(\mu_i + \mu_j\right)} \tag{8.41}$$

In these equations, R is the internuclear distance expressed in Angstroms, and μ is the quantity $(Z - S)/n$ from the Slater wave functions:

$$\psi = [R_{n,l}(r)] \exp\left[-(Z - S)r/a_0 n^*\right] Y(\theta, \varphi) \tag{8.42}$$

When the spherical harmonics are combined with the radial function $R(r) = r^{n^*-1}$, the resulting wave function is written as

$$\psi = Nr^{n^*-1} \exp\left(-\mu r/a_0\right) \tag{8.43}$$

where N is the normalization constant.

For atoms having $n \leq 3$, $n = n^*$, as was shown in the list of rules in Chapter 5 for constructing Slater wave functions. A brief list of μ values is shown in Table 8.3.

Table 8.3 Some values for μ for use in calculating overlap integrals

H	1.00	Li	0.65	Be	0.975
B	1.30	C	1.625	B	1.95
O	2.275	D	2.60	Na	0.733

If the calculations are for two orbitals on atoms that have different n values, the subscript i represents the orbital of smaller n. For cases where the orbitals have the same n, the i subscript represents the larger t value. The publication by Mulliken et al. includes extensive tables of values for the overlap integrals (Mulliken et al., 1949). The tables include the values for orbital integrals for orbital combinations of $1s$, $1s$; $1s$, $2s$; $2s$, $2s$; etc., with the integral being evaluated over a wide range of p and t values that cover a wide range of internuclear distances and types of atoms. These tables should be consulted if values for specific overlap integrals are needed. For a simple case such as H_2, the μ values are identical (each is 1.00), so $t=0$. Because the internuclear distance is 0.74 Å, the value for p is calculated to be 1.40. Consulting the appropriate table for $1s$–$1s$ orbital overlap shows a value of 0.753, which indicates very effective overlap of the $1s$ orbitals.

As shown earlier the energy of the bonding molecular orbital can be written as

$$E_b = \frac{H_{11} + H_{12}}{1 + S} \tag{8.44}$$

The value of H_{11} is obtained from the VSIP for the atom. If the molecular orbital holds two electrons, the bond energy is the difference between the energy of two electrons in the bonding state and their energy in the valence shells of the separated atoms. Therefore,

$$BE = 2H_{11} - 2\left(\frac{H_{11} + H_{12}}{1 + S}\right) \tag{8.45}$$

Therefore, the energy of the bond will be approximately $-2H_{12}/(1 + S)$, but it is necessary to have values for H_{12} and S. As discussed earlier, the integral H_{12} represents the exchange integral and is a negative quantity. The interaction that gives rise to this integral disappears at an infinite internuclear distance, so $H_{12} = 0$ at $R = \infty$. However, it is also apparent that the overlap of the orbitals also becomes 0 when the atoms are separated by an infinite distance. Thus, there are two quantities that have values of 0 at infinity, but, S becomes larger as the atoms get closer together, whereas H_{12} becomes more negative under these conditions. What is needed is a way to express H_{12} as a function of the overlap integral. However, one might also *suspect* that the exchange integral should be related to the energies with which electrons are bonded to atoms 1 and 2. These energies are the electron binding energies H_{11} and H_{22}. Three ways of connecting these properties are commonly encountered in describing bonds.

In a first attempt, the value of H_{12} will be expressed in terms of S and the *average* of H_{11} and H_{22}. As a result of greater binding energies, one could assume that increasing the average of H_{11} and H_{22} should also increase H_{12}. By taking these factors into consideration, Mulliken assumed many years ago that the so-called off-diagonal elements (H_{12}) should be proportional to the overlap integral (Mulliken et al., 1949). The function that results can be written as

$$H_{12} = -KS \left(\frac{H_{11} + H_{22}}{1 + S} \right) \tag{8.46}$$

where K is a proportionality constant that has a numerical value of about 1.75 (a rather wide range of values has been used). This relationship is known as the *Wolfsberg-Helmholtz approximation* and is one of the most widely used approximations for the exchange integrals in molecular orbital calculations (Wolfsberg and Helmholtz, 1952). Unfortunately, there is no clear agreement on the value to be used for K. Roald Hoffmann has given a detailed analysis of the factors involved in the choice of K value (Hoffmann, 1963).

Because atoms with considerably different ionization potentials must frequently be considered, the *arithmetic* mean used in the Wolfsberg-Helmholtz approximation may not be the best way to calculate an average. As was shown in Section 8.2, the geometric mean is preferable in some instances. Therefore, the exchange integral can be represented using the *geometric* mean of the ionization potentials to give

$$H_{12} = -KS(H_{11}H_{22})^{1/2} \tag{8.47}$$

This relationship is known as the *Ballhausen-Gray approximation* (Ballhausen and Gray, 1962). It was previously mentioned that the bond energy can be written as $-2H_{12}/(1 + S)$, so suddenly it can be seen why the relationship given in Eq. (8.2), which gives the bond energy in terms of the geometric mean of the ionization potential, works so well! The intuitive approach used earlier is equivalent to the Ballhausen-Gray approximation (Ballhausen and Gray, 1962).

A chemical bond between two atoms has an energy related to the bond length by a potential energy curve like that shown in Fig. 8.1. There is a minimum energy at the equilibrium distance. Neither of the expressions shown in Eqs. (8.46) and (8.47) goes through a minimum when the bond length varies. The value of H_{12} simply gets more negative as the bond length decreases because the value of S increases as the bond gets shorter. An approximation for H_{12} that *does* show a minimum as internuclear distance changes is that given by Cusachs (1965),

$$H_{12} = \frac{1}{2}S(K - |S|)(H_{11} + H_{22}) \tag{8.48}$$

This function passes through a minimum with respect to bond length because S is a function of bond length, and this function is a quadratic in S. As a result, the Cusachs approximation to H_{12} is the most nearly correct of the three presented. However, the Wolfsberg-Helmholtz and Ballhausen-Gray approximations are still widely used.

8.6 HETERONUCLEAR DIATOMIC MOLECULES

In the case of homonuclear diatomic molecules, the wave functions for the bonding and antibonding states were shown to be expressed as

$$\psi_b = \frac{1}{\sqrt{2}}(\varphi_1 + \varphi_2) \tag{8.49}$$

and

$$\psi_a = \frac{1}{\sqrt{2}}(\varphi_1 - \varphi_2) \tag{8.50}$$

The electron density contours of these MOs were found to be symmetrical about the center of the internuclear axis. In the case of heteronuclear diatomic molecules, this is not true because the bulk of the bonding orbital lies toward the atom having the higher electronegativity. Thus, the form of the molecular wave functions are changed by including a weighting factor to take this difference into account. Accordingly, the wave functions are written as

$$\psi_b = \varphi_1 + \lambda\varphi_2 \tag{8.51}$$

$$\psi_a = \varphi_1 - \lambda\varphi_2 \tag{8.52}$$

in which the parameter λ takes into account the difference in electronegativity of the two atoms.

The extent to which the atomic orbitals of two atoms will combine to produce molecular orbitals depends upon the relative energies of the orbitals. The closer the energies of the orbitals on the two atoms, the more complete the "mixing" and the atomic orbitals lose their individuality more completely.

In considering the bonding in diatomic molecules composed of atoms of different types, there is an additional factor that must be kept in mind. This additional factor arises from the fact that more than one Lewis structure can be drawn for the molecule. In valence bond terms, when more than one acceptable structure can be drawn for a molecule or ion, the structures

are called resonance structures, and the actual structure is said to be a resonance hybrid of all the structures with appropriate weightings for each. As
will soon be described, the *weighting coefficient* λ takes into account the
contribution of the ionic structure to the wave function. Thus, the adjustable coefficients in a wave function serve the same purpose as being able to
draw more than one valence bond structure for the molecule.

For a molecule AB, the possible structures can be shown as

$$A - B \underset{I}{\longleftrightarrow} A^{+} \cdots B^{-} \underset{II}{\longleftrightarrow} A^{-} \cdots B^{+}$$

Even for nonpolar molecules like H_2, the ionic structures contribute a substantial amount to the overall stability of the molecule. It can be shown that
the added stability in the case of H_2 amounts to about 0.24 eV molecule^{-1}
(23 kJ mol^{-1}). For the AB molecule described by the structures above, a
wave function can be written to take the three structures into account.
The result is

$$\psi_{molecule} = a\psi_{I} + b\psi_{II} + c\psi_{III} \tag{8.53}$$

where a, b, and c are constants and ψ_{I}, ψ_{II}, and ψ_{III} represent wave functions
corresponding to resonance structures I, II, and III, respectively. For a mol-
ecule such as H_2, two ionic structures contribute equally, but they contribute much less than does the covalent structure. so the coefficients in the wave
function are related by $a \gg b = c$. In the case of a heteronuclear molecule,
one of the ionic structures is usually insignificant because it is unrealistic
to expect the atom having the *higher* electronegativity to assume a positive
charge and the atom having the *lower* electronegativity to assume a negative
charge. For example, in HF the structures H-F and $H^{+}F^{-}$ would contribute
roughly equally, but the structure $H^{-}F^{+}$ is unrealistic owing to the much
higher electronegativity of fluorine. Accordingly, it is often possible to
neglect one of the structures of the most heteronuclear diatomic molecules
and write the wave function as

$$\psi_{molecule} = \psi_{covalent} + \lambda\psi_{ionic} \tag{8.54}$$

In this case ψ_{ionic} corresponds to the ionic structure having the negative
charge on the element of higher electronegativity. It is now important to
obtain the relative weightings of the covalent and ionic portions of the wave
function for the molecule. A procedure that makes use of dipole moments
provides the means to do so.

A purely covalent structure with equal sharing of the bonding electron
pair would result in a dipole moment of 0 for the molecule. Likewise, a

Table 8.4 Some data for HX molecules (X=a halogen)

Molecule	r, pm	μ_{obs}, D	μ_{ionic}, D	$100\mu_{obs}/\mu_{ionic}$	$\chi_X - \chi_H$
				% Ionic character	
HF	92	1.91	4.41	43	1.9
HCl	128	1.03	6.07	17	0.9
HBr	143	0.78	6.82	11	0.7
HI	162	0.38	7.74	5	0.4

1 Debye 10^{-18} esu cm. The electronegativities of atoms H and X are χ_H and χ_X.

completely ionic structure in which an electron is transferred would result in a dipole moment μ equal to $e \bullet r$, where e is the charge on the electron and r is the internuclear distance. The ratio of the observed dipole moment to that calculated for a completely ionic structure gives the relative ionic character of the bond (i.e., the percent ionic character). Consider the data shown in Table 8.4 for the hydrogen halides. The percent ionic character is given as

$$\% \text{ionic character} = 100\mu_{obs}/\mu_{ionic} \qquad (8.55)$$

It should be recalled at this point that it is the *square* of the wave function that is related to the weighting given to the structure described by the wave function. Consequently, λ^2 represents the weighting given to the ionic structure and $1 + \lambda^2$ gives the weighting of the contribution from both the covalent and the ionic structures. Therefore, the ratio of the weighting of the ionic structure to the total is $\lambda^2/(1 + \lambda^2)$. Thus,

$$\% \text{ionic character} = \frac{100\lambda^2}{\left(1 + \lambda^2\right)} \qquad (8.56)$$

from which it can be seen that

$$\frac{\lambda^2}{\left(1 + \lambda^2\right)} = \frac{\mu_{obs}}{\mu_{ionic}} \qquad (8.57)$$

For HF, the ratio of the observed dipole moment (1.91 D) to that for the assumed ionic structure (4.41 D) is 0.43. Therefore,

$$0.43 = \frac{\lambda^2}{\left(1 + \lambda^2\right)} \qquad (8.58)$$

From this equation, the value of λ can be found, and the result is $\lambda = 0.87$. Therefore, for HF, the molecular wave function can be written as

$$\psi_{\text{molecule}} = \psi_{\text{covalent}} + 0.87\psi_{\text{ionic}} \qquad (8.59)$$

Based on this model, HF appears to be 43% ionic, so the actual structure can be considered a hybrid consisting of 57% of the covalent structure and 43% of the ionic structure. This statement should not be taken too literally, but it does give an approach to the problem of wave functions for polar molecules. A similar procedure shows that HCl is 17% ionic with $\lambda = 0.45$, HBr is 11% ionic with $\lambda = 0.36$, and HI is 5% ionic with $\lambda = 0.23$, according to this model. It was mentioned earlier that including the ionic structures H^+H^- and H^-H^+ for H_2 results in increased stability of the molecule. This effect is also manifested for molecules like HF, as will now be described.

The difference between the *actual* bond dissociation energy and that predicted for the purely *covalent* bond provides a measure of this resonance stabilization energy, which is the "extra" energy provided as a result of the contribution from the ionic structure. Although the purely covalent bond does not exist, there are two commonly used methods of estimating its energy that were proposed by Linus Pauling. The first of these methods makes use of the postulate of the arithmetic mean, in which the energy of the hypothetical covalent bond between atoms A and B is taken as $(D_{AA} + D_{BB})/2$, where D_{AA} and D_{BB} are the bond dissociation energies of the molecules A_2 and B_2, respectively. Thus, the difference between the actual bond energy D_{AB} and $(D_{AA} + D_{BB})/2$ is the added stability of the bond Δ that results from the ionic resonance structure. Thus,

$$\Delta = D_{AB} - (D_{AA} + D_{BB})/2 \qquad (8.60)$$

The quantity Δ, which is the resonance energy, is always positive because the actual bond energy is greater than predicted for the purely covalent bond alone.[1] Pauling realized that the extent to which an ionic structure stabilizes a diatomic molecule is related to a fundamental difference in the ability of the atoms to attract electrons. He therefore related Δ to the difference in this property, which is now recognized as the electronegativity χ. Pauling's relationship can be shown as (Pauling, 1960)

$$\Delta = 23.06|\chi_A - \chi_B|^2 (\text{kcal mol}^{-1}) = 96.48|\chi_A - \chi_B|^2 (\text{kJ mol}^{-1}) \qquad (8.61)$$

The values of Δ can now be considered as experimentally determined quantities from which the electronegativities of A and B must be determined, but only the *difference* is known from Eq. (8.61). Obviously, the differences

---

[1] (Strictly, this is not entirely true. Pauling presents a discussion of cases like LiH, NaH, etc., for which Δ is negative (Pauling (1960), p. 82). However, such cases need not concern us now.)

$100 - 99$ and $3 - 2$ are exactly the same. Pauling *assigned* the value for fluorine as 4.0 so that all other atoms had electronegativities between 0 and 4.0. The constant 23.06 merely converts eV atom^{-1} to kcal mol^{-1}, because 1 eV molecule^{-1} is equivalent to 23.06 kcal mol^{-1}. Table 8.5 summarizes the results of similar calculations for all of the hydrogen halides.

Earlier in this chapter, the postulate of the geometric mean was used to determine the average ionization potential for two atoms. Pauling found also that the geometric mean gave better correlations of electronegativity and bond energy in cases where the atoms have considerably different electronegativities. In using the geometric mean, the bond energy for the hypothetical purely covalent structure D_{AB} was taken as $(D_{AA} \cdot D_{BB})^{1/2}$. Using this approximation, the ionic stabilization Δ' is given by

$$\Delta' = D_{AB} - (D_{AA} \cdot D_{BB})^{1/2} \tag{8.62}$$

The values of Δ' are not linearly related to differences in electronegativity, but $(\Delta')^{\frac{1}{2}}$ is approximately a linear function of $|\chi_A - \chi_B|$. Table 8.6 shows the resonance stabilization for the hydrogen halides obtained using the geometric mean approximation.

There are several other equations that are sometimes used to estimate the percent ionic character of bonds in heteronuclear molecules. These

Table 8.5 Resonance energies for hydrogen halides when the arithmetic mean is used to estimate the energy of the covalent AB bond

	HF	HCl	HBr	HI
D_{HX} (kJ mol^{-1})	563	431	366	299
$(D_{HH} + D_{XX})/2$ (kJ mol^{-1})	295	339	316	295
Δ (kJ mol^{-1})	268	92	50	4
Electronegativity difference	1.9	0.9	0.7	0.4

Note: The bond energies used are H$_2$, 436; F$_2$, 153; Cl$_2$, 243; Br$_2$, 193; and I$_2$, 151 kJ mol^{-1}.

Table 8.6 Resonance energies for hydrogen halides obtained using the geometric mean for calculating D_{HX}

	HF	HCl	HBr	HI
D_{HX} (kJ mol^{-1})	563	431	366	299
$(D_{HH} \cdot D_{XX})^{1/2}$ (kJ mol^{-1})	259	326	290	257
Δ' (kJ mol^{-1})	304	106	76	42
$\sqrt{\Delta'}$	17.4	10.3	8.7	6.5
Electronegativity difference	1.9	0.9	0.7	0.4

Note: The bond energies used are the same as those in Table 8.5.

are semiempirical equations that relate the percent ionic character to the difference in electronegativity for the atoms. Two of these equations are

$$\%\text{ionic character} = 16|\chi_A - \chi_B| + 3.5|\chi_A - \chi_B|^2 \qquad (8.63)$$

$$\%\text{ionic character} = 18|\chi_A - \chi_B|^{1.4} \qquad (8.64)$$

Although these equations appear quite different, the predicted percentage of ionic character is approximately the same when $|\chi_A - \chi_B|$ is in the range of 1–2. Using these equations in combination with Eq. (8.56) enables an estimate of the weighting constant λ to be made if the electronegativities of the atoms are known.

Not only is there a contribution of the ionic structure to the bond energy, but it also leads to a shortening of the bond. For diatomic molecules, the bond length is usually given as the sum of the covalent radii of the atoms. However, if the atoms have different electronegativities, the sum of the covalent radii does not accurately give the observed internuclear distance because of the contribution of the ionic structure. A more accurate value for the bond length of a molecule AB that has some ionic character is given by the Shoemaker-Stevenson equation,

$$r_{AB} = r_A + r_B - 9.0|\chi_A - \chi_B| \qquad (8.65)$$

where r_A and r_B are the covalent radii of atoms A and B, respectively, and χ_A and χ_B are their electronegativities. In the molecule ClF, the observed bond length is 163 pm. The covalent radii for Cl and F are 99 and 72 pm, respectively, which leads to an expected bond length of 171 pm for ClF. As a result of the electronegativities of Cl and F being 3.0 and 4.0, respectively, Eq. (8.65) predicts a bond length of 162 pm for ClF, in excellent agreement with the experimental value. The experimental bond energy for ClF is 253 kJ mol^{-1}, which is considerably greater than the covalent value of 198 kJ mol^{-1} predicted from the arithmetic mean or 193 kJ mol^{-1} predicted using the geometric mean.

Combinations of atomic orbitals from atoms having different electronegativities (and hence different valence orbital energies) produce molecular orbitals that have energies closer to that of the atomic orbital of lower energy. In fact, the greater the electronegativity difference between the atoms, the closer the bond comes to being ionic. The bonding molecular orbital in that case represents an *atomic* orbital on the atom having the higher electronegativity to which the electron is transferred. That is, the bond is essentially ionic. Figure 8.11 shows the change in energies of molecular orbitals from equal sharing to electron transfer as the difference in electronegativity between the two atoms increases.

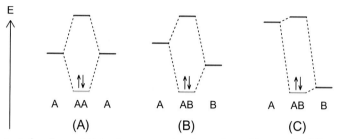

Fig. 8.11 Molecular orbital diagrams showing the effects of differences in electronegativity. (A) The two atoms have the same electronegativity. (B) Atom B has higher electronegativity. The molecular orbital has more of the character of an orbital of atom B (similar energy). (C) The difference in electronegativity is large enough so that the electron pair essentially resides in an orbital on atom B (ionic bond).

The concept of electronegativity provides one of the most useful principles available when it comes to dealing with chemical bonding. It provides a measure of the ability of an atom in a molecule to attract electrons to itself. Therefore, it is possible to predict bond polarities in most cases, although a few (e.g., CO) seem contradictory at first. However, there are several electronegativity scales with approximate values. Therefore, there are some difference in the values reported in different publications. Table 8.7 shows the electronegativities for atoms of the main group elements.

Because the basic idea regarding electron distribution in a bond is so useful, there have been numerous attempts to establish electronegativity scales based on atomic properties rather than on bond energies, as in the case of the Pauling scale. One such scale is the Mulliken scale, which predicts the electronegativity of an atom from its ionization potential and electron affinity. Both of these properties are measures of the ability of an atom to attract an electron, so it is natural to base electronegativity on them. Using this approach, the electronegativity, EN, is represented as

$$EN = (IP + EA)/2 \qquad (8.66)$$

in which IP is the ionization potential and EA is the electron affinity of the atom. The values for these properties are often expressed in electron volts, whereas bond energies are typically expressed as kcal mol^{-1} or kJ mol^{-1}. To convert the Mulliken electronegativity values to the equivalent Pauling values, the former are divided by 3.17. There have been numerous other scales devised to represent the electronegativities of atoms, but the Pauling scale is still the most widely used.

Table 8.7 Electronegativities of atoms

H 2.2									
Li 1.0	Be 1.6				B 2.0	C 2.6	N 3.0	O 3.4	F 4.0
Na 1.0	Mg 1.3				Al 1.6	Si 1.9	P 2.2	S 2.6	Cl 3.2
K 0.8	Ca 1.0	Sc 1.2	....	Zn 1.7	Ga 1.8	Ge 2.0	As 2.2	Se 2.6	Br 3.0
Rb 0.8	Sr 0.9	Y 1.1	....	Cd 1.5	In 1.8	Sn 2.0	Sb 2.1	Te 2.1	I 2.7
Cs 0.8	Ba 0.9	La 1.1	....	Hg 1.5	Tl 1.4	Pb 1.6	Bi 1.7	Po 1.8	At 2.0

8.7 SYMMETRY OF MOLECULAR ORBITALS

The formation of the H_2 molecule, which has a center of symmetry, gives rise to the combinations of atomic orbitals that can be written as $\varphi_1 + \varphi_2$ and $\varphi_1 - \varphi_2$. The topic of symmetry will be discussed in greater detail in Chapter 10. However, a center of symmetry is simply a point through which each atom can be moved to give the same orientation of the molecule. For a diatomic molecule like H_2, that point is the midpoint of the bond between the two atoms. It is equally valid to speak of a center of symmetry for wave functions. The first of the combinations of wave functions (as shown in Fig. 8.4A) possesses a center of symmetry, whereas the second does not. Therefore, the $\varphi_1 + \varphi_2$ molecular wave function corresponds to the orbital written as σ_g, whereas the $\varphi_1 - \varphi_2$ combination corresponds to σ_u^*. In these designations, "g" refers to the fact that the wave function retains the same sign when inflected through the center of symmetry, and "u" indicates that the wave function changes sign when it is inflected through the center of symmetry. It said that the bonding orbital is symmetric and antibonding orbital is antisymmetric. However, for π and π^* orbitals, g and u refer to symmetry with respect to a plane that contains the internuclear axis (see Fig. 8.8).

For diatomic molecules, the order of filling of molecular orbitals is σ, σ^*, (π, π), σ, (π^*, π^*), σ^* for the early part of the first long period and σ, σ^*, σ, (π, π), (π^*, π^*), σ^* for the latter part of the first long series. The designations (π, π)

and (π^*, π^*) indicate pairs of degenerate molecular orbitals. For the hydrogen molecule, the electron configuration can be shown as $(\sigma_g)^2$, whereas that for the C_2 molecule is designated as $(\sigma_g)^2 (\sigma^*_u)^2 (\pi_u)^2 (\pi_u)^2$ (see Fig. 8.9).

The molecular orbitals can be identified by applying labels that show the atomic orbitals that were combined to produce them. For example, the orbital of lowest energy is $1s\ \sigma_g$ or $\sigma_g\ 1s$. In this way, other orbitals would have designations like $2p_x\pi_u$, $2p_y\pi^*_g$, etc.

As a result of orbital mixing, a σ molecular orbital may not arise from the combination of pure s atomic orbitals. For example, it was shown in Section 8.4 that there is substantial mixing of $2s$ and $2p$ orbitals in molecules like B_2 and C_2. Therefore, labels like $2s\ \sigma_g$ and $2s\ \sigma^*_u$ may not be strictly correct. Because of this, the molecular orbitals are frequently designated as $1\sigma_g$, $1\sigma_u$, $2\sigma_g$, $2\sigma_u$, $3\sigma_u$, $1\pi_u$, $1\pi_g$, ...

In these designations, the leading digit refers to the order in which an orbital having that designation is encountered as the orbitals are filled. For example, $1\sigma_g$ denotes the first σ orbital having g symmetry (a bonding orbital), $3\sigma_u$ means the third σ orbital having u symmetry (an antibonding orbital), etc. The asterisks on antibonding orbitals are not really needed since a σ orbital having u symmetry is an antibonding orbital, and it is the antibonding π orbital that has g symmetry. Therefore, the g and u designations alone are sufficient to denote a bonding or antibonding character. These ideas elaborate on those discussed in Section 8.4.

8.8 ORBITAL SYMMETRY AND REACTIVITY

For many years, it has been recognized that symmetry plays a significant role in the reactions between chemical species. In simple terms, many reactions occur because electron density is transferred (or shared) between the reacting species as the transition state forms. In order to interact favorably (i.e., to give an overlap integral greater than 0), it is necessary for the interacting orbitals to have the same symmetry (see Section 4.4). Otherwise, orthogonal orbitals give an overlap integral equal to 0. The orbitals involved in the interactions of reacting species are those of higher energy, the so-called *frontier orbitals*. These are the highest occupied molecular orbital (HOMO) and the lowest unoccupied molecular orbital (LUMO). As the species interact, electron density flows from the HOMO on one species to the LUMO on the other. In more precise terms, it can be stated that the orbitals must belong to the same symmetry type or point group for the orbitals to overlap with an overlap integral greater than 0.

As was described earlier in this chapter, orbitals of similar energy interact (overlap) best. Therefore, it is necessary that the energy difference between the HOMO on one reactant and the LUMO on the other be less than some threshold value for effective overlap to occur. As a reaction takes place, a bond in one reactant molecule is broken as another is being formed. When both of the orbitals are *bonding* orbitals, the bond being broken (i.e., electron density has been donated from it) is the one representing the HOMO in one reactant and the bond being formed is represented by the LUMO in the other (which is empty and receives electron density as the molecules interact). When the frontier orbitals are antibonding in character, the LUMO in one reactant molecule corresponds to the bond broken, while the HOMO corresponds to the bond formed.

Consider the N_2 and O_2 molecules: The HOMOs of the N_2 molecule are π_u orbitals, which are antisymmetric, whereas the LUMOs of O_2 are half filled π_g (or π_g^*). Therefore, interaction between the frontier orbitals of these two molecules is forbidden by symmetry. Electron density *could* flow from the HOMO (having "g" symmetry) of the O_2 molecule to the π_g orbital on N_2, except for the fact that oxygen has a higher electronegativity than nitrogen. Therefore, transfer of electron density from O_2 to N_2 is excluded on the basis of their chemical properties. As a result, the reaction

$$N_2(g) + O_2(g) \rightarrow 2NO(g) \qquad (8.67)$$

is accompanied by a high activation energy and, in accordance with these observations, the reaction does not take place readily. This is fortunate, because a reacting atmosphere could be a serious environmental issue.

In the reaction

$$H_2(g) + I_2(g) \rightarrow 2HI(g) \qquad (8.68)$$

one would normally expect to find that electron density is transferred from hydrogen to iodine because of the difference in their electronegativities. However, the LUMO of I_2 is an antibonding orbital designated as σ_u (or σ_u^*), whereas the HOMO for H_2 is bonding and is designated as σ_g. Consequently, the overlap is zero for the HOMO of an H_2 molecule with the LUMO of an I_2 molecule, and the expected interaction is symmetry forbidden as shown in Fig. 8.12.

The transfer of electron density from filled molecular orbitals on I_2 to an empty molecular orbital on H_2 is not symmetry forbidden, but it is contrary to the difference in electronegativity. In view of these principles, it is not surprising that the reaction between H_2 and I_2 does not take place by a bimolecular

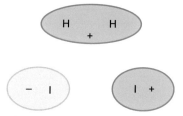

Fig. 8.12 The interaction of hydrogen T_g and iodine T_u orbitals to form a bimolecular complex is symmetry forbidden.

process involving *molecules*, although it was thought to do so for many years. A transition state such as

where the dots indicate breaking and forming bonds, is forbidden on the basis of the chemical nature of both the reactants and symmetry. This reaction actually occurs by the reaction of two iodine atoms with a hydrogen molecule, which is not symmetry forbidden. Sullivan showed that a mixture of hydrogen and iodine that was irradiated with 5780 Å light reacted with a rate that was proportional to the square of the concentration of iodine atoms (Sullivan, 1967).

8.9 TERM SYMBOLS

In Chapter 5, the spectroscopic states of atoms that result from the coupling of spin and orbital angular momenta were discussed. An analogous coupling of angular momenta occurs in molecules, and for diatomic molecules the coupling is similar to the Russell-Saunders scheme.

For a diatomic molecule, the internuclear axis is defined as the z-axis (this will be discussed in greater detail in Chapter 10). In the case of atoms, it was seen that the m value gave the projection of the l vector on the z-axis. Also, the value of s gives the spin angular momentum in units of $\hbar$. To determine the spectroscopic ground state (indicated by a *term symbol*) for an atom, the sum of the spin angular momenta gave a value S and the sum of the orbital contributions gave the value of L.

In a molecule, each electron has its own component of spin angular momentum and orbital angular momentum along the z-axis. These angular momenta couple as they do in atoms, and the resultants determine the *molecular* term symbol. However, to represent the angular momenta of electrons in molecules, different symbols are used. The orbital angular momentum for an electron in a molecule is designated as λ. Also, the atomic orbitals combined to produce a molecular orbital then have the same value for the z component of angular momentum. If the molecular orbital is a σ orbital, the orbital angular momentum quantum number is 0, so λ must equal 0 because that is the only projection on the z-axis that a vector 0 units long can have. This means that for an electron residing in a σ orbital, the only m value possible is 0. For an electron in a π orbital, m can have values of $+1$ and -1, which are both projections of a vector l having one unit in length. Therefore, if the molecular orbital is a π orbital, the λ value is 1. The total orbital angular momentum, which is 2, is then determined. For a molecule, the sum of the λ values is designated as Λ. The resultant spin angular momentum is given by the sum of the electron spins, as in the case of atoms. For atoms, the values $L=0$, 1, and 2 give rise to spectroscopic states designated as S, P, and D, respectively. For molecules, the values $\Lambda=0$, 1, and 2 give rise to spectroscopic states designated as Σ, Π, and Δ, respectively. Thus, coupling procedures for atoms and molecules are analogous, except for the fact that Greek letters are used to denote molecular term symbols. The term symbol for a molecule is expressed as $^{2S+1}\Lambda$.

After the resultant spin and orbital angular moment vectors have been determined, these vectors can couple to give a total angular momentum. For atoms, the vector was designated as J, but for molecules, it is usually designated as Ω, which has the possible values 0, 1, 2, ... As in the case of atoms, all filled shells have a total spin of 0 and the sum of the m values is also 0. Therefore, all of the lower-lying filled shells can be ignored in determining the ground term symbol for a diatomic molecule. In Section 8.8, it was shown that the molecular orbitals for homonuclear diatomic molecules are designated as g or u, depending on whether they are symmetric or antisymmetric with respect to a center of symmetry. Of course, heteronuclear diatomic molecules do not possess a center of symmetry. The overall g or u character for more than one occupied orbital can be determined by using g to represent a $+$ sign and u to represent a $-$ sign. Then the g or u character of each orbital is multiplied by that of each other orbital. As a result, $g \times g = g$, $g \times u = u$, and $u \times u = g$. The σ states are also designated as Σ^+ or Σ^- based on whether the wave function that represents the molecular orbital is symmetric or antisymmetric, with respect to reflection in any plane that contains the molecular axis.

For the H_2 molecule, the two electrons reside in the $1\sigma_g$ molecular orbital and the configuration is $(1\sigma_g)^2$. For a σ orbital, $\lambda=0$ so the sum of the values for the two electrons is 0, which is the value for Λ. Therefore, the ground state is a Σ state and with the two electrons being paired, the sum of the spins is 0. As a result, the multiplicity, which is $2S+1$, equals 1 so the ground state is $^1\Sigma$. Moreover, two electrons reside in the $1\sigma_g$ molecular orbital, and the product of $g \times g$ gives an overall symmetry of g. The molecular orbital wave function is symmetric with respect to a plane that contains the internuclear axis, so the superscript $+$ is appropriate. For the H_2 molecule, the correct term symbol is $^1\Sigma_g{}^+$, which expresses all of the information for the spectroscopic state.

When deriving the term symbol for the O_2 molecule, the starting point is the electron configuration $(1\sigma_g)^2 \, (1\sigma_u)^2 \, (2\sigma_g)^2 \, (1\pi_u)^4 \, (1\pi_g)^2$, and the outer two electrons are unpaired in degenerate $1\pi_g$ orbitals. The sum of spins could be either 1 or 0, depending on whether the spins are aligned or opposed. The first of these values would give $2S+1=3$, whereas the latter would give $2S+1=1$. As seen for atomic term symbols, the states of highest multiplicity correspond to the lowest energy. This would occur when one electron resides in each $1\pi_g$ orbital, with the two having parallel spins. However, this occurs when one electron is in each orbital so $m_{(1)}=+1$ and $m_{(2)}=-1$, so the sum is 0 and results in a Σ term. Therefore, the triplet state is a $^3\Sigma_g$ because both electrons reside in orbitals with g symmetry. Finally, the orbitals are symmetric with respect to a plane containing the internuclear axis, so the superscript $+$ is added to give $^3\Sigma_g{}^+$ as the term symbol for the O_2 molecule. By drawing all of the microstates as was illustrated in Section 5.4, we find that other terms exist, though they do not represent the ground state.

Removal of an electron from the O_2 molecule to produce $O_2{}^+$ (the dioxygenyl ion) leaves one electron in a $1\pi_g$ orbital. In this case, $S=\frac{1}{2}$ and $\Lambda=1$ and the molecular orbital has g symmetry, so the term symbol is $^2\Pi_g$. The cases that have been worked out here illustrate the procedures involved in finding the term of lowest energy for a particular configuration. However, as in the case of atoms, other terms are possible when all permissible microstates are considered.

LITERATURE CITED

Ballhausen, C.; Gray, H. B. The Electronic Structure of the Vanadyl Ion. *Inorg. Chem.* **1962**, *1*, 111–121.

Cusachs, L. C. Semiempirical Molecular Orbitals for General Polyatomic Molecules. II. One-Electron Model Prediction of H-O-H Angle. *J. Chem. Phys.* **1965**, *43*, S157–S159.

Hoffmann, R. An Extended Hückel Theory. I. Hydrocarbons. *J. Chem. Phys.* **1963**, *39*, 1397–1412.

House, J. E. *Inorganic Chemistry*; 2nd ed.; Academic Press/Elsevier: Amsterdam, 2013.

Mulliken, R. S.; Rieke, C. A.; Orloff, D.; Orloff, H. Formulas and Numerical Tables for Overlap Integrals. *J. Chem. Phys.* **1949**, *17*, 1248–1267.

Pauling, L. *The Nature of the Chemical Bond*; 3rd ed.; Cornell University Press: Ithaca, NY, 1960.

Sullivan, J. H. Mechanism of the "Bimolecular" Hydrogen–Iodine Reaction. *J. Chem. Phys.* **1967**, *46*, 73–78.

Wolfsberg, M.; Helmholtz, L. The Spectra and Electronic Structure of the Tetrahedral Ions MnO_4^-, CrO_4^-, and ClO_4^-. *J. Chem. Phys.* **1952**, *20*, 837.

PROBLEMS

1. For a molecule XY, a molecular wave function can be written as $\Psi_{molecule} = \Psi_{covalent} + 0.50\,\Psi_{ionic}$. Calculate the percent ionic character of the X—Y bond. If the bond length is 1.50 Å, what is the dipole moment?

2. For a homonuclear diatomic molecule, $H_{11} = H_{22}$ and $H_{12} = H_{21}$. This is not true for a heteronuclear molecule. Derive the expressions for the bonding and antibonding states for a heteronuclear diatomic molecule. You may still assume that $S_{12} = S_{21}$.

3. Suppose that the bond energies for A_2 and X_2 are 209 and 360 kJ mol^{-1}, respectively. If atoms A and X have electronegativities of 2.0 and 3.0, what will be the strength of the A—X bond? What will be the dipole moment if the internuclear distance is 1.25 Å?

4. Suppose that a diatomic molecule XZ contains a single σ bond. The binding energy of an electron in the valence shell of atom X is -10.0 eV. Spectroscopically, it is observed that the promotion of an electron to the antibonding state leads to an absorption band at 16,100 cm^{-1}. Using a value of 0.10 for the overlap integral, determine the value of the exchange integral. Sketch the energy level diagram and determine the actual energies of the bonding and antibonding states. What is the bond energy?

5. For the molecule ICl, the wave function can be written as $\Psi_{ICl} = \Psi_{covalent} + 0.33\,\Psi_{ionic}$. If the dipole moment for ICl is 0.65 D, what is the internuclear distance?

6. Write molecular orbital descriptions for NO, NO$^+$, and NO$^-$. Predict the relative bond energies of these species and account for any that are paramagnetic.

7. The covalent radii of F and Cl are 0.72 and 0.99 Å, respectively. Given that the electronegativities are 4.0 and 3.0, what would be the expected bond distance for ClF?

8. The H—S bond moment is 0.68 D and the bond length is 1.34 Å. What is the percent ionic character of the H—S bond?

9. Write the molecular orbital configurations for the following: (a) CO^+, (b) C_2^-, and (c) BO. Determine the bond order for these species.
10. Write out the molecular orbital configurations for Na_2 and Si_2 (omit the $n=1$ and $n=2$ states and represent them only as KK and LL). Give the bond order for each molecule. Describe the stabilities of these molecules when they are excited to the first excited state.
11. Calculate the percent ionic character for the following bonds: (a) HCl, (b) HC, (c) NH, and (d) LiH.
12. Give the term symbols for the boron atom in its ground electronic state, and arrange them in the order of increasing energy. What would be the spectroscopic state for boron in the first excited state?
13. A heteronuclear diatomic species has the molecular orbital electron configuration $1\sigma_g^2\, 1\sigma_u^2\, 2\sigma_g^2\, 2\sigma_u^2\, 1\pi_u^3$.
   (a) What is the bond order for the species?
   (b) Would the dissociation energy for the molecule be higher or lower if an electron is removed from the $1\pi_u$ orbital? Explain.
   (c) Would you expect this species to form a stable -1 ion?
14. What would be the symmetry designation (including g or u, as appropriate) for the following atomic orbitals with respect to the z-axis? (a) $2s$, (b) $2p_z$, (c) $2p_y$, and (d) $3d_{z^2}$.
15. Determine the ground state term symbol for the following: (a) Li_2, (b) C_2, (c) O_2^-, and (d) B_2^+
16. One term symbol for N_2^+ is $^2\Sigma_g^+$. From which orbital was the electron removed?
17. Explain the difference between the reactions

$$N_2 \rightarrow N_2^+ + e^- \,^2\Sigma_g^+$$
$$N_2 \rightarrow N_2^+ + e^- \,^2\Pi_u$$

18. The bond energy in C_2^- is 2.2 eV greater than that of C_2, but the bond energy of O_2^- is 1.1 eV less than that of O_2. Explain this difference.
19. Explain why the bond length in F_2 is 142 pm while that of F_2^+ is 132 pm.
20. The ground state term for B_2 is $^3\Sigma_g^-$. Explain how this fact gives information about the order of filling the molecular orbitals for diatomic molecules of the second-row elements.
21. For the O_2 molecule, draw all of the microstates that could result for the $1\pi_g^1\, 1\pi_g^1$ configuration. Determine which term each belongs to. What is the ground state term?

22. Write out the molecular orbital populations for NO^+, F_2, and O_2^+, then determine their spectroscopic ground states.

23. What would be the effect on the molecular orbital diagram be if a deuterium atom were substituted for a hydrogen atom in H_3^+? Why?

24. Oxygen molecules have an intense absorption at 1800 Å. Calculate the energy associated with this band in eV molecule^{-1} and kJ mol^{-1}. What transition could this correspond to?

25. By means of the molecular orbital energy level diagrams, arrange the following in the order of increasing force constant: (a) NO, (b) O_2, (c) N_2, and (d) B_2.

26. The force constant for bond stretching in the superoxide ion O_2^- is only about half that in the O_2 molecule, whereas that of the dioxygenyl cation O_2^+ is about 1.5 times that of O_2. Explain these observations.

27. For the NO molecule, the internuclear distance is 1.15 Å (115 pm) and the stretching force constant is 15.95 mdyn Å^{-1} (1595 N m^{-1}). Give estimates of these parameters for the NO^+ ion and explain the basis for your answers.

28. The oxygen molecule has a ground state configuration of $(\pi_u)^4 (\pi_g)^2$. What is the spectroscopic state for this configuration?

29. Although the oxygen molecule has an electronic ground state configuration of $(\pi_u)^4 (\pi_g)^2$, the excited state having the configuration $(\pi_u)^3 (\pi_g)^3$ lies 35,700 cm^{-1} higher in energy. What wave length of radiation would be necessary to cause a transition to the excited state? In what region of the spectrum would this absorption be observed?

CHAPTER 9

The Hückel Molecular Orbital Method

When the phrase "molecular orbital calculations" is first encountered, the mental image may well be one of hopelessly complicated mathematics and piles upon piles of computer output. It is interesting to note, however, that sometimes a relatively simple calculation may provide useful information that correlates well with experimental observations. Such is the case with the method known as the Hückel molecular orbital (HMO) calculation. This method was developed in 1931 by German physicist Erich Hückel, who was trying to understand the concept of aromaticity in benzene. Almost 50 years ago, the method was described in these words: "In spite of its approximate nature, Hückel molecular orbital (HMO) theory has proved itself extremely useful in elucidating problems concerned with electronic structures of π-electron systems" (Purins and Karplus, 1968). The method still has value, even though the procedures for performing the calculations are relatively simple and have become known as the "back of an envelope" calculations. In fact, a very popular book in the early 1960s describing such methods was John D. Roberts' *Notes on Molecular Orbital Calculations* (Roberts, 1962). That book did much to popularize the application of the Hückel method in organic chemistry. Arno Liberles' book, *Introduction to Molecular-Orbital Theory*, also provides a good introduction to Hückel methods (Liberles, 1966). The extended Hückel method (EHMO) will be described briefly, and the more sophisticated methods for dealing with the structure of molecules will be treated in Chapter 14.

9.1 THE HÜCKEL METHOD

In the Hückel method, the assumption is made that the σ and π parts of the bonding in molecules can be separated. Also, the overlap of orbitals on nonadjacent atoms is assumed to be zero. Additionally, the interaction energy between nonadjacent atoms is assumed to be zero. The energies to be considered are represented as H_{ii}, the valence state ionization potential (VSIP) for atom i, and H_{ij}, the exchange energy between atoms i and j. In the Hückel treatment, it is assumed that $H_{ij} = 0$ when $|i - j| = 2$ (nonadjacent atoms). When the overlap is not included in the calculations, it is assumed

Fundamentals of Quantum Mechanics
http://dx.doi.org/10.1016/B978-0-12-809242-2.00009-7

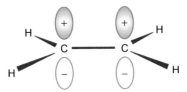

Fig. 9.1 The ethylene (ethene) molecule showing the p orbitals used in π bonding.

that $S_{12}=S_{21}=0$. The Hückel method will be illustrated by treating the π bond in the C_2H_4 molecule, which is shown in Fig. 9.1.

In the case of ethylene, it will be assumed that the carbon atoms use sp^2 hybrid orbitals to form the three σ bonds, and the p_z orbitals (which are perpendicular to the plane of the molecule) are left to form the π bonds. The wave functions for the bonding orbitals can be written as

$$\psi CH(\sigma) = a_1\varphi(1s) + a_2\varphi\left(sp^2\right) \tag{9.1}$$

$$\psi CC(\sigma) = a_1\varphi\left(sp^2\right) + a_2\varphi\left(sp^2\right) \tag{9.2}$$

$$\psi CC(\pi) = a_1\varphi(p_1) + a_2\varphi(p_2), \tag{9.3}$$

where $\psi CC(\sigma)$ corresponds to the σ bond between two carbon atoms, etc. Using the procedure developed earlier (see Chapter 8), we can write the secular determinant as

$$\begin{vmatrix} H_{11} - E & H_{12} - S_{12}E \\ H_{21} - S_{21}E & H_{22} - E \end{vmatrix} = 0 \tag{9.4}$$

The parameters representing energies are denoted as

$$\alpha = H_{11} = H_{22} = H_{33} = \cdots = H_{ii} = H_{jj} = \text{the Coulomb integral}$$

and

$$\beta = H_{12} = H_{21} = H_{23} = H_{32} = \cdots = H_{ij} = \text{the exchange integral}$$

Therefore, the secular determinant can be written as

$$\begin{vmatrix} \alpha - E & \beta \\ \beta & \alpha - E \end{vmatrix} = 0 \tag{9.5}$$

If each element in the determinant is divided by β, the result is

$$\begin{vmatrix} \dfrac{\alpha - E}{\beta} & 1 \\ 1 & \dfrac{\alpha - E}{\beta} \end{vmatrix} = 0 \tag{9.6}$$

It is customary to now let $x = (\alpha - E)/\beta$, which allows the determinant to be written as

$$\begin{vmatrix} x & 1 \\ 1 & x \end{vmatrix} = 0 \qquad (9.7)$$

Expansion of this determinant gives

$$x^2 - 1 = 0 \qquad (9.8)$$

The roots of this equation are $x = -1$ and $x = +1$. Therefore, we can write

$$\frac{\alpha - E}{\beta} = -1 \qquad (9.9)$$

so that $E = \alpha + \beta$, and

$$\frac{\alpha - E}{\beta} = 1 \qquad (9.10)$$

so that $E = \alpha - \beta$.

When we recall that both α and β are *negative* energies, it can be seen that $E = \alpha + \beta$ corresponds to the state having *lower* energy. An energy level diagram showing the states can be constructed, as shown in Fig. 9.2.

In ethylene, each carbon atom contributes one electron to the π bond so that the lowest level is filled with two electrons. If the two electrons were on isolated carbon atoms, their binding energies would be 2α. Therefore, the energy for the two electrons in the *molecular* orbital, as opposed to what the energy would be for two isolated atoms, is

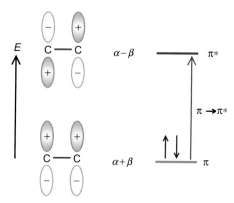

Fig. 9.2 Energy level diagram for ethylene. *(Modified with permission from House, J. E. Inorganic Chemistry, 2nd ed.; Academic Press/Elsevier: Amsterdam, 2013.)*

$$2(\alpha + \beta) - 2\alpha = 2\beta. \tag{9.11}$$

The fact that the molecular orbital encompasses the complete carbon structure whereas the atomic orbitals do not leads to the description of the *molecular orbital* as being *delocalized*. The energy 2β is called the *delocalization* or *resonance energy*. An average energy for a typical C—C bond is about 335–355 kJ mol^{-1}, whereas that for a typical C═C bond is about 600 kJ mol^{-1}. Therefore, the π bond adds about 250 kJ mol^{-1}, which means that this is the approximate value of 2β, so β is approximately 125 kJ mol^{-1}. The energies of the two molecular orbitals (bonding and antibonding) have been obtained, but a major objective of the calculations is to be able to obtain useful information from the wave functions. In order to do this, it is necessary to determine the values of the constants a_1 and a_2. These constants determine the weighting given to the atomic wave functions and thereby determine electron density, etc. The wave function for the bonding orbital in ethylene can be written as

$$\psi_b = a_1\varphi_1 + a_2\varphi_2, \tag{9.12}$$

and we know that for the normalized wave function,

$$\int \psi_b^2 \, d\tau = 1 = \int (a_1\varphi_1 + a_2\varphi_2)^2 \, d\tau \tag{9.13}$$

By expansion of the integral and letting S_{11}, S_{22}, and S_{12} represent the overlap integrals, we obtain

$$a_1{}^2 S_{11} + a_2{}^2 S_{22} + 2a_1 a_2 S_{12} = 1 \tag{9.14}$$

If we let $S_{11} = S_{22} = 1$ and neglect the overlap between adjacent atoms by letting $S_{12} = S_{21} = 0$, then

$$a_1{}^2 + a_2{}^2 = 1 \tag{9.15}$$

The secular equations can be written in terms of α and β as

$$a_1(\alpha - E) + a_2\beta = 0 \tag{9.16}$$

$$a_1\beta + a_2(\alpha - E) = 0 \tag{9.17}$$

Dividing both equations by β and letting $x = (\alpha - E)/\beta$, we find that

$$a_1 x + a_2 = 0 \tag{9.18}$$

$$a_1 + a_2 x = 0 \tag{9.19}$$

For the bonding state, we saw that $x = -1$, so it follows that $a_1{}^2 = a_2{}^2$. Therefore, we can write

$$a_1{}^2 + a_1{}^2 = 1 = 2a_1{}^2 \tag{9.20}$$

From this equation, the values of a_1 and a_2 can be found:

$$a_1 = \frac{1}{\sqrt{2}} = 0.707 = a_2$$

The wave function for the bonding state can be written in terms of the atomic wave functions as

$$\psi_b = 0.707\phi_1 + 0.707\phi_2 \tag{9.21}$$

For an atom in a molecule, it is the square of the coefficient of the atomic wave function in the molecular wave function that gives the probability (density) of finding an electron on that atom. Therefore, $a_1{}^2 = a_2{}^2 = \frac{1}{2}$, so $\frac{1}{2}$ of the electrons should be on each atom. Because there are two electrons in the bonding orbital, the electron density (ED) is $2(1/2) = 1$ and one electron resides on each atom. As expected, electrons are not transferred from one carbon atom to the other.

Another useful property for describing bonding in the molecule is the bond order, B. This quantity gives an electron population in terms of the number of π bonds between two bonded atoms. In this case, it is the product of the coefficients on the atomic wave functions that gives the electron density of the bond between them. However, we must also take into account the total number of electrons in the occupied molecular orbital(s). Therefore, the bond order between atoms X and Y can be written as B_{XY}, which is given by

$$B_{XY} = \sum_{i=1}^{n} a_X a_Y p_i \tag{9.22}$$

where a is the weighting coefficient, n is the number of populated orbitals, and p is the population (number of electrons) in that orbital. For the case of ethylene, there is only one orbital, and it is populated with two electrons, so we find that

$$B_{CC} = 2(0.707)(0.707) = 1 \tag{9.23}$$

which indicates that there is one π bond between the carbon atoms. Before considering larger molecules, we will show some of the mathematics necessary for the application of the Hückel method.

9.2 DETERMINANTS

From the example in the previous section, it should be apparent that formulating a problem using the HMO method to describe a molecule results in a *determinant.* A determinant represents a function in the form of an array that contains *elements* in *rows* and *columns.* A secular determinant was also encountered in dealing with diatomic molecules earlier (see Section 8.3). The number of rows or columns (they are equal) is called the rank (or *order*) of the determinant. Determinants are essential in the Hückel method, so we need to show how they are manipulated.

A determinant can be reduced to an equation known as the *characteristic equation.* Suppose we consider the 2×2 determinant,

$$\begin{vmatrix} a & b \\ c & d \end{vmatrix} = 0 \tag{9.24}$$

Simplifying this determinant to obtain the characteristic equation involves multiplying along one diagonal and subtracting the product obtained by multiplying along the other diagonal. This rule applied to the preceding determinant gives

$$\begin{vmatrix} a & b \\ c & d \end{vmatrix} = ad - bc = 0 \tag{9.25}$$

If a determinant can be written as

$$\begin{vmatrix} x & 1 \\ 1 & x \end{vmatrix} = 0 \tag{9.26}$$

then expanding the determinant gives the characteristic equation

$$x^2 - 1 = 0 \tag{9.27}$$

which is the equation that arises from the treatment of the π bond in ethylene (see Section 9.1).

If the molecule under consideration contains three atoms, we will obtain a 3×3 determinant such as

$$\begin{vmatrix} a & b & c \\ d & e & f \\ g & h & i \end{vmatrix} = 0 \tag{9.28}$$

Expansion of a 3×3 determinant is somewhat more elaborate than that of a 2×2 determinant. One method can be illustrated as follows: Initially, extend the determinant by writing the first two columns again to the right of the determinant:

$$\begin{vmatrix} a & b & c \\ d & e & f \\ g & h & i \end{vmatrix} \begin{matrix} a & b \\ d & e \\ g & h \end{matrix}$$

We now perform the multiplication along each of the three-membered diagonals, with multiplication to the *right* resulting in products that are *positive,* and multiplication to the *left* resulting in products that are *negative.* Therefore, in this case, we obtain the characteristic equation

$$aei + bfg + cdh - ceg - afh - bdi = 0 \qquad (9.29)$$

If the determinant under consideration is

$$\begin{vmatrix} x & 1 & 0 \\ 1 & x & 1 \\ 0 & 1 & x \end{vmatrix} = 0 \qquad (9.30)$$

then we expand the determinant as follows:

$$\begin{vmatrix} x & 1 & 0 \\ 1 & x & 1 \\ 0 & 1 & x \end{vmatrix} \begin{matrix} x & 1 \\ 1 & x \\ 0 & 1 \end{matrix}$$

This method results in the characteristic equation that reduces to

$$x^3 - 2x = 0, \qquad (9.31)$$

and the roots are $x = 0$, $-(2)^{1/2}$, and $+(2)^{1/2}$. However, this expansion method using diagonals works only for 2×2 and 3×3 determinants, so we need a more general method for the expansion of higher-order determinants.

Consider the 4×4 determinant:

$$\begin{vmatrix} a & b & c & d \\ e & f & g & h \\ i & j & k & l \\ m & n & o & p \end{vmatrix} = 0 \qquad (9.32)$$

Expansion of this determinant is accomplished by a procedure known as the *method of cofactors.* In this method, we begin with element a and remove the row and column that contain a. Then the rest of the determinant is multiplied by a to obtain

$$a \begin{vmatrix} f & g & h \\ j & k & l \\ n & o & p \end{vmatrix}$$

The portion of the determinant that is multiplied by a is called a *minor* and will have a rank $(n-1)$, where n is the rank of the original determinant. We now repeat this process, except that b is removed, the sign preceding this term is negative, and so on. Therefore,

$$\begin{vmatrix} a & b & c & d \\ e & f & g & h \\ i & j & k & l \\ m & n & o & p \end{vmatrix} = a \begin{vmatrix} f & g & h \\ j & k & l \\ n & o & p \end{vmatrix} - b \begin{vmatrix} e & g & h \\ i & k & l \\ m & o & p \end{vmatrix} + c \begin{vmatrix} e & f & h \\ i & j & l \\ m & n & p \end{vmatrix} - d \begin{vmatrix} e & f & g \\ i & j & k \\ m & n & o \end{vmatrix} \quad (9.33)$$

Expansion of each 3×3 determinant can now be continued as previously illustrated.

The cofactors are designated as C_{ij} and their general formula is

$$C_{ij} = (-1)^{i+j} M_{ij} \quad (9.34)$$

where M_{ij} is the minor having rank $(n-1)$. For the determinant

$$\begin{vmatrix} x & 1 & 0 & 0 \\ 1 & x & 1 & 0 \\ 0 & 1 & x & 1 \\ 0 & 0 & 1 & x \end{vmatrix} = 0 \quad (9.35)$$

we can show the expansion to find the characteristic equation as

$$\begin{vmatrix} x & 1 & 0 & 0 \\ 1 & x & 1 & 0 \\ 0 & 1 & x & 1 \\ 0 & 0 & 1 & x \end{vmatrix} = x \begin{vmatrix} x & 1 & 0 \\ 1 & x & 1 \\ 0 & 1 & x \end{vmatrix} - 1 \begin{vmatrix} 1 & 1 & 0 \\ 0 & x & 1 \\ 0 & 1 & x \end{vmatrix} + 0 \begin{vmatrix} 1 & x & 0 \\ 0 & 1 & 1 \\ 0 & 0 & x \end{vmatrix} - 0 \begin{vmatrix} 1 & x & 1 \\ 0 & 1 & x \\ 0 & 0 & 1 \end{vmatrix} \quad (9.36)$$

As a result of the last two terms being 0, the characteristic equation can be written as

$$x^4 - 3x^2 + 1 = 0 \quad (9.37)$$

Another useful property of determinants is illustrated by *Laplace's Expansion Theorem*, which relates to the expansion of a determinant in terms of smaller units or subdeterminants. Due to the symmetry of the determinant, it is sometimes possible to simplify the determinant, as illustrated by the following example.

Consider the determinant

$$\begin{vmatrix} x^2 & 4 & 0 & 0 \\ x & 3 & 0 & 0 \\ 0 & 0 & 12 & 2x \\ 0 & 0 & 5x & 2 \end{vmatrix} = f(x) \tag{9.38}$$

In this case, the function represented can be written as

$$\begin{vmatrix} x^2 & 4 \\ x & 3 \end{vmatrix} \cdot \begin{vmatrix} 12 & 2x \\ 5x & 2 \end{vmatrix} = f(x) \tag{9.39}$$

Expanding each of the 2×2 determinants is carried out as illustrated earlier to give the characteristic equation

$$f(x) = \left(3x^2 - 4x\right)\left(24 - 10x^2\right) \tag{9.40}$$

$$f(x) = -30x^4 + 40x^3 + 72x^2 - 96x$$

This technique is useful in simplifying the secular determinants that arise for some organic molecules having certain structures. Because only interactions between adjacent atoms are considered, several elements in the secular determinant can be set equal to 0.

Determinants also have other useful and interesting properties that will be only listed here. For a complete discussion of the mathematics of determinants, see the references listed at the end of this book.

1. Interchanging two rows (or columns) of a determinant produces a determinant that is the negative of the original.
2. If each element in a row (or column) of a determinant is multiplied by a constant, the result is the constant times the original determinant.
   If we consider the determinant

$$D = \begin{vmatrix} a & b \\ c & d \end{vmatrix} \tag{9.41}$$

it is possible now to multiply each member of the second column by k to obtain

$$D = \begin{vmatrix} a & kb \\ c & kc \end{vmatrix} \tag{9.42}$$

Expansion of this determinant gives

$$adk - bck = k(ad - bc) = k\begin{vmatrix} a & b \\ c & d \end{vmatrix}$$

which shows that the result is indeed the constant times the determinant.

3. If two rows (or columns) in a determinant are identical, the determinant has the value of 0. For example,

$$\begin{vmatrix} a & a & b \\ c & c & d \\ e & e & f \end{vmatrix} = acf + ade + bce - bce - ade - acf = 0 \qquad (9.43)$$

4. If a determinant has one row (or column) where each element is zero, the determinant evaluates to 0. This is very easy to verify, so an example will not be provided here. As we shall see later in this chapter, the evaluation of determinants is an integral part of the HMO method.

9.3 SOLVING POLYNOMIAL EQUATIONS

It has just been shown that the expansion of determinants leads to polynomial equations that we will need to solve. Before the widespread availability of sophisticated calculators and computers, such equations were solved graphically (Roberts, 1962).

The roots of the polynomial equations can often be found by graphing the functions if they cannot be factored directly. As we shall see, the equation

$$x^4 - 3x^2 + 1 = 0 \qquad (9.44)$$

arises from the expansion of the secular determinant for butadiene. To find the values of x where this function $y = f(x)$ crosses the x-axis, we let

$$y = x^4 - 3x^2 + 1 = 0 \qquad (9.45)$$

and we make the graph by assigning values to x. It is possible to get a rough estimate of the range of x values to consider by realizing that for large values of x, x^4, and $3x^2$ are both much greater than 1. Therefore, for y to have a value near 0, the terms x^4 and $3x^2$ must be approximately equal:

$$x^4 \approx 3x^2$$

$$x^2 \approx 3$$

$$x \approx \pm\sqrt{3}$$

Therefore, it is appropriate to choose the range of values $2 > x > -2$, which should encompass the roots. The values of y when x is assigned values in this range are shown as follows:

x	y	x	y
−2.0	5.0	0.4	0.546
−1.6	−0.126	0.8	−0.510
−1.2	−1.254	1.2	−1.254
−0.8	−0.510	1.6	−0.126
−0.4	0.546	2.0	5.0
0.0	1.000		

The graph of these values is shown in Fig. 9.3. It is apparent from the data or the graph that the four roots are approximately 0.6 and 1.6. In this case, the actual roots of Eq. (9.45) are $x = -1.62$, -0.618, $+0.618$, and $+1.62$.

The advent of the graphing calculator has made this a very rapid procedure to carry out. In many situations, calculus is now taught in such a way as to require students to use these electronic marvels. Solving the polynomial equations resulting from Hückel MO calculations is indeed very easy. Also, many calculators that are not graphing ones have a built in "solve" function. This lets the user enter the expression to be solved and obtain the roots of polynomial equations in a convenient manner. Most of these "hardwired" root finding capabilities make use of the *Newton-Raphson* or *secant* methods that are standard techniques in numerical analysis. For details of how these methods work, see the references on numerical analysis listed at the end of this book. Electronic calculators and computer software have made possible the routine use of numerical analysis procedures.

Another numerical technique useful in certain types of problems is *iteration*. In an iterative technique, some operation is used repetitively in order to solve a problem. A programmable calculator or computer is ideally

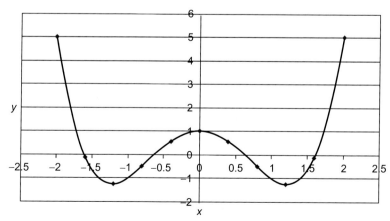

Fig. 9.3 A graph of the data shown in the text for $y = x^4 - 3x^2 + 1$.

suited to this type of calculation. Finding the roots of polynomial equations, like those arising from Hückel calculations, is essentially finding the zeros of functions (values of x where $y=0$). The roots are the values of x that satisfy the equation

$$a_n x^n + a_{n-1} x^{n-1} + \cdots + a_1 x + a_0 = 0 \qquad (9.46)$$

From the quadratic formula, an equation that can be written

$$a x^2 + b x + c = 0 \qquad (9.47)$$

has the roots

$$x = \frac{-b \pm \sqrt{b^2 - 4ac}}{2a} \qquad (9.48)$$

If $b^2 - 4ac$ (called the *discriminant*) is negative, the roots of the equation are complex. If $b^2 - 4ac$ is positive, the roots of the equation will be real. Most polynomial equations arising from secular determinants will be higher than second order, so an analytical method for their solution is not nearly as simple as it is for a quadratic equation. Suppose we wish to solve the equation

$$x^2 + 3.6x - 16.4 = 0 \qquad (9.49)$$

By applying the quadratic formula, we obtain

$$x = \frac{-3.6 \pm \sqrt{3.6^2 - 4(-16.4)}}{2} = +2.63, \ -6.23 \qquad (9.50)$$

After writing Eq. (9.49) as $x(x+3.6)-16.4$, x can be written as

$$x = \frac{16.4}{x + 3.6} \qquad (9.51)$$

Suppose we call $x=2$ a "first guess" at a solution, so we let it equal x_0. The next value, which is x_1, can be written in terms of x_0 as

$$x_1 = \frac{16.4}{x_0 + 3.6} = \frac{16.4}{2 + 3.6} = 2.93 \qquad (9.52)$$

We will now use this value as an "improved" guess and calculate a new value x_2:

$$x_2 = \frac{16.4}{x_1 + 3.6} = \frac{16.4}{2.93 + 3.6} = 2.51$$

Repeating the process with each new value of x_{n+1}, being given in terms of x_n,

$$x_3 = \frac{16.4}{x_2 + 3.6} = \frac{16.4}{2.51 + 3.6} = 2.68$$

$$x_4 = \frac{16.4}{x_3 + 3.6} = \frac{16.4}{2.68 + 3.6} = 2.61$$

$$x_5 = \frac{16.4}{x_2 + 3.6} = \frac{16.4}{2.61 + 3.6} = 2.64, \text{ etc.}$$

In this process, we are using the formula (known as a *recursion formula*)

$$x_{n+1} = \frac{16.4}{x_n + 3.6} \tag{9.53}$$

As was illustrated by Eq. (9.50), one root is $x = 2.63$, so it is apparent that this *iterative* process is converging to a correct root.

If we write Eq. (9.53) in a general form as

$$x = f(x) \tag{9.54}$$

it can be shown that the iterative process will converge if $|f'(x)| < 1$. In the preceding case,

$$f(x) = \frac{16.4}{x + 3.6} \tag{9.55}$$

and taking the derivative gives

$$f'(x) = \frac{-16.4}{(x + 3.6)^2} \tag{9.56}$$

For the root where $x = 2.63$, $|f'(x)| = 0.423$, which is less than 1 and convergence is thereby achieved. In the general case using the equation

$$x^2 + bx + c = 0$$

$$x = \frac{-c}{x + b} \tag{9.57}$$

and

$$f'(x) = \frac{c}{(x + b)^2} \tag{9.58}$$

If $|(x + b)^2| \approx c$, convergence will be slow. Verify that this is indeed true using the equation

$$x^2 - 4x + 3.99 = 0 \tag{9.59}$$

For an equation such as

$$x^2 - 5.0x + 7.5 = 0 \qquad (9.60)$$

the roots are complex because b^2-4ac is negative, and the iterative method does not work. Iterative methods are well suited to solving equations like

$$x - (1 + \cos x)^{1/2} = 0 \qquad (9.61)$$

It is easy to see that for *some* value of x, the two sides of the equation must be equal:

$$x = (1 + \cos x)^{1/2} \text{(when } x \text{ is in radians)} \qquad (9.62)$$

A graph showing $y = f(x)$ can now be prepared for each of the two sides of the equation (Fig. 9.4). From the graphs we can see that x is approximately 1, so we can begin an iterative process using a trial value of $x_0 = 1$. Then

$$x_1 = (1 + \cos 1)^{1/2} = 1.241$$
$$x_2 = (1 + \cos 1.241)^{1/2} = 1.151$$
$$x_3 = (1 + \cos 1.151)^{1/2} = 1.187$$
$$x_4 = (1 + \cos 1.187)^{1/2} = 1.173$$
$$x_5 = (1 + \cos 1.173)^{1/2} = 1.178$$
$$x_6 = (1 + \cos 1.178)^{1/2} = 1.176$$
$$x_7 = (1 + \cos 1.176)^{1/2} = 1.177$$
$$x_8 = (1 + \cos 1.177)^{1/2} = 1.1764.$$

Therefore, the root of Eq. (9.61) is 1.176.

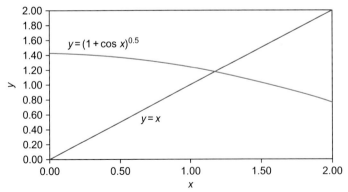

Fig. 9.4 The graphs showing $y = x$ and $y = (1 + \cos x)^{1/2}$.

Sometimes, problems in the sciences and engineering may lead to some very interesting equations in order to be solved. Although such transcendental equations as those just solved do not arise very often, it is useful to know that a procedure for solving them easily exists.

9.4 HÜCKEL CALCULATIONS FOR LARGER MOLECULES

The system containing three carbon atoms in a chain shown in Fig. 9.5 is that of the allyl species, which includes the neutral radical as well as the carbocation and the anion. The combination of atomic wave functions will be constructed using $2p$ wave functions from the carbon atoms. In this case, the Coulomb integrals will be equal, so $H_{11} = H_{22} = H_{33}$, and the exchange integrals between adjacent atoms will be set equal, as $H_{12} = H_{21} = H_{23} = H_{32}$. However, in this approximate method, interactions between nonadjacent atoms are ignored. Thus, $H_{13} = H_{31} = 0$. All overlap integrals of the type S_{ii} are set equal to 1 and all overlap between adjacent atoms is neglected ($S_{ij} = 0$). The secular determinant is written as follows when the usual substitutions are made:

$$\begin{vmatrix} H_{11} - E & H_{12} & 0 \\ H_{21} & H_{22} - E & H_{23} \\ 0 & H_{32} & H_{33} - E \end{vmatrix} = \begin{vmatrix} \alpha - E & \beta & 0 \\ \beta & \alpha - E & \beta \\ 0 & \beta & \alpha - E \end{vmatrix} = \begin{vmatrix} x & 1 & 0 \\ 1 & x & 1 \\ 0 & 1 & x \end{vmatrix} = 0$$

(9.63)

By utilizing the techniques shown in Section 9.2, the characteristic equation can be written as

$$x^3 - 2x = 0,$$

(9.64)

which has the roots $x = 0$, $-(2)^{1/2}$, and $(2)^{1/2}$. Therefore, we obtain the energies of the molecular orbitals from these roots as

$$\frac{\alpha - E}{\beta} = -\sqrt{2} \qquad \frac{\alpha - E}{\beta} = 0 \qquad \frac{\alpha - E}{\beta} = \sqrt{2}$$

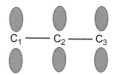

$C_1 \;\!\!—\!\!\; C_2 \;\!\!—\!\!\; C_3$

Fig. 9.5 The structure of the allyl group.

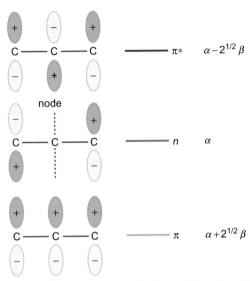

Fig. 9.6 The allyl model showing the *p* orbitals used in π bonding. *(Modified with permission from House, J. E. Inorganic Chemistry, 2nd ed.; Academic Press/Elsevier: Amsterdam, 2013.)*

$$E = \alpha + \sqrt{2}\beta \qquad\qquad E = \alpha \qquad\qquad E = \alpha - \sqrt{2}\beta$$

Both α and β are negative quantities, so the lowest energy is $E = \alpha + (2)^{1/2}\beta$. Figure 9.6 shows the molecular orbitals and energy levels for this system.

The molecular orbital diagram for the neutral, positive, and negative allyl species can be shown, as in Fig. 9.7. The electrons have been placed in the orbitals to show the radical, the cation, and the anion. It is now possible to evaluate the coefficients, the electron densities, and the bond orders. The elements in the secular determinant represent the coefficients in secular equations. Therefore, the determinant of the coefficients can be written directly:

$$\begin{vmatrix} a_1 x & a_2 & 0 \\ a_1 & a_2 x & a_3 \\ 0 & a_2 & a_3 x \end{vmatrix} = 0 \qquad\qquad (9.65)$$

From Eq. (9.65), we see that the three equations are represented as

$$a_1 x + a_2 = 0 \qquad\qquad (9.66)$$

$$a_1 + a_2 x + a_3 = 0 \qquad\qquad (9.67)$$

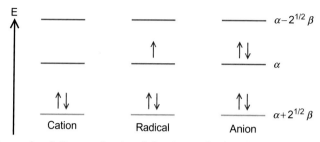

Fig. 9.7 Energy level diagram for the allyl cation, radical, and anion.

$$a_2 + a_3x = 0 \tag{9.68}$$

Taking first the root $x = -(2)^{1/2}$, which corresponds to the molecular orbital of lowest energy, substitution of that value in the equations above gives

$$-a_1\sqrt{2} + a_2 = 0 \tag{9.69}$$

$$a_1 - a_2\sqrt{2} + a_3 = 0 \tag{9.70}$$

$$a_2 - a_3\sqrt{2} = 0 \tag{9.71}$$

From Eq. (9.69) it can be seen that $a_2 = (2)^{1/2}a_1$, and from Eq. (9.71), it can be seen that $a_2 = (2)^{1/2}a_3$. Therefore, it is clear that $a_1 = a_3$, and substituting this value in Eq. (9.70) gives

$$a_1 - a_2\sqrt{2} + a_1 = 0 \tag{9.72}$$

Solving this equation for a_2, we find that $a_2 = (2)^{1/2}a_1$. We now make use of the normalization requirement that

$$a_1^2 + a_2^2 + a_3^2 = 1 \tag{9.73}$$

and by substitution in terms of a_1,

$$a_1^2 + 2a_1^2 + a_1^2 = 4a_1^2 = 1 \tag{9.74}$$

It is therefore found that $a_1 = 0.5 = a_3$ and $a_2 = (2)^{1/2}/2 = 0.707$.

Using these coefficients for the atomic orbitals in the expression for the molecular wave function gives

$$\psi_1 = 0.500\varphi_1 + 0.707\varphi_2 + 0.500\varphi_3 \tag{9.75}$$

Next we use the root $x = 0$ with the secular equations, which leads to

$$0a_1 + a_2 + 0 = 0 \tag{9.76}$$

$$a_1 + 0a_2 + a_3 = 0 \tag{9.77}$$

$$0 + a_2 + 0a_3 = 0 \tag{9.78}$$

From these equations, it is easy to verify that $a_2 = 0$ and $a_1 + a_3 = 0$, so $a_1 = -a_3$. Therefore, the requirement that the sum of the squares of the coefficients is equal to 1 gives

$$a_1^2 + a_3^2 = 1 = 2a_1^2 \tag{9.79}$$

from which we find $a_1 = 1/(2)^{1/2}$ and $a_3 = -1/(2)^{1/2}$. These coefficients obtained from the $x = 0$ root lead to the wave function

$$\psi_2 = 0.707\varphi_1 - 0.707\varphi_3. \tag{9.80}$$

When the root $x = (2)^{1/2}$ is used to evaluate the constants by the procedures above, the wave function obtained can be written as

$$\psi_3 = 0.500\varphi_1 - 0.707\varphi_2 + 0.500\varphi_3 \tag{9.81}$$

However, this orbital remains unpopulated regardless of whether the radical, the cation, or the anion is considered (see Fig. 9.7). It is now possible to calculate the electron densities and bond orders in the cation, radical, and anion species. For the allyl radical, there are three electrons in the π system with only one electron in the state having $E = \alpha$. Because the square of the coefficients of the atomic wave functions multiplied by the occupancy of the orbitals gives the ED, we see that

$$\text{ED at } C_1 = 2(0.500)^2 + 1(0.707)^2 = 1.00$$
$$\text{ED at } C_2 = 2(0.707)^2 + 1(0)^2 = 1.00$$
$$\text{ED at } C_3 = 2(0.500)^2 + 1(-0.707)^2 = 1.00$$

As expected, the three electrons are distributed equally on the three carbon atoms. The allyl carbocation has only the orbital of lowest energy $(E = \alpha + \sqrt{2}\beta)$ populated with two electrons, so the electron densities are

$$\text{ED at } C_1 = 2(0.500)^2 = 0.50$$
$$\text{ED at } C_2 = 2(0.707)^2 = 1.00$$
$$\text{ED at } C_3 = 2(0.500)^2 = 0.50.$$

For the anion, the two molecular orbitals of lowest energy are occupied with two electrons in each and the electron densities are

$$\text{ED at } C_1 = 2(0.500)^2 + 2(0.707)^2 = 1.50$$
$$\text{ED at } C_2 = 2(0.707)^2 + 2(0)^2 = 1.00$$
$$\text{ED at } C_3 = 2(0.500)^2 + 2(-0.707)^2 = 1.50.$$

From these results, we see that the additional electron that the anion contains compared to the radical resides in a molecular orbital centered on the terminal carbon atoms. It is to be expected that the negative regions would be separated in this way.

The bond orders can now be obtained, and in this case the two ends are identical for the three species, so we expect to find $B_{12} = B_{23}$. Using the populations of the orbitals shown in Fig. 9.7 and the coefficients of the wave functions, we find π bond orders as

Cation $(B_{12} = B_{23})$: $2(0.500)(0.707) = 0.707$

Radical $(B_{12} = B_{23})$: $2(0.500)(0.707) + 1(0)(0.707) = 0.707$

Anion $(B_{12} = B_{23})$: $2(0.500)(0.707) + 2(0)(0.707) = 0.707.$

Because the difference in electron populations for the three species involves an orbital of energy α, it is an orbital having the same energy as the atomic orbital and it is therefore nonbonding. Thus, the number of electrons in this orbital (0, 1, or 2 for the cation, radical, or anion, respectively) does not affect the bond orders of the species. The bond orders are determined by the population of the bonding orbital, which has $E = \alpha + \sqrt{2}\beta$.

If the structures

$$C = C - C \quad \text{and} \quad C \cdots C \cdots C$$

are considered, we find that the first structure has one π bond between adjacent carbon atoms, as did ethylene. If it is assumed that the bond is localized, then the orbital energy for two electrons would be $2(\alpha + \beta)$, as it was for the ethylene case. For the second structure, the energy that we found for the orbital having the lowest energy is that for the bonding orbital in the allyl system. Therefore, for two electrons populating that orbital, the energy would be $2[\alpha + \sqrt{2}\beta]$. The difference between these energies is known as the *delocalization* or *resonance* energy and amounts to -0.828β. Therefore, the structure showing the delocalized π bond represents a lower energy. If three carbon atoms are placed in a ring structure, carbon atom 1 is bonded to carbon atom 3 and the secular determinant must be modified to take into account that bond:

Therefore, the elements H_{13} and H_{31}, rather than being zero, are set equal to β, and the secular determinant is written as

$$\begin{vmatrix} x & 1 & 1 \\ 1 & x & 1 \\ 1 & 1 & x \end{vmatrix} = 0 \qquad (9.82)$$

Therefore, expansion of the determinant leads to the characteristic equation

$$x^3 - 3x + 2 = 0 \qquad (9.83)$$

The roots of this equation are $x = -2$, 1, and 1, which correspond to the energies $E = \alpha + 2\beta$ and $E = \alpha - \beta$ (twice). The result is the molecular orbital diagram can be represented as in Fig. 9.8. The coefficients of the wave functions, electron densities, and bond orders can be calculated by the procedures employed for the allyl system. However, we will leave that as an exercise and progress directly to the determination of the resonance energy. For the cation, two electrons populate the lowest state with a total energy of $E = 2(\alpha + 2\beta)$, compared with two electrons in an ethylene (localized) bond that have an energy of $E = 2(\alpha + \beta)$. Therefore, the resonance energy is -2β. For the anion, the two degenerate orbitals having an energy of $\alpha - \beta$ are singly occupied, and the total energy for four electrons is $E = 2(\alpha + 2\beta) + 2(\alpha - \beta)$. For one localized bond as in ethylene and two single electrons on carbon atoms $(E = \alpha)$, the total energy would be $E = 2(\alpha + \beta) + 2\alpha$. Therefore, the resonance energy for the anion would be 0. It is correctly predicted that the cyclopropenyl cation (resonance energy of 2β) would be more stable than the anion.

Having solved the problems of three carbon atoms in a chain or ring structure, we could use the same methods to examine a totally different chemical system. Suppose a simple calculation is carried out for the H_3^+ species to determine which of the following structures is more stable.

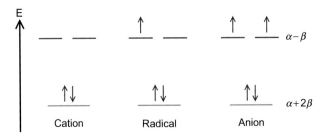

Fig. 9.8 Energy level diagrams for cyclopropenyl species.

$$[H_1 - H_2 - H_3]^+ \quad \text{or} \quad \left| \begin{array}{c} H_3 \\ \diagup \quad \diagdown \\ H_1 \underline{\quad\quad} H_2 \end{array} \right|^+$$

(A)   (B)

It is still possible to use α as the binding energy of an electron in an atom, but it represents the binding energy of an electron in a *hydrogen* atom rather than a carbon atom. Therefore, the secular determinants for the two cases will look exactly as they did for the allyl and cyclopropene systems. For structure A, the roots of the secular determinant lead to $E = \alpha + (2)^{1/2} \beta$, $E = \alpha$, and $E = \alpha - 2^{1/2} \beta$. For structure B, the roots of the secular determinant lead to the energies $E = \alpha + 2\beta$, $E = \alpha - \beta$, and $E = \alpha - \beta$. Therefore, with two electrons to place in the sets of molecular orbitals, we place them in the lowest energy level. For the ring structure that corresponds to $E = 2(\alpha + 2\beta)$, but for the linear structure the energy would be $E = 2[\alpha + 2^{1/2} \beta]$. The energy is more favorable for the ring structure, and we predict that it would be more stable by -1.2β.

As an additional example of the use of the Hückel method, the calculation will be performed for 1,3-butadiene. The Hückel calculation for butadiene begins with the determinant

$$\begin{vmatrix} \alpha - E & \beta & 0 & 0 \\ \beta & \alpha - E & \beta & 0 \\ 0 & \beta & \alpha - E & \beta \\ 0 & 0 & \beta & \alpha - E \end{vmatrix} = \begin{vmatrix} x & 1 & 0 & 0 \\ 1 & x & 1 & 0 \\ 0 & 1 & x & 1 \\ 0 & 0 & 1 & x \end{vmatrix} = 0 \qquad (9.84)$$

which results in the characteristic equation

$$x^4 - 3x^2 + 1 = 0 \qquad (9.85)$$

The roots of this equation are ± 1.62 and ± 0.62, giving energies of $E_1 = \alpha + 1.62\beta$, $E_2 = \alpha + 0.62\ \beta$, $E_3 = \alpha - 0.62\ \beta$, and $E_4 = \alpha - 1.62\beta$. With there being only four electrons in the π system, only the first two energy levels are populated. Therefore, the resonance energy amounts to 0.48β in magnitude.

Following the procedures illustrated earlier, it is possible to obtain the explicit form of the four wave functions, which can now be written as

$$\psi_1 = 0.372\phi_1 + 0.602\phi_2 + 0.602\phi_3 + 0.372\phi_4$$
$$\psi_2 = 0.602\phi_1 + 0.372\phi_2 - 0.372\phi_3 - 0.602\phi_4$$
$$\psi_3 = 0.602\phi_1 - 0.372\phi_2 - 0.372\phi_3 + 0.602\phi_4$$
$$\psi_4 = 0.372\phi_1 - 0.602\phi_2 + 0.602\phi_3 - 0.372\phi_4$$

in which ϕ indicates a carbon p wave function. The graphs of the four wave functions and the orbital diagrams for butadiene are shown in Fig. 9.9.

Hückel calculations can be performed on cyclic systems with the only modification being that the secular determinant has an element equal to one in the top right and bottom left positions where the atoms that would be in terminal positions bond to each other. Thus, for cyclobutadiene, the secular determinant would be

$$\begin{vmatrix} \alpha-E & \beta & 0 & \beta \\ \beta & \alpha-E & \beta & 0 \\ 0 & \beta & \alpha-E & \beta \\ \beta & 0 & \beta & \alpha-E \end{vmatrix} = \begin{vmatrix} x & 1 & 0 & 1 \\ 1 & x & 1 & 0 \\ 0 & 1 & x & 1 \\ 1 & 0 & 1 & x \end{vmatrix} = 0 \qquad (9.86)$$

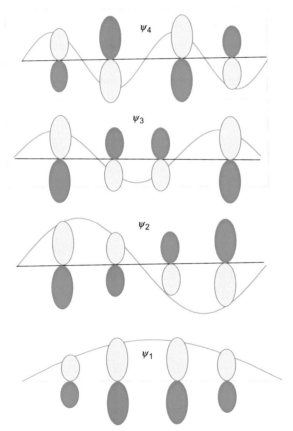

ψ_4

ψ_3

ψ_2

ψ_1

Fig. 9.9 The orbitals and wave function plots for butadiene. As in numerous previous cases, green is used to indicate the lobes have a positive mathematical sign, whereas the lobes shaded yellow have a negative sign.

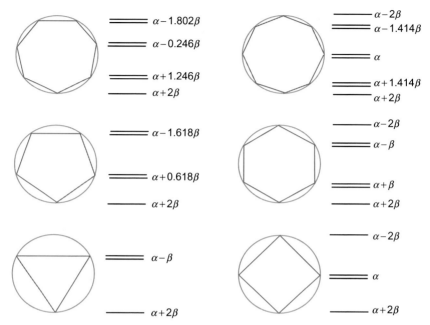

Fig. 9.10 A Frost-Musulin diagram for cyclic hydrocarbons (Frost and Musulin, 1953).

which leads to the characteristic equation $x^4 - 4x^2 = 0$. Solving this equation leads to the energy levels $\alpha + 2\beta$, α, α, and $\alpha - 2\beta$.

Many years ago, a useful mnemonic was developed by Frost and Musulin to illustrate the results for cyclic systems (Frost and Musulin, 1953). This device is shown in Fig. 9.10, along with the actual energies of the levels. Figure 9.10 makes it simple to see the energy levels that result from Hückel calculations for cyclic systems. However, the numerical results of such calculations for cyclic molecules are also shown in Table 9.1.

9.5 CALCULATIONS INCLUDING HETEROATOMS

In the treatment of organic molecules by the Hückel method, the two energy parameters are H_{11} or α and H_{12} or β. These energies refer to carbon atoms and specifically to the p orbitals of the carbon atom. If atoms other than carbon are present, the Hückel method can still be used, but the energy parameters must be adjusted to reflect the fact that an atom other than carbon has a different binding energy for its electron. Pauling and Wheland developed a procedure to adjust for the heteroatom by relating the α and β parameters for that atom to those for carbon. For example, if a nitrogen atom is

Table 9.1 Energy levels for cyclic systems

No. of Atoms =	3	4	5	Orbital energies 6	7	8
E_8	–	–	–	–	–	$\alpha - 2\beta$
E_7	–	–	–	–	$\alpha - 1.802\beta$	$\alpha - 1.414\beta$
E_6	–	–	–	$\alpha - 2\beta$	$\alpha - 1.802\beta$	$\alpha - 1.414\beta$
E_5	–	–	$\alpha - 1.618\beta$	$\alpha - \beta$	$\alpha - 0.246\beta$	α
E_4	–	$\alpha - 2\beta$	$\alpha - 1.618\beta$	$\alpha - \beta$	$\alpha - 0.246\beta$	α
E_3	$\alpha - \beta$	α	$\alpha + 0.618\beta$	$\alpha + \beta$	$\alpha + 1.246\beta$	$\alpha + 1.414\beta$
E_2	$\alpha - \beta$	α	$\alpha + 0.618\beta$	$\alpha + \beta$	$\alpha + 1.246\beta$	$\alpha + 1.414\beta$
E_1	$\alpha + 2\beta$	$\alpha + 2\beta$	$\alpha + 2\beta$	$\alpha + 2\beta$	$\alpha + 2\beta$	$\alpha + 2\beta$

present and donates one electron to the π system, the atom is represented in the secular determinant by $\alpha_{N1} = \alpha + 0.5\beta$, where α and β are the values used for a carbon atom, and the subscript "1" indicates the number of electrons used in π bonding. When an atom like nitrogen or oxygen donates more than one electron to the π system, the parameters are adjusted to reflect this difference. A series of values have been adopted for various atoms other than carbon and their adjusted values are shown in Table 9.2.

Calculations carried out for the pyrrole molecule lead to electron densities on the atoms that can be shown as

The electron densities are in accordance with the fact that nitrogen has an electronegativity of 3.0, whereas that of carbon is 2.5. However, the chemical behavior of pyrrole with regard to substitution reactions does not always agree with the predicted electron densities, so mixtures of products are often obtained.

Table 9.2 Values of coulomb and exchange integrals for heteroatoms

$$\alpha_{N1} = \alpha + 0.5\beta$$
$$\alpha_{N2} = \alpha + 1.5\beta$$
$$\alpha_{O1} = \alpha + 1.5\beta$$
$$\alpha_{O2} = \alpha + 2.5\beta$$

9.6 SOME TRIATOMIC INORGANIC MOLECULES

Studies dealing with matter in interstellar space and as transient molecules in irradiated matrices have resulted in a great deal of interest in species such as CCN, CNC, NNC, NCN, and several others. It would be interesting to determine what a Hückel calculation on some of these species might reveal. Accordingly, such calculations will be illustrated in this section.

Earlier, a discussion of the allyl system was presented as an example of a molecule containing three atoms. To illustrate the type of calculation possible for systems containing atoms other than carbon, the N–C–N molecule will be considered, in which it is assumed that each atom contributes one electron to the π system. In this case, the value for α_{N1} will be taken as $\alpha + 0.5\beta$, and without introducing any other changes, the secular determinant can be written as

$$\begin{vmatrix} \alpha + 0.5\beta - E & \beta & 0 \\ \beta & \alpha - E & \beta \\ 0 & \alpha + 0.5\beta - E \end{vmatrix} = 0 \qquad (9.87)$$

As before, each element will be divided by β and we will let $x = (\alpha - E)/\beta$. The secular determinant then becomes

$$\begin{vmatrix} x + 0.5 & 1 & 0 \\ 1 & x & 1 \\ 0 & 1 & x + 0.5 \end{vmatrix} = 0 \qquad (9.88)$$

This determinant can be expanded to give the polynomial equation

$$x(x + 0.5)^2 - (x + 0.5) - (x + 0.5) = 0 \qquad (9.89)$$

or

$$x^3 + x^2 - 1.75x - 1 = 0 \qquad (9.90)$$

The roots of this equation are $x = -1.686$, -0.500, and 1.186, which lead to the energy levels shown in Fig. 9.11.

The coefficients in the wave functions can be evaluated in the usual way because from the secular determinant we can write the equations

$$a_1(x + 0.5) + a_2 = 0 \qquad (9.91)$$

$$a_1 + a_2 x + a_3 = 0 \qquad (9.92)$$

$$a_2 + a_3(x + 0.5) = 0 \qquad (9.93)$$

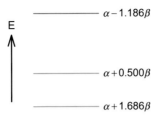

Fig. 9.11 Energy level diagram for the NCN molecule, assuming that each nitrogen atom donates *one* electron.

From Eqs. (9.91) and (9.93), it can be seen that $a_1 = a_3$. Starting with the root $x = -1.686$, we find that $x + 0.5 = -1.186$ and after substituting $a_1 = a_3$, the equations become

$$-1.186a_1 + a_2 = 0 \qquad (9.94)$$

$$a_1 - 1.686a_2 + a_1 = 0 \qquad (9.95)$$

$$a_2 - 1.186a_1 = 0 \qquad (9.96)$$

From these equations, it is easy to see that $a_2 = 1.186a_1 = 1.186a_3$, and from the normalization condition, it is found that

$$a_1^2 + a_2^2 + a_3^2 = 1 = a_1^2 + (1.186a_1)^2 + a_1^2 \qquad (9.97)$$

Therefore, it is found that $a_1 = (1/3.41)^{1/2} = 0.542 = a_3$, which results in $a_2 = 1.186a_1 = 0.643$. Accordingly, the wave function that corresponds to the root $x = -1.686$ is

$$\psi_1 = 0.542\varphi_1 + 0.643\varphi_2 + 0.542\varphi_3 \qquad (9.98)$$

Using the $x = -0.500$ root in an analogous procedure leads to the wave function

$$\psi_2 = 0.707\varphi_1 - 0.707\varphi_3 \qquad (9.99)$$

As a result of these two orbitals for which we have obtained wave functions being able to hold the three electrons in the π system, it is not necessary to determine the coefficients for the third molecular orbital.

A convenient way to determine whether the calculated energy expressions are correct is to substitute the wave function in the equation

$$E = \int \psi_1^* \hat{H} \psi_1 \, d\tau \qquad (9.100)$$

If the calculations are correct, the integral must evaluate to the energy that was obtained for that orbital, which is $E = \alpha + 0.5\,\beta$. Substituting the wave function ψ_2 into this expression gives

$$E = a_1^2(\alpha + 0.5\beta) + a_2^2\alpha + a_3^2(\alpha + 0.5\beta) + 2a_1 a_2\beta + 2a_2 a_3\beta \qquad (9.101a)$$

In this expression the last two terms will be equal to zero because a_2 is zero.

$$E = 0.707^2(\alpha + 0.5\beta) + (-0.707)^2(\alpha + 0.5\beta) \qquad (9.101b)$$

Simplifying Eq. (9.101b) gives $E_2 = 1.00\alpha + 0.5\beta$, which is the energy calculated earlier for the second energy state. Therefore, the calculated energy is correct. It is also possible to verify the value of E_1 by a similar procedure.

The electron densities for $N_1-C_2-N_3$, where subscripts indicate atomic positions in the chain, are

$$\text{ED at } N_1 = 2(0.542)^2 + 1(0.707)^2 = 1.09$$

$$\text{ED at } C_2 = 2(0.643)^2 = 0.827$$

$$\text{ED at } N_3 = 2(0.542)^2 + 1(-0.707)^2 = 1.09.$$

Note that within the round-off errors encountered, the sum of the electron densities is 3.0. As expected based on the relative electronegativities of nitrogen and carbon, it is found that the ED is higher on the nitrogen atoms than it is on the carbon atoms. The bond orders $B_{12} = B_{23}$ can be calculated as illustrated earlier:

$$B_{12} = B_{23} = 2(0.542)(0.643) + 1(0.707)(0) = 0.697 \qquad (9.102)$$

If it is assumed that the nitrogen atoms each donate two electrons to the π system, the value $\alpha_{N2} = \alpha + 1.5\beta$ is used. After making the usual substitutions, the secular determinant can be written as

$$\begin{vmatrix} x + 1.5 & 1 & 0 \\ 1 & x & 1 \\ 0 & 1 & x + 1.5 \end{vmatrix} \qquad (9.103)$$

This results in the characteristic equation

$$x(x + 1.5)^2 - (x + 1.5) - (x + 1.5) = 0 \qquad (9.104)$$

This equation has the roots $x = -2.35, -1.50$, and $+0.85$. These roots are then set equal to $(\alpha - E)/\beta$, and the calculated energies lead to the molecular orbital diagram shown in Fig. 9.12.

Using the procedures developed earlier, we find the coefficients to be $a_1 = 0.606$, $a_2 = 0.515$, and $a_3 = 0.606$, when the root $x = -2.35$ (corresponding to the lowest energy) is used. Therefore, the lowest lying molecular orbital has the wave function

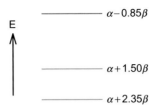

Fig. 9.12 Energy level diagram for the NCN molecule, assuming that each nitrogen atom donates *two* electrons.

$$\psi_1 = 0.606\varphi_1 + 0.515\varphi_2 + 0.606\varphi_3.$$

From the root $x=-1.50$, we find the coefficients $a_1=0.707$, $a_2=0$, and $a_3=-0.707$, which lead to the wave function

$$\psi_2 = 0.707\varphi_1 - 0.707\varphi_3.$$

The coefficients $a_1=0.365$, $a_2=-0.857$, and $a_3=0.365$ are found from the root $x=0.85$. That root gives rise to the wave function

$$\psi_3 = 0.365\varphi_1 - 0.857\varphi_2 + 0.365\varphi_3,$$

which corresponds to the orbital of highest energy and is populated by only one electron. When the coefficients of the three wave functions are used to calculate the electron densities on the atoms, the results are

$$\text{ED at } N_1 = 2(0.606)^2 + 2(0.707)^2 + 1(0.365)^2 = 1.87$$
$$\text{ED at } C_2 = 2(0.515)^2 + 2(0)^2 + 1(0.857)^2 = 1.26$$
$$\text{ED at } N_3 = 2(0.606)^2 + 2(-0.707)^2 + 1(0.365)^2 = 1.87.$$

As expected, the total electron density is equivalent to a total of five electrons, and the electron density is higher on the nitrogen atoms than it is on the carbon atom. In fact, the values obtained are of reasonable magnitude based on the electronegativities of the atoms. These calculations are qualitative, of course, but the results are in agreement with expectations based on electronegativities.

It should be mentioned that the cyanamide ion, CN_2^{2-}, has the structure $N=C=N$ and constitutes a logical extension of the preceding problem. It is thus apparent that possibilities exist for using the HMO calculations for inorganic species as well. The main use of this method has, of course, been in the area of organic chemistry.

9.7 KERNELS, REPULSION, AND STABILITY

As a result of their presence in interstellar matter, numerous small molecules have been identified spectroscopically. For example, species such as C_2N

and CN_2 have been the subject of a great deal of study, both experimentally and theoretically (Martin et al., 1994; Hensel and Brown, 1996; Curtis et al., 1988; Mebel and Kaiser, 2002). Calculations show that the structure CNN has an energy that is approximately 31 kcal mol^{-1} higher than that of the NCN structure. This obeys the general rule that in triatomic molecules, the atom having lowest electronegativity usually occupies the middle position. One of the most familiar molecules that illustrates this principle is nitrous oxide, N_2O for which the structure is NNO rather than NON. Although experimental evidence makes this clear now, that was not the case when Pauling and Hendricks predicted the correct structure in 1926 (Pauling and Hendricks, 1926).

A simple and interesting approach to stability of such species described by Pauling (1975) as "… a simple of way of remembering which structure is to be assigned to the more stable isomer," was to consider the core $1s^2$ electrons and the nucleus as constituting a *kernel*, with a residual positive charge of $(N-2)$. In this way, the structure is determined by the minimum repulsion of the kernels in the structure. For example, when considering the structures of cyanate (NCO^-) and fulminate (CNO^-), the following situation results.

$$C^{4+}\text{------}N^{5+}\text{------}O^{6+} \quad N^{5+}\text{------}C^{4+}\text{------}O^{6+}$$

The actual bond lengths in these structures do not vary greatly, so it was assumed that a bond length of 1.15 Å (115 pm) could be assumed without invalidating the method. In this way, the repulsions can be expressed in terms of e^2/r. The energies due to repulsion of the kernels can be summarized as follows:

$$C^{4+}\text{.........}N^{5+}\text{.........}O^{6+} \quad N^{5+}\text{.........}C^{4+}\text{.........}O^{6+}$$
$$20e^2/r \; + \; 30e^2/r \qquad\qquad 20e^2/r \; + \; 24e^2/r$$
$$+ \, 24e^2/2r \qquad\qquad\qquad + \, 30e^2/2r$$

In these structures, the lower quantity represents the repulsion of the terminal kernels at a distance of $2r$. The difference is $3e^2/r$ with the cyanate structure having the lower energy. The valence shell electrons are distributed so as to reduce the repulsion. However, when the difference of $3e^2/r$ is converted to an equivalent on a molar basis, it amounts to approximately 3600 kJ mol^{-1}. On the basis of the heat of the reaction of N_2O with CO, Pauling estimated that the effect produced by the electrons reduced this by about 90%, but the difference shows that the fulminate ion is considerably less stable than is the cyanate ion. This is in accord with the chemical of these classes of compounds.

The same approach can be taken with the C_2N, with the results as follows:

$$C^{4+}.........C^{4+}.........N^{5+} \quad C^{4+}.........N^{5+}.........C^{4+}$$
$$16e^2/r \ + \ 20e^2/r \qquad\quad 20e^2/r \ + \ 20e^2/r$$
$$+ 20e^2/2r \qquad\qquad\quad + 16e^2/2r$$

It can be seen that there is a difference in the kernel repulsion of $2e^2/r$, with the CCN structure having lower repulsion, which amounts to approximately 2400 kJ mol^{-1}. This does not take into account the "cushioning" effect of the valence shell electrons, but as in the case of the fulminate and cyanate ions, it is unlikely that the kernel repulsion is negated.

For the CN_2 molecule, the repulsion of the kernels can be represented as follows:

$$N^{5+}.........C^{4+}.........N^{5+} \quad N^{5+}.........N^{5+}.........C^{4+}$$
$$20e^2/r \ + \ 20e^2/r \qquad\quad 25e^2/r \ + \ 20e^2/r$$
$$+ 25e^2/2r \qquad\qquad\quad + 20e^2/2r$$

It can be seen that the NCN structure has $2.5e^2/r$ lower repulsion than of that the CNN structure.

There are numerous triatomic molecules that have been either identified or suspected of interstellar existence. Several such species include those having 13, 14, or 15 valence shell electrons. Among the most extensively studied, both experimentally and theoretically, are those containing carbon and nitrogen atoms. Some of the most interesting are those that are isomers of C_2N and CN_2. On the basis of *ab initio* calculations, Martin et al. (1994) concluded that the CNN isomer (in the $^3\Pi$ state, which does not have a center of symmetry) is higher in energy than the NCN isomer (in its $^3\Pi_g$ state, which has a center of symmetry) by approximately 31.5 kcal mol^{-1}. It thus appears that a simple approach developed by Pauling many years ago leads to correct conclusions, at least on a qualitative basis, regarding the stability of molecules unknown at the time and the subject of a great deal of study currently.

The values for heat of atomization for CN_2 and C_2N were reported to be 288.6 and 294.1 kcal mol^{-1}, respectively (Martin et al., 1994), which indicates that C_2N is more stable than CN_2, which would be expected. Mebel and Kaiser have reported the vibrational frequencies for the C_2N molecules (Mebel and Kaiser, 2002). The CNC structure has a ground state of $^2\Pi_g$, whereas that of the CCN is $^2\Pi$ (there is no center of symmetry). Each of these molecules gives rise to three infrared active vibrations (see Chapter 11). From these and other studies, it is apparent that the isomeric species of C_2N and CN_2 have been extensively studied.

9.8 BAND THEORY OF METALS

One view of a solid metal is that it is composed of a collection of ions having positive charges around which is the collection of mobile electrons. When viewed in this way, the properties of electrical conductivity and malleability of metals follow logically. Alloys consist of solid solutions of two or more metals, so substitution of one type of ion for another would result in an alloy. It is well known that when metals are combined with hydrogen (hydrides), oxygen (oxides), nitrogen (nitrides), or carbon (carbides), the metallic properties mentioned above are greatly diminished.

Even though some electrons may be mobile, it is largely those in the valence shell that are capable of being moved. The most common structures of metal lattices are cubic closest packed and hexagonal closest packed. In these structures, each metal atom (except those on a surface or edge) has 12 nearest neighbors. Bonding between metal atoms consists of a sharing of the electrons, which are mobile as a result of their being held in *conduction bands*. The result is that although metals are malleable, there is generally a high cohesion and strength of the lattices.

Even though metal atoms may not move *through* a solid lattice, there is motion *within* the lattice. There is some vibration of the lattice members about the average positions. The resistivity of a metal results from the fact that metal atoms hinders the motion of the electrons through the lattice. When a solid is heated, the vibration of the metal atoms becomes more extensive. This increase in vibrational amplitude leads to greater impedance to the flow of electrons, so the resistivity of the metal increases with temperature. Conductivity is the reciprocal of resistivity, so the conductivity of the metal decreases with an increase in temperature.

The molecular orbital approach is a suitable way to describe bonding in many systems. When applied to bonding in a metal, it is assumed that the atomic orbitals of metal atoms combine to form molecular orbitals that extend over a larger number of atoms. At least for atoms in the interior of the metal, each atom can be presumed to contribute an orbital for the combinations. As shown in Section 9.1, when two *atomic* orbitals combine, then two *molecular* orbitals result, with one designated as a *bonding* orbital and the other an *antibonding* orbital. The interaction of three atoms leads to the formation of three molecular orbitals (see Section 9.4). From the molecular orbital energy diagrams shown earlier for ethylene, allyl, and butadiene molecules, the calculations described for those molecules indicate that as the number of molecular orbitals *increases*, the difference in energy between

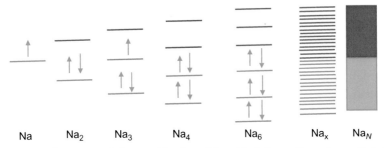

Fig. 9.13 An illustration of the combination of 3s orbitals for sodium atoms. Note that the combination of two atomic orbitals gives rise to the same type of energy level diagram that applied to ethylene. Note also that as the number of atomic orbitals increases, the molecular orbitals get closer together in energy. When x atoms (where x is a large number) combine, the energy levels become very close together and if a very large number of atoms, one mole of atoms (N) combine, the levels essentially form a continuum or band.

them *decreases*. Figure 9.13 illustrates this principle as the number of atoms increases to Avogadro's number N.

In the Hückel approach, the difference in energy between *adjacent* molecular orbitals decreases as the number of orbitals increases to the point where the term *band* is applied. Thermal energy as a function of temperature can be represented by kT, where k is Boltzmann's constant and T is the temperature (K). For a collection of molecular orbitals that contains a large number of atoms, the separation between adjacent orbitals is smaller than kT. As a result, numerous closely spaced molecular orbitals may be populated to form a conduction band. Because vibrations of metal atoms in a lattice increase with temperature, the conductivity of metals decreases as the temperature increases as a result of reduced electron mobility.

For a set of molecular orbitals in the Hückel approximation, the energy of the nth orbital is given by

$$E = \alpha + 2\beta \cos \frac{n\pi}{N+1} \qquad (9.105)$$

In this equation, α represents a Coulomb integral (H_{ii}), β is a resonance integral (H_{ij}), and N approaches infinity for a metal. As a result, there are N energy levels that comprise an energy *band* with an overall width approaching 4β. Within this band, the difference in energy between n and $n+1$ levels approaches zero and N increases. This description of the energy levels in a metal gives rise to the *band theory* descriptor, which is illustrated in Fig. 9.13.

If the metal being considered is presumed to be sodium (as is the case shown in Fig. 9.13), the electron configuration is $3s^1$, so the set of bonding orbitals constructed from these orbitals is only half filled. It is sometimes considered that all of the sodium orbitals can form bands, but the gaps between the bands limit electrical conductance to the upper band. For purposes of illustration, it will be assumed that the core electrons that reside in the $1s$ and $2s$ orbitals are not involved in bonding. The resulting band structure can be illustrated, as shown in Fig. 9.14. When N atoms are considered, electron population of the bands can be expressed as $2(2\,l+1)N$, in which l is the orbital quantum number. Both the $2p$ and $3s$ orbitals interact to form bands because both types of orbitals are occupied, and therefore a band is formed from the combination of each type of orbital.

For sodium, the band of highest energy is only half filled because each sodium atom has a single electron in the $3s$ orbital. As a result, it is possible for electrons to enter and leave the band. This action can be caused by light in the visible region, which gives rise to an absorption and emission process at the surface of the metal. This process causes the metal to have the shiny appearance known as metallic luster. With electrons being able to move in a partially filled band, the metal is a good conductor of electricity.

Materials that behave as semiconductors or insulators have greater energy differences between the bands (sometimes referred to as the *band gap*). In the case of insulators, the movement of electrons requires energies that are of a magnitude similar to the binding energies of electrons in atoms (up to 10–12 eV). Semiconductors have band gaps that typically range from 1–2.5 eV. For example, some representative values are as follows: Ge, 0.67; Si, 1.14; and CdS, 2.42 eV (Serway and Jewett, 2014). At room temperature, thermal energy per mole is RT or about 2.49 kJ mol^{-1} = 596 cal mol^{-1} = 0.596 kcal mol^{-1}.

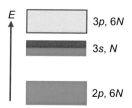

Fig. 9.14 Bands formed by the combination of orbitals on sodium atoms. A green color signifies that a band or portion of the band is filled, whereas a red color signifies a portion of the band that is empty. The band arising from the $3s$ orbitals is only half filled, because there is only one electron in the $3s$ state of each sodium atom. The band from combination of $3p$ states is empty. *(Modified with permission from House, J. E. Inorganic Chemistry, 2nd ed.; Academic Press/Elsevier: Amsterdam, 2013.)*

This is equivalent to 0.026 eV molecule^{-1} so increasing the temperature increases the number of electrons that can populate orbitals of higher energy, as is the case with a semiconductor (see Section 6.7). For a much more complete discussion of semiconductors, see the books of Serway and Jewett (2014) and Blinder (2004).

The HMO theory is adequate to deal with many significant problems in molecular structure and reactivity. In view of its gross approximations and very simplistic approach, it is surprising how many qualitative aspects of molecular structure and reactivity can be dealt with using the Hückel approach. For a more complete discussion of this topic, consult the references listed at the end of this book. As was stated at the beginning of this chapter "In spite of its approximate nature, Hückel molecular orbital (HMO) theory has proved itself extremely useful in elucidating problems concerned with electronic structures of π-electron systems" (Purins and Karplus, 1968). It is still true.

LITERATURE CITED

Blinder, S. M. *Introduction to Quantum Mechanics*. Academic Press/Elsevier: Amsterdam, 2004; (chapter 12).

Curtis, M. C.; Levick, A. P.; Sarre, P. J. Laser-Induced Fluorescence Spectrum of the CNN Molecule. *Laser Chem.* **1988**, *9*, 359–368.

Frost, A. A.; Musulin, B. A Mnemonic Device for Molecular Orbital Energies. *J. Chem. Phys.* **1953**, *21*, 572.

Hensel, K. D.; Brown, J. M. The ν_3 Band of the NCN Radical Studied by LMR. *J. Mol. Spectrosc.* **1996**, *180*, 170–174.

Liberles, A. *Introduction to Molecular-Orbital Theory*. Holt, Rinehart, and Winston, Inc.: New York, 1966.

Martin, J. M. L.; Taylor, P. R.; Francois, J. P.; Gijbels, R. Ab Initio Study of the Spectroscopy and Thermochemistry of the C_2N and CN_2 Molecules. *Chem. Phys. Lett.* **1994**, *226*, 475–483.

Mebel, A. M.; Kaiser, R. I. The Formation of Interstellar C_2N Isomers in Circumstellar Envelopes of Carbon Stars: An Ab Inito Study. *Astron J.* **2002**, *564*, 787–791.

Pauling, L.; Hendricks, S. B. The Prediction of the Relative Stabilities of Isosteric Isomeric Ions and Molecules. *J. Am. Chem. Soc.* **1926**, *48*, 641–651.

Pauling, L. The Relative Stability of Isosteric Ions and Molecules. *J. Chem. Educ.* **1975**, *52*, 577.

Purins, D.; Karplus, M. Methyl Group Inductive Effect in the Toluene Ions. Comparison of Hückel and Extended Hückel Theory. *J. Am. Chem. Soc.* **1968**, *90*(23), 6275–6281.

Roberts, J. D. *Notes on Molecular Orbital Calculations*. Benjamin: New York, 1962.

Serway, R. E.; Jewett, J. W. *Physics for Scientists and Engineers*; 9th ed.; Brooks/Cole: Boston, MA, 2014.

PROBLEMS

1. Find the roots of these equations using graphical or numerical methods:
   (a) $x^3 - 2x^2 - 2x + 3 = 0$
   (b) $x^4 - 5x^2 + 2x - 5 = 0$
   (c) $x^3 + 2x^2 - 6x - 8 = 0$
   (d) $x^6 - 6x^4 + 9x^2 - 4 = 0$.
2. Find the solution to $e^x + \cos x = 0$, where x is in radians.
3. Solve $\exp(-x^2) - \sin x = 0$ if x is in radians.
4. Find two roots for the equation $e^{2t} - 2 \cos 2\,t = 0$.
5. In the text, the HMO calculations were carried out for H_3^+. Perform a similar analysis for H_3^-.
6. Two possible structures of I_3^+ are

   Use a HMO calculation to determine which structure is more likely.
7. The tri-iodide ion I_3^- forms when I_2 reacts with I^- in aqueous solutions of KI. Perform the same calculations from Problem 6 for the I_3^- to determine the preferred structure.
8. Perform HMO calculations for the molecules N–C–C and C–N–C. In each case, assume that the nitrogen atom is a two-electron donor to the π system. Determine the energy levels, the coefficients for the wave functions, and the charge densities on the atoms.
9. Perform HMO calculations for bicyclobutadiene,

$$\begin{array}{c}C\\C \overline{\quad\quad} C\\C\end{array}$$

10. Carry out a HMO calculation for 1,3-butadiene, $CH_2=CH-CH=CH_2$. Determine the energy levels, the coefficients of the wave functions, the bond orders, and the electron density at each carbon atom.
11. Using the HMO approach, determine the resonance or delocalization energy for cyclobutadiene, Complete the calculations by determining the coefficients of the wave functions, the electron density on each atom, and the bond orders.

$$
\begin{array}{ccc}
C & = & C \\
| & & | \\
C & = & C
\end{array}
$$

12. Repeat the calculations of Problem 11 for the ring structure in which each nitrogen atom is assumed to contribute one electron to the π system.

$$
\begin{array}{ccc}
C & = & N \\
| & & | \\
N & = & N
\end{array}
$$

13. Repeat the calculations of Problem 11 for a ring structure like that shown below, assuming that each nitrogen atom contributes one electron to the π system.

$$
\begin{array}{ccc}
N & = & N \\
| & & | \\
C & = & C
\end{array}
$$

14. Consider the molecule

$$
\begin{array}{ccc}
C - C - C \\
\| \quad \| \quad \| \\
C - C - C
\end{array}
$$

Show how the symmetry of this molecule could be used to simplify the Hückel calculations.

15. Assume that a linear structure having a -1 charge is composed of one atom of I, one of Cl, and one of Br. Based on the chemical nature of the atoms, what should be the arrangement of atoms in the structure? Consider only the p orbitals and assume that the same α value can be used for each of these atoms (a very crude approximation). Perform Hückel calculations to determine the electron density at each atom and the bond orders.

16. Because Cl, Br, and I atoms have different electronegativities and electron binding energies (ionization potentials), the same value for α should not be used for each atom. Use the values α for I, $\alpha+0.2\beta$ for Br, and $\alpha+0.4\beta$ for Cl to compensate for the difference in properties of the atoms. Perform Hückel calculations for the three possible linear arrangements of atoms and calculate the electron densities on the atoms for each structure. Explain how the results of the calculations support or contradict the structure that you would expect based on the chemical nature of the atoms.

CHAPTER 10

Molecular Structure and Symmetry

In Chapter 3, it was shown that the symmetry (dimensions) of a box can affect the nature of the energy level diagram for a particle in a three-dimensional box model. Symmetry is, in fact, one of the most important and universal aspects of structures of all types. In chemistry and physics, symmetry and its application through group theory have a direct relationship to structures of molecules, combinations of atomic orbitals that form molecular orbitals, the types of vibrations possible for molecules, and several other facets of the chemical sciences. For this reason, it is imperative that a brief introduction to symmetry and its relationship to quantum mechanics is presented at an early stage in the study of molecules.

10.1 VALENCE BOND DESCRIPTION OF MOLECULAR STRUCTURE

In earlier chapters, it has been shown that the wave equation for the hydrogen atom can be solved to obtain wave functions. These functions describe regions of probability (in one interpretation of quantum mechanics) in which the electron may be found. Geometrical descriptions of the regions were shown, the same orbital features were applied to other atoms, and the directional attributes of the wave functions were also shown. However, due to the fact that the water molecule has a bond angle of 104.5 degrees, it can be inferred that the orbitals used by oxygen are *not* $2p$ orbitals, which are directed at 90 degrees from each other. The bonding orbitals involved must be different from p orbitals, despite the fact that the oxygen atom has a valence shell configuration of $2s^2\,2p^4$. Moreover, carbon has a valence shell configuration of $2s^2\,2p^2$, but methane (CH_4) has a regular tetrahedral structure.

The issue of *molecular* structure based on *atomic* orbitals leads to the conclusion that something must change the nature of the latter when a molecule is formed. The *valence bond* approach to describing molecular structure resulted from the explanation of this issue. In the valence bond description,

Fundamentals of Quantum Mechanics
http://dx.doi.org/10.1016/B978-0-12-809242-2.00010-3

231

atomic orbitals are somehow "mixed" to produce a new set of *bonding* orbitals that differ in directional properties and size. As a result, overlap of such orbitals can lead to stronger bonds and less repulsion of peripheral atoms. The interpretation of the overlap integral for two atomic orbitals φ_1 and φ_2,

$$S_{12} = \int \varphi_1 {}^* \varphi_2 \, d\tau \tag{10.1}$$

is based on the effectiveness of the overlap. In simplest terms, the orbitals describe probability regions, and the overlap of atomic orbitals leads to the reinforcement or cancelation of the probability. One of the basic ideas regarding hybrid orbitals is that the combination of *atomic* orbitals leads to the same number of *hybrid* orbitals, with each being able to hold a pair of electrons.

The simplest combination of atomic orbitals is that of one s orbital and one p orbital to yield two sp hybrid orbitals. Combining the two atomic wave functions leads to two wave functions for the hybrid orbitals. The combinations can be shown as

$$\psi_{(1)} = \frac{1}{\sqrt{2}} \left(\varphi_s + \varphi_{p_z} \right) \quad \text{and} \quad \psi_{(2)} = \frac{1}{\sqrt{2}} \left(\varphi_s - \varphi_{p_z} \right) \tag{10.2}$$

The reason for choosing the p_z orbital, as shown later in this chapter, is the axis of highest symmetry is designated as the z-axis. For a linear molecule, that axis is coincident with the internuclear axis. The combination of one s and one p orbital can be shown graphically as illustrated in Fig. 10.1. As a result of the two sp hybrid orbitals situated at 180 degrees to each other, a molecule that involves such orbitals should be linear. This is the case for BeH_2.

The most common halides of boron have the formula BX_3, and they have trigonal planar structures. With boron having a $2s^2\,2p^1$ electron configuration, the s and p orbitals must combine to produce a set of hybrid orbitals that have the appropriate directional characteristics. Although the mathematical details will not be illustrated, the resulting combinations of wave functions can be shown as follows:

Fig. 10.1 The formation sp hybrid orbitals.

$$\Psi_{(1)} = \frac{1}{\sqrt{3}}\varphi_s + \sqrt{\frac{2}{3}}\varphi_{p_x} \tag{10.3}$$

$$\Psi_{(2)} = \frac{1}{\sqrt{3}}\varphi_s - \frac{1}{\sqrt{6}}\varphi_{p_x} + \frac{1}{\sqrt{2}}\varphi_{p_y} \tag{10.4}$$

$$\Psi_{(3)} = \frac{1}{\sqrt{3}}\varphi_s - \frac{1}{\sqrt{6}}\varphi_{p_x} - \frac{1}{\sqrt{2}}\varphi_{p_y} \tag{10.5}$$

It can be shown mathematically that the resulting three orbitals are directed toward the corners of an equilateral triangle. Pictorially, the combination of orbitals to produce sp^2 hybrids is shown in Fig. 10.2.

The structure of the methane molecule (CH_4) is tetrahedral, so there must be hybrid orbitals that have lobes pointing toward the corners of that structure. A combination of atomic orbitals that does is composed of one s orbital and three p orbitals to give a set of four sp^3 hybrids. The combinations of wave functions are as follows:

$$\Psi_{(1)} = \frac{1}{2}\left(\varphi_s + \varphi_{p_x} + \varphi_{p_y} + \varphi_{p_z}\right) \tag{10.6}$$

$$\Psi_{(2)} = \frac{1}{2}\left(\varphi_s + \varphi_{p_x} - \varphi_{p_y} - \varphi_{p_z}\right) \tag{10.7}$$

$$\Psi_{(3)} = \frac{1}{2}\left(\varphi_s - \varphi_{p_x} + \varphi_{p_y} - \varphi_{p_z}\right) \tag{10.8}$$

$$\Psi_{(4)} = \frac{1}{2}\left(\varphi_s - \varphi_{p_x} - \varphi_{p_y} + \varphi_{p_z}\right) \tag{10.9}$$

A pictorial representation of a set of four sp^3 orbitals is shown in Fig. 10.3.

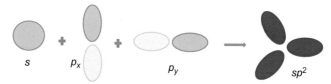

Fig. 10.2 The formation of a set of sp^2 hybrid orbitals.

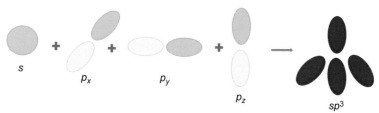

Fig. 10.3 The formation of a set of sp^3 hybrid orbitals.

Molecules such as PF_5 have a structure that is described as a trigonal bipyramid,

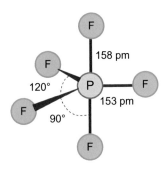

so a hybridization scheme must product five orbitals directed with three in a plane, as well as one directed above and one below that are perpendicular to the plane. Such a set of orbitals is the dsp^3 or sp^3d combination that in reality, can be described more accurately as $(sp^2 + dp)$. As shown later, three sp^2 orbitals are utilized in forming the three bonds in the plane (called the *equatorial* positions), and the linear dp orbitals form bonds to the atoms in *axial* positions. The two sets of orbitals have been shaded differently to illustrate that they are not identical.

sp^3d

Note that the PF_5 molecule has two different bond lengths, with those in the plane being shorter than those in the axial positions. This is evidence that there are two types of hybrid orbitals used to form bonds in such a molecule.

There are numerous molecules in which six groups are bonded to a central atom in an octahedral arrangement. Although such a molecule is SF_6, there are also numerous ions, such as PF_6^-, SbF_6^-, and SnF_6^{2-}, that have octahedral structures. The hybrid orbital type is either sp^3d^2 or d^2sp^3, depending on the nature of the species. All of the orbitals are equivalent in this case, so the structure is a regular octahedron. The lobes are shaded different colors to emphasize the geometry, but they are equivalent with respect to bonding.

sp^3d^2

A large number of species have structures in which the central atom utilizes hybrid orbitals. Of course, in some cases (e.g., H_2O, NH_3, SF_4, etc.), there are one or more unshared pairs of electrons (sometimes called lone pairs). These electrons exert an effect that alters the structure of such molecules as a result of differences in repulsion. For example, an unshared pair of electrons is bound to only one atom, whereas a shared pair is more or less localized between the two atoms sharing them. The result is that bond angles are altered. For example, CH_4 has bond angles of 109.5 degrees, but the bond angles in NH_3 are approximately 107.5 degrees because the repulsion between a shared pair and an unshared pair is greater than that between two shared pairs, which are more localized. This rationale is the basis for the valence shell electron pair repulsion (VSEPR) theory, and invoking it makes it possible to rationalize why bond angles frequently deviate from those predicted for the various types of hybrid orbitals (House, 2013). Table 10.1 illustrates the structures of many species that involve the hybrid orbitals described earlier. The symmetry labels will be explained in later sections of this chapter.

10.2 WHAT SYMMETRY MEANS

One of the most efficient ways to describe the spatial arrangement of atoms in a molecule is to describe the symmetry of the molecule. The symmetry of a molecule is denoted by a symbol that succinctly conveys the necessary information about how the atoms are arranged. The symbol used describes the *point group* to which the molecule belongs. Thus, the symbol O_h is used to describe a molecule like SF_6 having octahedral symmetry. However, C_{2v} is used to describe a bent molecule like H_2O. These symbols indicate the structures of the molecules having these symmetries. When they are encountered, these symbols denote to the reader a particular arrangement of atoms. For example, H_2O can have two orientations, as shown in Fig. 10.4. One orientation can be changed to the other by rotation of the molecule by 180 degrees.

Table 10.1 A summary of molecular structures

Number of pairs on central atom and hybrid type	Number of unshared pairs on the central atom			
	0	1	2	3
2 sp	$D_{\infty h}$ Linear $BeCl_2$			
3 sp^2	D_{3h} Trig. planar BCl_3	C_{2v} Bent $SnCl_2$		
4 sp^3	T_d Tetrahedral CH_4	C_{3v} Trig. pyramid NH_3	C_{2v} Bent H_2O	
5 sp^3d	D_{3h} Trig. bipyramid PCl_5	C_{2v} Irreg. tetrahred. $TeCl_4$	C_{2v} "T" shaped ClF_3	$D_{\infty h}$ Linear ICl_2^-
6 sp^3d^2	O_h Octahedral SF_6	C_{4v} Sq. base bipyr. IF_5	D_{4h} Square planar ICl_4^-	

(Modified with permission from House, J. E. *Inorganic Chemistry*, 2nd ed.; Academic Press/Elsevier: Amsterdam, 2013.)

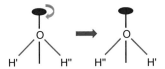

Fig. 10.4 Two orientations of the water molecule. The oval is used to indicate rotation of the molecule by a half circle or 180 degrees.

On the other hand, the molecule HCl can have the two orientations shown in Fig. 10.5. However, a rotation of the HCl molecule by 180 degrees around the axis shown in the figure does *not* lead to an orientation that is indistinguishable from the original, as it does in the case of H_2O. Thus, H_2O and HCl do not have the same *symmetry*. The details of the structure

Fig. 10.5 Two orientations of the HCl molecule.

of a molecule can be adequately described if the symmetry properties are known and vice versa. The rules by which symmetry operations are applied and interpreted (group theory) give certain useful results for considering combinations of atomic wave functions as well.

The rules of group theory govern the permissible ways in which the wave functions can be combined. Thus, some knowledge of molecular symmetry and group theory is essential in order to understand molecular structure and molecular orbital theory. The discussion provided in this chapter will serve as only a brief introduction to the subject, and the references at the end of the book should be consulted for a more exhaustive treatment. As we have seen, a molecule may have two or more orientations in space that are indistinguishable from each other. Certain parts of the molecule can be interchanged in position by performing some operation that changes the relative positions of atoms. Such operations are called *symmetry operations*, and they include rotations about axes and reflections through planes. The imaginary axes about which the rotations are carried out are called *rotation axes*. The planes through which reflections of atoms occur are called *mirror planes*. Symmetry *elements* are the lines, planes, and points that relate the objects in the structure spatially. A symmetry *operation* is performed by making use of a symmetry element. Thus, symmetry elements and operations are not the same, but a rotation axis will be indicated by the symbol C and the operation of actually rotating the molecule around this axis by the same symbol. As a result, the *axis* is the C axis and the *rotation* about that axis is the C operation.

10.3 SYMMETRY ELEMENTS

Center of Symmetry or Inversion Center (*i*)

If the inversion of each atom through this point results in an identical arrangement of atoms, then a molecule possesses a center of symmetry. Thus, in XeF_4, which has a square planar structure, there is a center of symmetry:

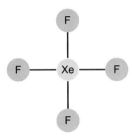

Of course, each atom must be moved through the center the same distance that it was initially situated from the center. The Xe atom is at the center of symmetry, and inversion of each fluorine atom through the Xe gives exactly the same arrangement as the original. However, for a tetrahedral CH_4 molecule, inversion through the *geometric center* of the molecule gives a different result.

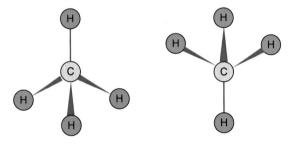

Thus, the *geometric center* (i.e., where the C atom is located) is *not* a center of symmetry. Similarly, the linear CO_2 molecule has a center of symmetry, whereas the bent SO_2 molecule does not:

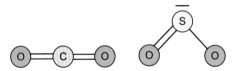

The (Proper) Rotation Axis (C_n)

If a molecule can be rotated around an imaginary axis to produce an equivalent orientation, the molecule possesses a *proper rotation axis*. There is also a symmetry element known as an improper rotation axis, which is designated as S_n and will be described later. Consider the boron trifluoride molecule BF_3, as shown in Fig. 10.6.

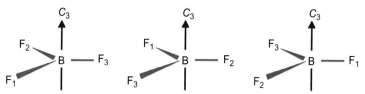

Fig. 10.6 Rotation of a BF_3 molecule around the C_3 axis. Note that a triangle is used on the axis of rotation to indicate a rotation by one-third of a revolution.

It can be seen that rotation around the z-axis, which passes through the boron atom and is perpendicular to the plane of the molecule, produces an indistinguishable structure when the angle of rotation is 120 degrees. In this case, the rotations producing indistinguishable orientations are 120 degrees, or 360 degrees/3, so that the rotation axis is a threefold or C_3 axis. Three such rotations return the fluorine atoms in the molecule to their original positions. For a C_n axis, the n value is determined by dividing 360 degrees by the angle through which the molecule must be rotated to give an equivalent orientation. For the BF_3 molecule, there are three other axes about which the molecule can be rotated by 180 degrees to arrive at the same orientation. These axes are shown in Fig. 10.7, and they lie along the B—F bonds. Therefore, there are three C_2 axes (one lying along each B—F bond) in addition to the C_3 axis that are perpendicular to the C_3 axis.

Although there are three C_2 axes, the C_3 axis is designated as the *principal* axis. The principal axis is designated as the axis of highest-fold rotation (the axis about which the *smallest* rotation produces an indistinguishable orientation of the molecule). This provides the customary way of assigning the z axis, which is the axis of highest symmetry, in setting up an internal coordinate system for a molecule.

Multiple rotations are indicated as $C_3{}^2$, which means two rotations of 120 degrees around the C_3 axis. This *clockwise* rotation of 240 degrees produces the same orientation as a counterclockwise rotation of 120 degrees. Such a

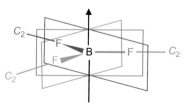

Fig. 10.7 The trigonal planar BF_3 molecule showing the rotation axes and planes of symmetry.

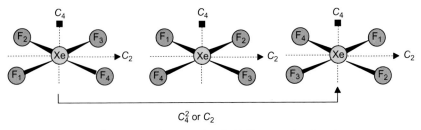

Fig. 10.8 Rotation of a square planar molecule XeF_4 around a C_4 axis.

rotation in the opposite direction is sometimes indicated as C_3^{-1}. It is readily apparent that the same orientation of BF_3 results from C_3^2 and C_3^{-1}.

Certain other "rules" can be seen when considering a square structure such as that of the XeF_4 molecule, which has four identical groups at its corners, as shown in Fig. 10.8. Through the center of the xenon atom, there is a C_4 axis perpendicular to the plane generated by the four fluorine atoms. If we perform four C_4 rotations, we arrive at a structure identical to the initial structure. Thus, C_n^n produces the *identity*, a structure indistinguishable from the original. The operation for the identity is designated as E, so $C_n^n = E$. It is also readily apparent that two rotations of 90 degrees produce the same result as a single rotation of 180 degrees, so $C_4^2 = C_2$.

Mirror Plane (Plane of Symmetry) (σ)

If a molecule has a plane that divides it into two halves that are mirror images, the plane is known as a mirror plane (plane of symmetry). Consider the H_2O molecule shown in Fig. 10.9: There are two mirror planes in this case, the xz and the yz planes. Reflection of the hydrogen atoms through the yz plane interchanges the locations of H′ and H″. Reflection through the yz plane simply interchanges the halves of the hydrogen atoms that are bisected by the xz plane. Both planes are designated as vertical planes because they encompass the z-axis, which is taken to be the vertical axis.

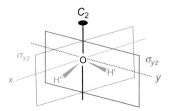

Fig. 10.9 The water molecule showing two mirror planes (outlined in *blue* and *red*) with the two hydrogen atoms lying in the σ_{xz} plane (*blue*). The intersection of these two planes generates a C_2 axis.

Improper Rotation Axis (S_n)

An improper rotation axis is an axis about which rotation followed by reflection through a plane perpendicular to the rotation axis produces an indistinguishable orientation. For example, the symbol S_6 means rotating the structure clockwise by 60 degrees and reflecting each atom through a plane perpendicular to the S_6 axis of rotation. This can be illustrated with an example: Suppose we consider the six objects illustrated on the coordinate system shown in Fig. 10.10. A red circle indicates a point below the xy plane, whereas a blue circle indicates a point lying above that plane (i.e., the plane of the page). The S_6 axis lies along the z-axis, which is perpendicular to the x- and y-axes at the point of intersection.

Rotation of the structure around the z-axis by 60 degrees followed by reflection through the xy plane moves the object at position 1 to position 2. Likewise, the object at position 2 moves to position 3, etc. Therefore, performing an S_6 operation has converted the original structure to another having the same orientation.

It should be apparent that the zigzag or puckered structure shown in Fig. 10.10 is the same as that of cyclohexane in the "chair" configuration. The S_6 axis is also a C_3 axis in this case because rotation by 120 degrees around the z-axis gives the same configuration:

From Fig. 10.10, it can be seen that the following relationships exist for symmetry operations for this structure:

$$S_6{}^2 = C_6 \cdot \sigma_h \cdot C_6 \cdot \sigma_h = C_6{}^2 \cdot \sigma_h{}^2 = C_6 \cdot E = C_3$$

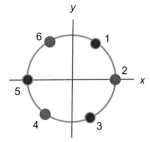

Fig. 10.10 A structure possessing an S_6 (and C_3) axis perpendicular to the plane of the page.

and that

$$S_6{}^4 = C_6 \cdot \sigma_h \cdot C_6 \cdot \sigma_h \cdot C_6 \cdot \sigma_h \cdot C_6 \cdot \sigma_h = C_6{}^4 \cdot E = C_3{}^2$$

Consider a tetrahedral molecule as represented in Fig. 10.11. In this case, it can be seen that rotation of the molecule by 90 degrees around the z-axis followed by reflection of each atom through the xy plane produces the structure in an unchanged orientation. The z-axis is thus an S_4 axis for this molecule. It is easily seen that both the x- and the y-axes are S_4 axes also, so a tetrahedral molecule has three S_4 axes as part of its symmetry elements.

The Identity (E)

The identity operation can be carried out for all molecules of all symmetries, because it leaves the orientation of the molecule unchanged. It is necessary to have the identity operation because an operation like $C_n{}^n$ returns the molecule to its original orientation. Thus,

$$C_n{}^n = E.$$

Any symmetry operation (B) of the point group has an inverse operation (B^{-1}), such that

$$B \cdot B^{-1} = B^{-1} \cdot B = E.$$

The importance of the identity operation will be considered in Section 10.4 when elementary group theory is considered.

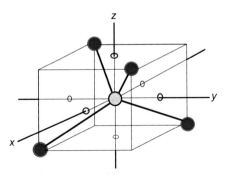

Fig. 10.11 A tetrahedral molecule. Each of the x-, y-, and z-axes are both S_4 and C_2 axes.

The assignment of a molecule to a specific symmetry type (also called the *point group*) requires the various symmetry elements present to be recognized. This may not always be obvious, but with practice most molecules can readily be assigned to a symmetry type.

10.4 WHAT POINT GROUP IS IT?

Determining the symmetry elements present in a molecule and then deducing the point group to which the molecule belongs begins with drawing the *correct* structure. For example, H_2O is sometimes shown on printed pages as

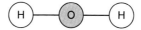

whereas the correct angular or bent structure is

The incorrect linear structure appears to have a C_∞ axis, because rotation of the molecule by any angle around a line along the bonds produces the same orientation. Also, a linear structure would possess a center of symmetry located at the center of the oxygen atom. Moreover, any line passing through the center of the oxygen atom and perpendicular to the C_∞ axis would be a C_2 axis. A plane perpendicular to the C_∞ axis cutting the oxygen atom in half would leave one hydrogen atom on either side and would be a mirror plane. Finally, a linear structure would have an infinite number of planes that bisect the molecule into equal fragments by cutting each atom in half, and the planes would intersect along the C_∞ axis. A molecule possessing all these elements of symmetry is designated as having $D_{\infty h}$ symmetry or belonging to the $D_{\infty h}$ point group. It is easily seen from Fig. 10.9 that the correct structure (bent or angular) has a C_2 axis and two mirror (vertical) planes that intersect along it, which are the only symmetry elements present. A molecule that has precisely these symmetry elements is called a C_{2v} molecule and belongs to the C_{2v} point group.

Methane (CH_4) is sometimes shown on printed pages as

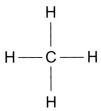

This structure would have a C_4 axis perpendicular to the plane of the page through the carbon atom, with four vertical planes intersecting along it. Moreover, it would have a horizontal plane of symmetry, a center of symmetry, and four C_2 axes perpendicular to the C_4 axis. The point group for such a structure is D_{4h}. This is the correct structure for the planar XeF_4 molecule. Methane is actually tetrahedral,

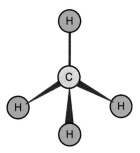

and there are four C_3 axes and six mirror planes (not 12, because each plane cleaves the molecule along two bonds). It has already been illustrated that this molecule has three S_4 axes (see Fig. 10.11). The point group for such a tetrahedral molecule is designated as T_d.

Table 10.1 summarizes the most common point groups of molecules and provides drawings showing the structures of the various types. Also, the molecular geometry is related to the hybrid orbital type of the central atom and the number of unshared pairs of electrons, if any. Table 10.2 provides a complete listing of the symmetry elements present in most of the commonly encountered point groups. Studying the structures in Table 10.1 shown earlier in this chapter and the listing of symmetry elements in Table 10.2 should make it easy to assign the point group for most molecules, assuming that the structure if the molecule has been drawn correctly!

Table 10.2 Common point groups and their symmetry elements

Point group	Structure	Symmetry elements	Examples
C_1	–	None	CHFClBr
C_s	–	One plane	ONCl, $OSCl_2$
C_2	–	One C_2 axis	H_2O_2
C_{2v}	Bent AB_2 or planar XAB_2	One C_2 axis and two σ_v planes at 90 degrees	NO_2, H_2CO
C_{3v}	Pyramidal AB_3	One C_3 axis and three σ_v planes	NH_3, SO_3^{2-}, PH_3
C_{nv}	–	One C_n axis and n σ_v planes	BrF_5 (C_{4v})
$C_{\infty v}$	Linear ABC	One C_∞ axis and ∞ σ_v planes	OCS, HCN, HCCH
D_{2h}	Planar	Three C_2 axes, one σ_h, two σ_v planes, and i	C_2H_4, N_2O_4
D_{3h}	Planar AB_3 or AB_5 trig. bipy.	One C_3 axis, three C_2 axes, three σ_v, and one σ_h	BF_3, NO_3^-, CO_3^{2-}, PCl_5
D_{4h}	Planar AB_4	One C_4 axis, four C_2 axes, four σ_v, one σ_h, and i	XeF_4, IF_4^-, $PtCl_4^{2-}$
$D_{\infty h}$	Linear AB_2	One C_∞ axis, one σ_h, ∞ σ_v planes, and i	CO_2, XeF_2, NO_2^+
T_d	Tetrahedral AB_4	Four C_3, three C_2, three S_4, and six σ_v planes	CH_4, BF_4^-, NH_4^+
O_h	Octahedral AB_6	Three C_4, four C_3, six C_2, four S_6 axes, nine σ_v and i	SF_6, PF_6^-, $Cr(CO)_6$
I_h	Icosohedral	Six C_5, 10 C_3, 15 C_2, 20 S_6 axes, and 15 planes	B_{12}, $B_{12}H_{12}^{2-}$

10.5 GROUP THEORY

The mathematical apparatus for treating combinations of symmetry operations lies in the branch of mathematics known as group theory. A group is a set of elements and corresponding operations that obey the following rules:

(1) The combination of any two members of a group must yield another member of the group (*closure*).

(2) The group contains the identity, E, multiplication by which commutes with all other members of the group ($EA = AE$) (*identity*).

(3) The associative law of multiplication must hold so that (AB) $C = A(BC) = (AC)B$ (*associative*).

(4) Every member of the group has a reciprocal such that $B \cdot B^{-1} = B^{-1} \cdot B = E$, where the reciprocal is also a member of the group (*inverse*).

The illustration of these rules will be initiated by considering the structure of the water molecule shown earlier in Fig. 10.9. First, it should be apparent that reflection of the hydrogen atoms through the yz plane, indicated by σ_{yz}, transforms H′ into H″. More precisely, it is appropriate to say that H′ and H″ are *interchanged* by a reflection through the yz plane. Because the z-axis coincides with a C_2 rotation axis, rotation by 180 degrees about the z-axis of the molecule will take H′ into H″ and H″ into H′, but with the "halves" of each interchanged with respect to the xz plane. The same result would follow from a reflection of the hydrogen atoms through the xz plane followed by reflection through the yz plane. Therefore, in terms of operations

$$\sigma_{xz} \cdot \sigma_{yz} = C_2 = \sigma_{yz} \cdot \sigma_{xz},$$

where C_2 is rotation around the z-axis by 360 degrees/2. This establishes that σ_{xz} and σ_{yz} are both members of the symmetry group for this molecule. Thus, it is illustrated that in accordance with Rule 1, the combination of two members of the group has produced another member of the group, which is C_2. If reflection through the xz plane is followed by repeating that operation, the molecule ends up with the arrangement shown in Fig. 10.9. Symbolically,

$$\sigma_{xz} \cdot \sigma_{xz} = E.$$

Also, it is easy to see from Fig. 10.9 that

$$\sigma_{yz} \cdot \sigma_{yz} = E$$

and

$$C_2 \cdot C_2 = E.$$

Further examination of Fig. 10.9 shows that a reflection of the molecule through the xz plane, σ_{xz}, will cause the "halves" of the H′ and H″ atoms lying on either side of the xz plane to be interchanged. If that operation is performed and the molecule is then rotated by 360 degrees/2 around the C_2 axis, the result obtained is exactly the same as that produced by reflection through the yz plane. This can be expressed as

$$\sigma_{xz} \cdot C_2 = \sigma_{yz} = C_2 \cdot \sigma_{xz}.$$

In a similar way, it is can be seen that a reflection through the yz plane followed by a C_2 operation gives the same result as that performed by the σ_{xz} operation. Finally, it can be seen from Fig. 10.9 that reflections through σ_{xz} and σ_{yz} in either order give the same orientation as results from the C_2 operation. Therefore, these operations can be summarized as

$$\sigma_{xz} \cdot \sigma_{yz} = C_2 = \sigma_{yz} \cdot \sigma_{xz}.$$

The associative law, Rule 3, has also been demonstrated in these operations. Additional relationships are provided by

$$E \cdot E = E$$

$$C_2 \cdot E = C_2 = E \cdot C_2$$

$$\sigma_{yz} \cdot E = \sigma_{yz} = E \cdot \sigma_{yz}, etc.$$

All of these combinations of operations can be summarized in a *group multiplication table*.

The multiplication table, shown in Table 10.3 for the C_{2v} group, is constructed so that the combination of operations follows the four rules presented at the beginning of this section. Obviously, a molecule having a structure other than C_{2v} (symmetry elements and operations) would require a different table.

To provide further illustration, of the use of symmetry elements and operations, the ammonia molecule NH_3, which has the structure shown in Fig. 10.12 will be considered. Figure 10.12 shows that the NH_3 molecule has a C_3 axis passing through the nitrogen atom and the three reflection planes containing that C_3 axis. The identity operation, E, and the C_3^2 complete the list of symmetry operations for the NH_3 molecule.

Table 10.3 Multiplication of symmetry operations for H_2O (C_{2v})

	E	C_2	σ_{xz}	σ_{yz}
E	E	C_2	σ_{xz}	σ_{yz}
C_2	C_2	E	σ_{yz}	σ_{xz}
σ_{xz}	σ_{xz}	σ_{yz}	E	C_2
σ_{yz}	σ_{yz}	σ_{xz}	C_2	E

To use this table, start with the operation in the left-hand column and proceed to the desired operation at the top of a column. Then, read down that column to obtain the desired product.

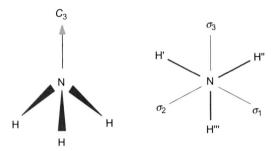

Fig. 10.12 The pyramidal ammonia (C_{3v}) molecule. In the right-hand structure, the C_3 axis is perpendicular to the page at the nitrogen atom.

Using the procedures illustrated above, the following relationships are now established:

$$C_3 \cdot C_3 = C_3{}^2$$

$$C_3{}^2 \cdot C_3 = C_3 \cdot C_3{}^2 = E$$

$$\sigma_1 \cdot \sigma_1 = E = \sigma_2 \cdot \sigma_2 = \sigma_3 \cdot \sigma_3.$$

A reflection through σ_2 does not change H″, but it does interchange H′ and H‴. Reflection through σ_1 leaves H′ in the same position but interchanges H″ and H‴. We can summarize these operations as

$$H' \overset{\sigma_2}{\leftrightarrow} H'''$$

$$H''' \overset{\sigma_1}{\leftrightarrow} H''$$

However, operation $C_3{}^2$ would move H′ to H‴, H″ to H′, and H‴ to H″, which is exactly the same orientation as that produced when σ_2 is followed by σ_1. It follows, therefore, that

$$\sigma_2 \cdot \sigma_1 = C_3{}^2.$$

This process could be continued so that all the combinations of symmetry operations would be worked out. Table 10.4 shows the resulting

Table 10.4 The multiplication table for the C_{3v} point group

	E	C_3	$C_3{}^2$	σ_1	σ_2	σ_3
E	E	C_3	$C_3{}^2$	σ_1	σ_2	σ_3
C_3	C_3	$C_3{}^2$	E	σ_3	σ_1	σ_2
$C_3{}^2$	$C_3{}^2$	E	C_3	σ_2	σ_3	σ_1
σ_1	σ_1	σ_2	σ_3	E	C_3	$C_3{}^2$
σ_2	σ_2	σ_3	σ_1	$C_3{}^2$	E	C_3
σ_3	σ_3	σ_1	σ_2	C_3	$C_3{}^2$	E

Table 10.5 Character table for the C_{2v} point group

	E	C_2	σ_{xz}	σ_{yz}
A_1	1	1	1	1
A_2	1	1	-1	-1
B_1	1	-1	1	-1
B_2	1	-1	-1	1

multiplication table for the C_{3v} point group, which is the point group to which a pyramidal molecule like NH_3 belongs.

Multiplication tables can be constructed for a combination of symmetry operations for a large number of other point groups. However, it is not the multiplication table as such that is of interest. Let us return to the multiplication table for the C_{2v} point group given in Table 10.3. The symbols at the left in Table 10.5 give the symmetry properties of the irreducible *representation* of the C_{2v} group. The meaning of these symbols will now be discussed in an elementary way.

Suppose a vector of unit length is lying coincident with the x-axis, as shown in Fig. 10.13. The identity operation does not change the orientation of the vector. A reflection in the xz plane leaves the vector unchanged, but a reflection through the yz plane changes it to a unit vector in the $-x$ direction. Likewise, the C_2 operation around the z-axis changes the vector so it points in the negative direction. Therefore, the vector is said to *transform* as +1 for the operations E and σ_{xz}, but it will transform as −1 for the operations C_2 and σ_{yz}. Table 10.5 shows the row containing the numbers +1, −1,+1, and −1 under the operations E, C_2, σ_{xz}, and σ_{yz}, respectively, labeled as B_1. It is easy to show how the numbers in the other rows can be obtained in a similar manner. The four representations, A_1, A_2, B_1, and B_2, are known as the *irreducible representations* of the C_{2v} group. It can be shown that these four irreducible representations cannot be separated or decomposed into other representations.

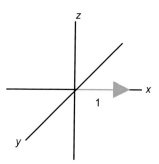

Fig. 10.13 A unit vector lying on the x axis.

For a given molecule belonging to a particular point group, it is possible to consider the A_1, A_2, B_1, and B_2 symmetry species as indicating the behavior of the molecule under symmetry operations. As will be shown later, similar species also determine the ways in which the atomic orbitals can combine to produce molecular orbitals because the combinations of atomic orbitals must satisfy the character table of the group. We need to give some meaning that is based on the molecular structure for the species A_1, B_2, etc.

The following conventions are used to label species in the character tables corresponding to the various point groups:

(1) The symbol A is used to designate a nondegenerate species symmetric about the principle axis.

(2) The symbol B is used to designate a nondegenerate species antisymmetric about the principle axis.

(3) The symbols E and T represent doubly and triply degenerate species, respectively.

(4) If a molecule possesses a center of symmetry, the letter g indicates symmetry with respect to that center (*gerade*), and the letter u indicates antisymmetry with respect to that center of symmetry (*ungerade*).

(5) For a molecule that has a rotation axis other than the principal one, symmetry or antisymmetry with respect to that axis is indicated by a subscript 1 or 2, respectively. When no rotation axis other than the principal one is present, these subscripts are sometimes used to indicate symmetry or antisymmetry with respect to a vertical plane σ_h.

(6) The marks "and" are sometimes used to indicate symmetry or antisymmetry with respect to a horizontal plane σ_h.

It should now be apparent how the species A_1, A_2, B_1, and B_2 arise. Character tables have been worked out and are tabulated for all of the common point groups. Presenting all the tables here would go beyond the scope of this introductory book.

We have barely scratched the surface of the important topic of symmetry. However, a brief discussion such as that presented here serves to introduce the concepts and nomenclature, as well as making one able to recognize the more important point groups. Thus, a symbol such as T_d or D_{4h} takes on precise meaning in the language of group theory. The applications of group theory include, among others, coordinate transformations, analysis of molecular vibrations, and the construction of molecular orbitals. Only the last of these uses will be illustrated here.

10.6 SYMMETRY OF MOLECULAR ORBITALS

The formation of the H_2 molecule, which has a center of symmetry, gives rise to the combinations of atomic orbitals that can be written as $\varphi_1 + \varphi_2$ and $\varphi_1 - \varphi_2$. The types of symmetry elements were discussed earlier in this chapter, and they also apply in a general way to orbitals. A center of symmetry is simply a point through which each atom can be moved to give the same orientation of the molecule. For a diatomic molecule like H_2, that point is the midpoint of the bond between the two atoms. It is equally valid to speak of a center of symmetry for wave functions. The first of the combinations of wave functions (as shown in Fig. 8.4) possesses a center of symmetry, whereas the second does not. Therefore, the $\varphi_1 + \varphi_2$ molecular wave function corresponds to the orbital written as σ_g, whereas the combination $\varphi_1 - \varphi_2$ combination corresponds to σ_u^*. In these designations, "g" refers to the fact that the wave function retains the same sign when inflected through the center of symmetry, and "u" indicates that the wave function changes sign when it is inflected through the center of symmetry. It is generally stated that the bonding orbital is *symmetric* and antibonding orbital is *antisymmetric*. However, for π and π^* orbitals, g and u refer to symmetry with respect to a plane that contains the internuclear axis (see Fig. 9.6).

For diatomic molecules, the order of filling of molecular orbitals is σ, σ^*, (π, π), σ, (π^*, π^*), σ^* for the early part of the first long period and σ, σ^*, σ, (π, π), (π^*, π^*), σ^* for the latter part of the first long series. The designations (π, π) and (π^*, π^*) indicate pairs of degenerate molecular orbitals. For the hydrogen molecule, the electron configuration can be shown as $(\sigma_g)^2$, whereas that for the C_2 molecule is designated as $(\sigma_g)^2 (\sigma_u^*)^2 (\pi_u)^2 (\pi_u)^2$ (see Fig. 8.9).

The molecular orbitals can be identified by applying labels that show the atomic orbitals that were combined to produce them. For example, the orbital of lowest energy is $1s\ \sigma_g$ or $\sigma_g\ 1s$. In this way, other orbitals would have designations like $2p_x\ \pi_u$, $2p_y\ \pi_g^*$, etc.

As a result of orbital mixing, a σ molecular orbital may not arise from the combination of pure s atomic orbitals. For example, it was described in Section 8.4 that there is substantial mixing of $2s$ and $2p$ orbitals in molecules like B_2 and C_2. Therefore, labels like $2s\ \sigma_g$ and $2s\ \sigma_u^*$ may not be strictly correct. Because of this, the molecular orbitals are frequently designated as $1\sigma_g$, $1\sigma_u$, $2\sigma_g$, $2\sigma_u$, $3\sigma_u$, $1\pi_u$, $1\pi_g$,....

In these designations, the leading digit refers to the order in which an orbital having that designation is encountered as the orbitals are filled.

For example, $1\sigma_g$ denotes the first σ orbital having g symmetry (a bonding orbital), $3\sigma_u$ means the third σ orbital having u symmetry (an antibonding orbital), etc. The asterisks on antibonding orbitals are not really needed, because a σ orbital having u symmetry is an antibonding orbital, and it is the antibonding π orbital that has g symmetry. Therefore, the g and u designations alone are sufficient to denote bonding or antibonding character. These ideas elaborate on those discussed in Section 8.4.

10.7 MOLECULAR ORBITAL DIAGRAMS

The application of symmetry concepts and group theory greatly simplifies the construction of molecular wave functions from atomic wave functions. For example, it can be shown that the combination of two hydrogen $1s$ wave functions $(\varphi_{1s}(1) + \varphi_{1s}(2))$ transforms as A_1 (sometimes written as a_1 when *orbitals* are considered), and the combination $(\varphi_{1s}(1) - \varphi_{1s}(2))$ transforms as B_1 (sometimes written as b_1). According to the description of species given above, we see that the A_1 combination is a singly degenerate state symmetric about the internuclear axis. Also, the B_1 combination represents a singly degenerate state that is antisymmetric about the internuclear axis. The states represented by the combinations $(\varphi_{1s}(1) + \varphi_{1s}(2))$ and $(\varphi_{1s}(1) - \varphi_{1s}(2))$ describe the bonding (A_1) and antibonding (B_1) molecular states, respectively, as shown in Fig. 10.14, where a_1 and b_1 are used to denote the orbitals.

It can be shown that for any group, the irreducible representations must be orthogonal. Therefore, only interactions (combinations) of orbitals having the same irreducible representations lead to nonzero elements in the secular determinant. It remains, then, to determine how the various orbitals transform under different symmetry groups. For the H_2O molecule, the

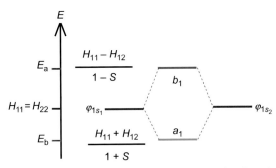

Fig. 10.14 Two combinations of $1s$ wave functions giving molecular orbitals of different symmetry. See Section 8.3 for a discussion of the energies of the states.

coordinate system was shown earlier in Fig. 10.9. Performing any of the four operations possible for the C_{2v} group (E, C_2, σ_{xz}, and σ_{yz}) leaves the $2s$ orbital unchanged. Therefore, that orbital transforms as A_1 (values of $+1$, $+1$, $+1$, and $+1$). Likewise, the p_x orbital does not change sign under E or σ_{xz} operations, but it does change signs under C_2 and σ_{yz} operations. This orbital thus transforms as B_1 ($+1$, -1, $+1$, and -1). In a like manner, we find that p_z transforms as A_2 (does not change signs under C_2, E, σ_{xz}, or σ_{yz} operations). Although it may not be readily apparent, the p_y orbital transforms as B_2. Using the four symmetry operations for the C_{2v} point group, we find that the valence shell orbitals of oxygen behave as follows:

Orbital Summary

$2s$	A_1
$2p_z$	A_2
$2p_x$	B_1
$2p_y$	B_2

The possible wave functions constructed for the molecular orbitals in molecules are those constructed from the irreducible representations of the groups giving the symmetry of the molecule. These are readily found in the character table for the appropriate point group for the molecule. For the water molecule, which has the point group C_{2v}, the character table (see Table 10.5) shows that only A_1, A_2, B_1, and B_2 representations occur for a molecule having C_{2v} symmetry. We can use this information to construct a qualitative molecular orbital scheme for the H_2O molecule, as shown in Fig. 10.15.

In constructing the molecular orbital diagram, it must be recognized that there are *two* hydrogen $1s$ orbitals, and the orbitals from the oxygen atom must interact with *both* of them. Therefore, it is not each hydrogen $1s$ orbital *individually* that is used, but rather a combination of the two. These combinations of atomic orbitals are called *group orbitals*, and in this case the linear combinations of atomic wave functions can be written as $(\varphi_{1s}(1) + \varphi_{1s}(2))$ and $(\varphi_{1s}(1) - \varphi_{1s}(2))$. The $2s$ and $2p_z$ orbitals having A_1 symmetry mix with the combination of hydrogen $1s$ orbitals, which have A_1 symmetry to produce three molecular orbitals having A_1 symmetry (one bonding, one nonbonding, and one antibonding). The $2p_x$ orbital having B_1 symmetry combines with the combination of hydrogen orbitals having same symmetry, and that combination can be written as $(\varphi_{1s}(1) - \varphi_{1s}(2))$. The $2p_y$ orbital, which has B_2 symmetry, remains as a π orbital, but it does not have the correct symmetry to interact with either of the combinations of hydrogen

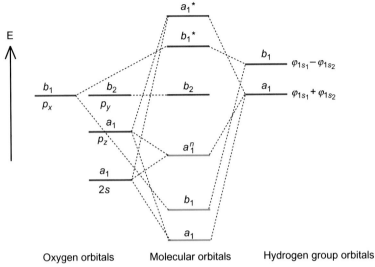

Fig. 10.15 The molecular orbital diagram for the water molecule.

orbitals. In the case of the H_2O molecule, the four molecular orbitals of lowest energy will be populated because the atoms have a total of eight valence shell electrons. Therefore, the bonding in the H_2O molecule can be represented as

$$(a_1)^2 (b_1)^2 (a_1{}^n)^2 (b_2)^2.$$

As in the case of atomic orbitals and spectroscopic states (see Chapter 5), we use *lower case letters to denote orbitals or configurations* and *upper case letters to indicate states*.

Having considered the case of the C_{2v} water molecule, we would like to be able to use the same procedures to construct the qualitative molecular orbital diagrams for molecules having other structures. To do this requires that we know how the orbitals of the central atom transform when the symmetry of the molecule may have any one of the common symmetry designations. Table 10.6 shows how the s and p orbitals are transformed in several common point groups, and more extensive tables can be found in the comprehensive books listed at the end of this book.

If we now consider a trigonal planar molecule such as BF_3 (D_{3h} symmetry), the z axis is defined as the C_3 axis. One of the B—F bonds lies along the x-axis, as shown in Fig. 10.16, and the C_3 axis passes through the boron atom perpendicular to the plane of the molecule. The symmetry elements present for this molecule include the C_3 axis, three C_2 axes (coincident with the B—F bonds), three mirror planes each containing a C_2 axis and a C_3 axis,

Table 10.6 Central atom s and p orbital transformations under different symmetries

Point group	Structure	Orbital			
		s	p_x	p_y	p_z
C_{2v}	Bent triatomic	A_1	B_1	B_2	A_1
C_{3v}	Pyramidal	A_1	E	E	A_1
D_{3h}	Trigonal planar	A'_1	E'	E'	A''_2
C_{4v}	Pyramidal	A_1	E	E	A_1
D_{4h}	Square plane	A'_1	E_u	E_u	A_{2u}
T_d	Tetrahedral	A_1	T_2	T_2	T_2
O_h	Octahedral	A_1	T_{1u}	T_{1u}	T_{1u}
$D_{\infty h}$	Linear	Σ_g	Σ_u	Σ_u	$\Sigma_g{}^+$

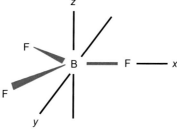

Fig. 10.16 The coordinate system for BF_3.

and the identity. Thus, there are 12 symmetry operations that can be performed with this molecule. It can be shown that the p_x and p_y orbitals both transform as E', and the p_z orbital transforms as A_2''. The s orbital is A_1' (the prime indicating symmetry with respect to σ_h). Similarly, it can be shown that the fluorine p_z orbitals are A_1, E_1, and E_1 for the three atoms. The qualitative molecular orbital diagram can then be constructed as shown in Fig. 10.17.

It is readily apparent from the molecular orbital diagram that the bonding molecular orbitals are capable of holding the six bonding electrons (in three σ bonds) in this molecule. The possibility of some π bonding can be illustrated in the molecular orbital diagram due to the presence of the a_2'' orbital that is not part of the σ bonding scheme, and in fact, there is some evidence for this type of interaction. The sum of the covalent radii of boron and fluorine atoms is about 1.52 Å (152 pm), but the experimental B—F bond distance in BF_3 is about 1.295 Å (129.5 pm). Part of this "bond shortening" may be due to partial double bonds that result from the π bonding. In valence bond terms, there is overlap between the empty p orbital that is not used in forming a set of sp^2 hybrids and the filled p orbitals on the fluorine

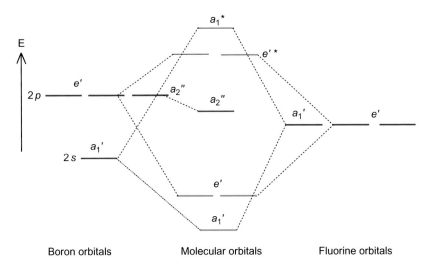

Fig. 10.17 A molecular orbital diagram for the BF_3 molecule.

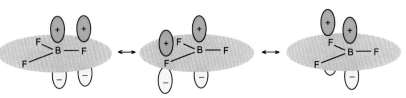

Fig. 10.18 The p_z orbital on boron forms a π bond with p orbitals on all three fluorine atoms. *(Modified with permission from House, J. E. Inorganic Chemistry, 2nd ed.; Academic Press/Elsevier: Amsterdam, 2013.)*

atoms. Showing how the double bonds arise leads to three resonance structures of the valence bond type that can be shown as in Fig. 10.18. From these resonance structures, which show one π bond spread in three directions toward the fluorine atoms, a bond order of 1.33 would result. This partial multiple bonding would predict that the observed bond length should be shorter than expected for a single bond. However, as discussed in Chapter 8, partial ionic character of the B—H bonds can also be invoked to explain the bond lengths.

Having seen the development of the molecular orbital diagrams for AB_2 and AB_3 molecules, we will now consider tetrahedral molecules like CH_4, SiH_4, or SiF_4. In this symmetry, the valence shell s orbital on the central atom transforms as A_1, while the p_x, p_y, and p_z orbitals transform as T_2 (see Table 10.6). For the methane molecule, a linear combination of hydrogen orbitals (sometimes referred to as *group orbitals*) that transforms as A_1 is

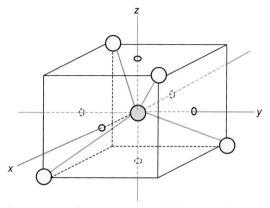

Fig. 10.19 Coordinate system for the tetrahedral CH_4 molecule.

$$\varphi_{1s}(1) + \varphi_{1s}(2) + \varphi_{1s}(3) + \varphi_{1s}(4),$$

and the combination transforming as T_2 is

$$\varphi_{1s}(1) + \varphi_{1s}(2) - \varphi_{1s}(3) - \varphi_{1s}(4),$$

where the coordinate system is as shown in Fig. 10.19. Using the orbitals on the carbon atom and combining them with the group orbitals from the four hydrogen atoms (the linear combination of atomic orbitals having appropriate symmetry to match those of the orbitals on the carbon atom), it is possible to arrive at the molecular orbital diagram shown in Fig. 10.20. The hydrogen group orbitals are referred to as symmetry adjusted linear combinations (SALC) because they have symmetry that matches the orbitals of the carbon atom. The molecular orbital diagrams for other tetrahedral molecules are similar to that for CH_4.

For an octahedral AB_6 molecule of which SF_6 is an example,

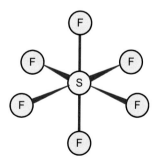

the valence shell orbitals of the central atom are considered to be the s, p, and d orbitals. The spatial orientations of the d orbitals are shown in Fig. 10.21.

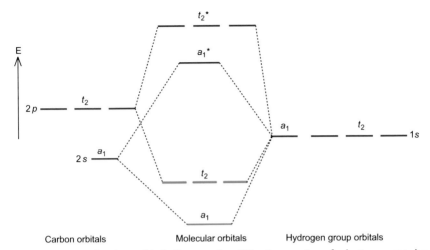

Fig. 10.20 A molecular orbital diagram for CH₄. Four pairs of electrons can be accommodated in the bonding orbitals.

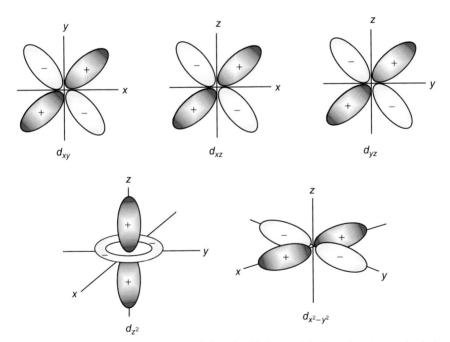

Fig. 10.21 The spatial orientation of the d orbitals used in bonding in octahedral molecules such as SF₆. Note that the d_{z^2} and $d_{x^2-y^2}$ orbitals have lobes lying *along* the axes, whereas lobes of the d_{xy}, d_{yz}, and d_{xz} orbitals lie *between* the axes. Thus, it is the d_{z^2} and $d_{x^2-y^2}$ orbitals that are used in forming bonding orbitals.

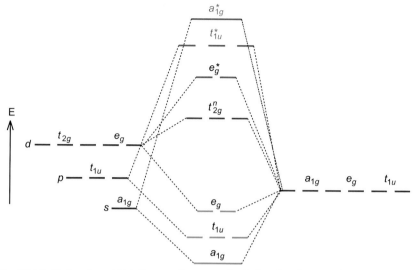

Fig. 10.22 Molecular orbital diagram for an octahedral molecule. Note that there are six bonding orbitals that can accommodate six pairs of electrons.

It is easy to see that a regular octahedron has a center of symmetry so that "g" and "u" designations must be used as part of the symmetry label to designate symmetry or asymmetry with respect to that center. It is clear that the s orbital transforms as A_{1g}. The three p orbitals, being directed toward the corners of the octahedron, are degenerate and change sign upon reflection through the center of symmetry. Thus, they constitute a set of three orbitals that can be designated as T_{1u}. Of the set of d orbitals, the d_{z^2} and $d_{x^2-y^2}$ orbitals are directed toward the corners of the octahedron, and they do not change sign upon inversion through the center of symmetry. Therefore these orbitals are designated as E_g. The remaining d_{xy}, d_{yz}, and d_{xz} orbitals form a triply degenerate set designated as T_{2g}.

If only σ bonding is considered, it is found that T_{1u}, E_g, and A_{1g} orbitals are used in bonding to the six groups attached. The resulting energy level diagram is shown in Fig. 10.22.

10.8 THE THREE-CENTER BOND

When Mg_3N_2 reacts with water, NH_3 is produced. However, when Mg_3B_2 reacts with water, the product is not BH_3 (borane), but rather B_2H_6 (diborane). If BH_3 were to form, its structure would be planar, with boron utilizing three sp^2 hybrid orbitals.

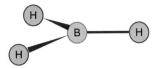

A structure of this type would leave the boron atom surrounded by only six valence shell electrons, making it an electron deficient molecule. Boron halides are also electron deficient, and as a result they behave as Lewis acids (electron pair acceptors) to form many complexes with electron pair donors (Lewis bases).

Bonding in B_2H_6 involves bonds that are not all of the usual shared electron pair types. Instead, each boron atom is surrounded by four hydrogen atoms that are located at the corners of an irregular tetrahedron. Four of the hydrogen atoms are in terminal positions and lie in the same plane as the boron atoms. However, the other two hydrogen atoms form bridges between the boron atoms. The structure of diborane is shown in Fig. 10.23.

The structure of diborane contains two bonds that are referred to as *three-center bonds* as a result of one pair of electrons being shared by three atoms that reside in an orbital that involves the simultaneous overlap of three atomic orbitals. This can be shown using a valence bond structure as follows:

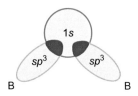

From the standpoint of molecular orbitals, a rationalization for a three-center bond can be provided by considering the boron atoms to utilize hybrid orbitals that combine with the hydrogen $1s$ orbital. Although the

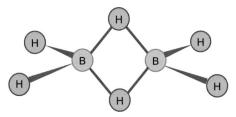

Fig. 10.23 The structure of diborane.

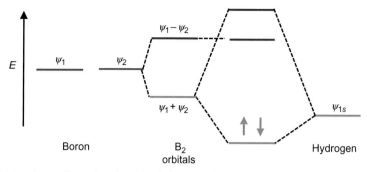

Fig. 10.24 A simple molecular orbital diagram for a three-center bond holding two electrons.

bond angles deviate considerably from those expected for complete sp^3 hybridization, it is convenient to consider them as such. In that way, the two boron wave functions combine to form two new wave functions that have symmetric and antisymmetric character. Those wave functions then combine with a hydrogen $1s$ wave function to produce two molecular orbitals, one bonding and the other antibonding. The result is illustrated by the molecular orbital diagram shown in Fig. 10.24. Essentially, the bonding molecular orbital is distributed over both of the boron atoms and the hydrogen atom, forming the bridge between them.

10.9 ORBITAL SYMMETRY AND REACTIVITY

For over half a century, it has been recognized that symmetry plays a significant role in the reactions between chemical species. In simple terms, many reactions occur because electron density is transferred (or shared) between the reacting species as the transition state forms. In order to interact favorably (to give an overlap integral greater than 0), it is necessary for the interacting orbitals to have the same symmetry (see Section 4.3). Otherwise, orthogonal orbitals give an overlap integral equal to 0. The orbitals involved in the interactions of reacting species are those of higher energy, the so-called *frontier orbitals*. These are the highest occupied molecular orbital (HOMO) and the lowest unoccupied molecular orbital (LUMO). As the species interact, electron density flows from the HOMO on one species to the LUMO on the other. In more precise terms, it can be stated that the orbitals must belong to the same symmetry type or point group for the orbitals to overlap to give an overlap integral greater than 0.

As has been described earlier, orbitals of similar energy interact (overlap) best. Therefore, it is necessary that the energy difference between the HOMO on one reactant and the LUMO on the other be less than some threshold value for effective overlap to occur. As a reaction takes place, a bond in one reactant molecule is broken as another is being formed. When both of the orbitals are *bonding* orbitals, the bond being broken (electron density is being donated from it) is the one representing the HOMO in one reactant and the bond being formed is represented by the LUMO in the other (which is empty and receives electron density as the molecules interact). When the frontier orbitals are antibonding in character, the LUMO in one reactant molecule corresponds to the bond broken and the HOMO to the bond formed.

In Chapter 8, the applications of orbital symmetry to the formation of transition states that form between reacting diatomic molecules were described. Having now shown in Chapter 9 the applications of Hückel molecular orbital theory to the structures of organic molecules, it is now possible to apply some of the principles to reactions of organic compounds. One type of reaction that can be described in terms of orbital symmetry is the ring-closing reaction of *cis*-1,3-butadiene to produce cyclobutene. This type of reaction is known as an *electrocyclic reaction*, and it could conceivably take place by two different pathways, which are shown in Fig. 10.25. In the first mechanism, known as *disrotatory*, the two CH_2 groups rotate in opposite directions. In the second mechanism, known as *conrotatory*, the terminal CH_2 groups rotate in the same direction. However, the stereochemistry of the product will be different for the two mechanisms.

If the reaction follows a disrotatory pathway, the rotation of terminal CH_2 groups is as shown in Fig. 10.25A, whereas that for the conrotatory pathway is shown in Fig. 10.25B. In order to show the stereochemistry of the product for each of the two pathways, one hydrogen atom on each terminal CH_2 group is labeled as H'. It is clear that the hydrogen atoms labeled as H' would be on the same side of the ring in a conrotatory process and on opposite sides of the ring in a disrotatory pathway. Let us now show how orbital symmetry can be used to predict which transition state will be formed.

A pictorial representation of the HOMO for butadiene is shown in Fig. 10.26. From this figure, due to the orientation of the lobes having

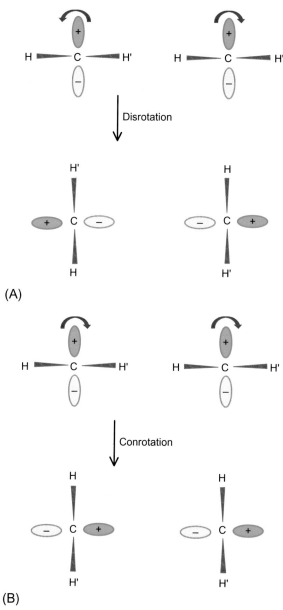

Fig. 10.25 An illustration of (A) disrotation and (B) conrotation of terminal methylene groups. *(Modified with permission from House, J. E.* Principles of Chemical Kinetics, *2nd ed.; Academic Press/Elsevier: Amsterdam, 2007 (chapter 9).)*

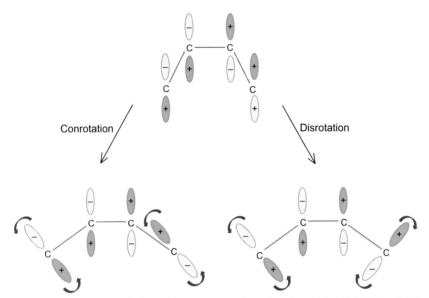

Fig. 10.26 Symmetry of the highest occupied molecular orbital (HOMO) of 1,3-butadiene showing changes in orbital orientation during conrotation and disrotation. *(Modified with permission from House, J. E.* Principles of Chemical Kinetics, *2nd ed.; Academic Press/Elsevier: Amsterdam, 2007 (chapter 9).)*

positive sign, it can be seen that the rotation of the CH_2 groups in a con-rotatory pathway leads to the overlap of orbitals having the same symmetry (bond formation). Also, the product formed has the H′ atoms on opposite sides of the ring. In the disrotatory process, the overlap of orbitals of opposite symmetry results, which gives an overlap of 0. In accordance with these observations, the product of the electrocyclic reaction that results when 1,3-butadiene is heated consists of 100% of that in which the H′ atoms are on opposite sides of the ring. Of course, the reaction can also be studied when one of the hydrogen atoms on the terminal CH_2 groups is replaced by a different atom. In that case, the substituted atoms are found on opposite sides of the ring in the cyclobutene produced.

When 1,3-butadiene is excited photochemically, an electron is moved from the HOMO to the LUMO, which has different symmetry (see Fig. 10.27). Therefore, it is the disrotatory motion that brings orbital lobes having the same symmetry (sign) in contact as the bond forms during the cyclization reaction of the excited state of 1,3-butadiene. As a result, the

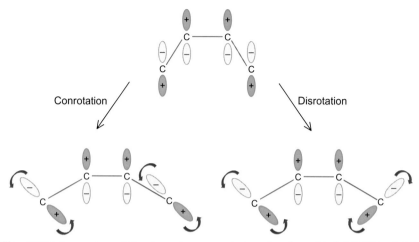

Fig. 10.27 Symmetry of the lowest unoccupied molecular orbital (LUMO) of 1,3-butadiene showing changes in orbital orientation during conrotation and disrotation. *(Modified with permission from House, J. E.* Principles of Chemical Kinetics, *2nd ed.; Academic Press/Elsevier: Amsterdam, 2007 (chapter 9).)*

product of this reaction of excited 1,3-butadiene is found to have the H′ atoms on the same side of the ring.

The electrocyclic ring closure of *cis*-1,3,5-hexatriene leads to the formation of 1,3-cyclohexadiene. Although the hexatriene molecule is planar, the cyclic product has two CH_2 groups in which the four hydrogen atoms are found with two of the atoms above the ring and two of the atoms below the ring. Therefore, in the transition state, the terminal CH_2 groups undergo a rotation that could be either conrotatory or disrotatory, as shown in Fig. 10.28. The rotations require breaking of a π bond formed from *p* orbitals on two carbon atoms to form a σ bond. In order to give a positive overlap to form the σ bond, the orbitals must match in symmetry, which is provided by the disrotatory pathway. Therefore, in disrotatory motion, the rotation of the terminal CH_2 groups leads to the formation of a bond between the two carbon atoms resulting in ring closure. In the conrotatory motion of the CH_2 groups, the resulting orbital overlap would be zero.

A guiding principle that can be used to predict how electrocyclic reactions take place was provided by R. B. Woodward and R. Hoffmann (1970). This rule is based on the number of electrons in the π bonding system of the molecule. The number of electrons in the π bonding system can be

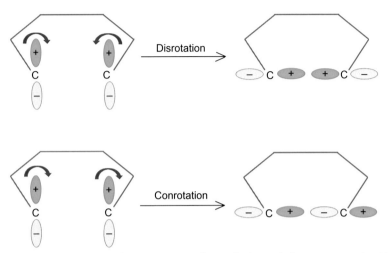

Fig. 10.28 Disrotation and conrotation of terminal methylene groups in 1,3,5-hexatriene. The remainder of the molecule is indicated by lines. *(Modified with permission from House, J. E. Principles of Chemical Kinetics, 2nd ed.; Academic Press/ Elsevier: Amsterdam, 2007 (chapter 9).)*

expressed as $4n$ or $4n+2$, where $n=0, 1, 2, \ldots$ The rule predicts the mechanism of electrocyclization as

$$4n = 4, 8, 12, \ldots, \text{thermal mechanism is conrotatory.}$$

$$4n + 2 = 2, 6, 10, \ldots, \text{thermal mechanism is disrotatory.}$$

It can be seen that the electrocyclization reactions of 1,3–butadiene (which has four electrons residing in π orbitals) and 1,3,5–hexatriene (which has six electrons residing in π orbitals) discussed earlier give products in accordance with these rules. However, it must be mentioned that the predictions given above are for the reactions that are thermally induced. If the reactions are carried out photochemically where excited states are produced, the predictions are reversed.

The discussion presented here has provided only an introduction to how symmetry of molecular orbitals can be used to predict reaction mechanisms. It should be apparent the requirement that the transition state involves interaction of orbitals of like symmetry is a powerful tool for predicting reaction pathways. Chemists doing synthetic work as well as theoretical chemists need to be familiar with these aspects of molecular orbital theory. However, the knowledge of only Hückel molecular orbital theory is required in order to deal with many significant problems in molecular structure and reactivity. In view of its gross approximations and very simplistic approach, it is surprising how many qualitative aspects of molecular structure and reactivity can be dealt with using the Hückel approach. For a more complete discussion of this topic, consult the book by J. P. Lowe cited in the suggested readings listed at the end of the book.

In this chapter, it has been shown how symmetry considerations are used to arrive at qualitative molecular orbital diagrams for molecules having several common structural types. The number of molecules and ions that have C_{2v}, C_{3v}, T_d, and O_h symmetry is indeed large. Energy level diagrams such as those shown in this chapter are widely used to describe structural, spectroscopic, and other properties of molecules. There has been no attempt, however, to set about calculating anything. In Chapter 9, we presented an overview of one of the simplest types of molecular orbital calculations that can be carried out easily for some simple systems. However, the more sophisticated mathematical treatments of molecular orbital calculations and the accompanying energy level diagrams are beyond the intended scope of this introductory book.

LITERATURE CITED

House, J. E. *Inorganic Chemistry*, 2nd ed.; Academic Press/Elsevier: Amsterdam, 2013.
Woodward, R. B., Hoffman, R. *The Conservation of Orbital Symmetry*. Verlag Chemie: Weinheim, 1970.

PROBLEMS

1. Make sketches of the listed species showing approximately correct geometry and all valence shell electrons. Identify all symmetry elements present and determine the point group for the species.
   (a) OCN^-
   (b) IF^+
   (c) ICl^-
   (d) SO_3^{2-}
   (e) SF_6
   (f) IF_5
   (g) ClF_3
   (h) SO_3

(i) ClO^-

(j) NSF

2. Make sketches of the listed species showing approximately correct geometry and all valence shell electrons. Identify all symmetry elements present and determine the point group for the species.

(a) CN_2^{2-}

(b) PH_3

(c) PO_3^-

(d) $B_3N_3H_6$

(e) SF_2

(f) ClO_3^-

(g) SF_4

(h) C_3O_2

(i) AlF_6^{3-}

(j) F_2O.

3. Consider the molecule AX_3Y_2, which has no unshared electron pairs on the central atom. Sketch the structures for all possible isomers of this compound and determine the point group to which each belongs.

4. Use the symmetry of the valence shell atomic orbitals of the central atom to construct (using appropriate hydrogen group orbitals) the molecular orbital diagrams for the following:

(a) BeH_2

(b) HF_2^-

(c) CH_2

(d) H_2S

5. Use the symmetry of the valence shell atomic orbitals of the central atom to construct (using appropriate group orbitals of terminal atoms) the molecular orbital diagrams for the following:

(a) AlF_3

(b) BH_4^-

(c) SF_6

(d) NF_3

6. Consider the molecule $Cl_2B{-}BCl_2$:

(a) If the structure is planar, what is the point group of the molecule?

(b) Draw a structure for $Cl_2B{-}BCl_2$ that would have an S_4 axis.

7. Use the procedure outlined in the text to obtain the multiplication table for the C_{4v} point group.

8. Follow the procedure used in the text in obtaining the character table for the C_{2v} point group and develop the character table for the C_{3v} point group.

9. Determine all of the symmetry elements possessed by the CH_4 molecule and give the point group for this molecule. In succession, replace hydrogen atoms with fluorine, chlorine, and bromine and determine what symmetry elements are present for each and determine the point group to which each product belongs.

10. There are three isomers possible for the N_2O molecule. Draw their structures, then determine the point group to which each structure belongs. The most stable of the isomers has an energy far below that of the other two. Which isomer is it? Explain your answer.

CHAPTER 11

Molecular Spectroscopy

Most of what we know about the structure of atoms and molecules has been obtained from studies involving their interactions with electromagnetic radiation. The general term applied to such studies is *spectroscopy*, and these studies involve changes in energy levels. Because different regions of the electromagnetic spectrum may be utilized in spectroscopic studies, it is not surprising that terms are used to describe the type of radiation interacting with the sample. Thus, types of spectroscopy include visible, ultraviolet, infrared, microwave, etc. In most cases, a different instrument is employed to carry out each type of spectroscopic study. In general, studies involving visible spectroscopy give information about the differences in electronic energy levels in the sample. Energy differences between vibrational states generally correspond to energies in the infrared region of the spectrum. In this chapter, some types of molecular spectroscopy will be described utilizing the quantum mechanical principles described in earlier chapters.

11.1 VISIBLE AND ULTRAVIOLET SPECTROSCOPY

In Chapters 6 and 7 the nature of molecular vibrations and rotations was discussed based on the harmonic oscillator and rigid rotor models. Although the discussion at that point did not include broad application of the results to molecular spectra, additional details will be presented later in this chapter. Before we progress to a discussion of additional details of infrared spectra, attention will be given to applications of visible and ultraviolet spectroscopy.

The basis for absorption of electromagnetic radiation in the visible and ultraviolet regions of the spectrum lies in electronic transitions (see Fig. 7.6). In general, a spectrophotometer that utilizes these regions of the spectrum is employed in studying electronic transitions. A schematic diagram of such a spectrophotometer is shown in Fig. 11.1.

As the wavelength of the radiation is continuously varied, the instrument measures the absorbance of the sample and a spectrum is recorded. The amount of radiation absorbed (the *absorbance*) is plotted versus wavelength to generate the spectrum. A sample absorbs radiation to give an absorbance

Fundamentals of Quantum Mechanics
http://dx.doi.org/10.1016/B978-0-12-809242-2.00011-5

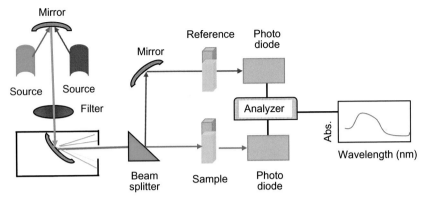

Fig. 11.1 A schematic diagram of a double-beam spectrophotometer. The sources used are most often a deuterium or xenon arc or a tungsten lamp. The reference cell often contains just the solvent. Note that by the use of two beams, the solvent is utilized as a reference and makes it possible to subtract the absorbance of the solvent, thus simplifying interpretation of the spectrum.

(*A*) that depends on three quantities that are related by a rule known as Beer's Law (sometimes Lambert–Beer Law),

$$A = \varepsilon l c \tag{11.1}$$

In this relationship, c is the concentration of the solute in the sample, l is the path length (thickness of the sample), and ε is a quantity that depends on the inherent nature of the sample to absorb radiation. It is sometimes called the extinction coefficient, but if the concentration of the sample is in mole per liters and the path length in centimeters, ε has the units of liters per mole per centimeter. The term *molar absorptivity* is often used to define ε. As can be seen from Eq. (11.1), a plot of absorbance versus concentration should be linear, and if the path length (l) is known, the value of ε can be determined. Measuring the absorbance of a sample has long been used to determine its concentration when the relationship given by Beer's law has been established.

11.2 ELECTRONIC TRANSITIONS IN MOLECULES

In many cases, it is possible to examine the molecular orbital diagram for a molecule and predict the type of electronic transition(s) that will occur. For example, it was shown in Chapter 9 that the orbital arrangement for the ethene molecule can be illustrated as

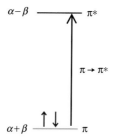

Therefore, it is easy to see that the electronic transition will be of the $\pi \rightarrow \pi^*$ type. As the Hückel calculation showed, the difference between the energy levels is 2β. Ethene, C_2H_4, shows an absorption band in the ultraviolet region at 165 nm, which corresponds to an energy of 173 kcal mol^{-1}. Therefore, the value of β is ~85 kcal mol^{-1}, a value that is at least *somewhat* realistic.

It is interesting to note that the orbital energies in 1,3-butadiene are $E_1 = \alpha + 1.62\beta$, $E_2 = \alpha + 0.62\beta$, $E_3 = \alpha - 0.62\beta$, and $E_4 = \alpha - 1.62\beta$. With there being only four electrons in the π system, only the lowest two levels are occupied. Note that the difference in energy between these levels and the third level is only 1.24β. In accord with this observation, the band in the ultraviolet spectrum corresponding to excitation of an electron from the highest occupied molecular orbital (HOMO) to the lowest unoccupied molecular orbital (LUMO) occurs at a wave length of 254 nm. As expected, this is a longer wavelength (lower energy) than in the case of ethene. Thus, although the Hückel molecular orbital method cannot be considered quantitative, it is "…useful in elucidating problems concerned with the electronic structure of π-electron systems" (Purins and Karplus, 1968).

It is interesting to note that for molecules that have cyclic structures, the excitation of an electron occurs at a longer wavelength (lower energy) the greater the number of atoms in the ring. Figure 9.10 shows the mnemonic of Frost and Musulin, which illustrates this graphically (Frost and Musulin, 1953; Baker and Baker, 1984). As a result of the number of vertices of the polygon increasing with the number of carbon atoms, the vertices are closer together, and thus the energy levels are closer together.

Some of the most common types of electronic transitions in molecules are summarized in Fig. 11.2. However, transitions between states having π and σ designations are symmetry forbidden.

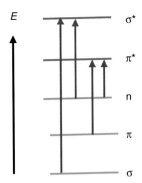

Fig. 11.2 Some types of electronic transitions that occur in molecules or functional groups.

Many functional groups give rise to absorption bands for which the positions do not change greatly regardless of the structure of the rest of the molecule. For example, the C=O group in ketones has nonbonding electrons on the oxygen atom that can be promoted to the next higher unoccupied level, so the $n \rightarrow \pi^*$ transition results in an absorption at $\sim$280 nm. Similarly, the —N=N— group gives an absorption at $\sim$340 nm as a result of the $n \rightarrow \pi^*$ transition. As a result, it is sometimes possible to employ electronic spectroscopy in the ultraviolet/visible region in much the same way as will be described in Section 11.6 for infrared spectroscopy.

11.3 PHOTOELECTRON SPECTROSCOPY

Spectroscopic studies of several types have provided the basis for much of the knowledge of atomic and molecular structure that is now considered so basic that it is taught in general chemistry courses. Consequently, it is virtually impossible to overemphasize the importance of spectroscopic methods. As has been shown in Chapters 6 and 7, spectroscopy, in which molecular vibrations and rotations are studied, has resulted in an understanding of molecular structure, bond lengths, and the types of bonds present in molecules. Electronic transitions give rise to absorptions primarily in the visible and ultraviolet regions of the spectrum, and the analysis of such spectra has revealed the nature of the energy levels in which electrons reside. One technique, known as photoelectron spectroscopy (PES), has been particularly important in elucidating electronic energy states in molecules.

Photoelectron spectroscopy has as its basis the analysis of the energies of electrons that are ejected by high-energy photons. One source of high-energy photons is that produced when He(I) (a helium atom) emits electromagnetic energy as an electron falls from the $2s^1 2p^1$ to the ground state ($1s^2$ ground state). For many molecules the first ionization potential is comparable to that of atoms. For example, the ionization potentials of O_2, NO, $(CH_3)_2CO$, and SO_2 are 12.06, 9.23, 9.69, and 11.7 eV, respectively. The hydrogen atom has an ionization potential of 13.6 eV, whereas that of the H_2 molecule is 15.43 eV.

Photoelectron spectroscopy is a type of emission spectroscopy the basis for which is the relaxation that occurs when electrons fall from high-energy states to those having lower energy. The excitation source most often used is He(I) (which is the spectroscopic designation of the helium atom) that emits photons of high energy, 21.22 eV. Such high-energy photos eject electrons from the target. The electron will leave with a kinetic energy of

$$\tfrac{1}{2} mv^2 = h\nu - I \qquad (11.2)$$

in which I is the ionization potential. Electrons having different energies are detected by using an analyzer that can employ varying voltages. As a result of counting electrons that have different energies, a spectrum can be produced. Ejection of electrons occurs on a rapid time scale, and most diatomic molecules are in their lowest vibrational state (see Chapters 6 and 7). However, if a series of vibrational states are populated, the spectrum will exhibit a series of closely spaced peaks. The absorption that corresponds to both the initial and ionized states of the molecule involving no change in vibrational level ($V=0$ to $V'=0$) is known as an *adiabatic transition* (see Fig. 7.6). When the ground and ionized states of the molecule have the same bond length, the transition is referred to as a *vertical ionization*. As can be seen in Fig. 7.6, this type of transition will result in the ionized molecule having excited vibrational states populated.

It is through the use of PES that a great deal of what we know about energy levels in molecules has been obtained. For diatomic molecules of the second row elements, combination of the $2p$ orbitals leads to a σ_{2p} and two π_{2p} orbitals. It was shown in Chapter 8 that the molecular orbital configuration of the O_2 molecule results in the σ orbital lying lower in energy that the two degenerate π orbitals. On the other hand, it is known that the reverse is true for the N_2 molecule and others comprised of atoms early in the second period. These possibilities are shown in Fig. 11.3.

The ionization process can be illustrated as

$$h\nu + M \rightarrow M^+ + e^- \qquad (11.3)$$

To illustrate a simple case in which PES has proven useful, the oxygen molecule will be considered. The σ_{2p} orbital in O_2 is filled with two electrons

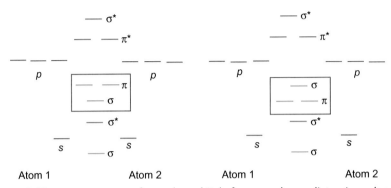

Fig. 11.3 The arrangements of σ and π orbitals for second row diatomic molecules with the boxed orbitals showing the differences in orbital energies that lead to differences in spectra obtained by PES.

(having spins of $+\frac{1}{2}$ and $-\frac{1}{2}$), and the two π^* orbitals of higher energy contain one electron each that has a spin of $+\frac{1}{2}$. When an electron is ejected from the σ orbital, the remaining electron has a spin that is the opposite of that of the ejected electron. Suppose that the ejected electron has a spin of $-\frac{1}{2}$. The two electrons in the π^* orbitals also have spins of $+\frac{1}{2}$, so there is an interaction with the electron remaining in the σ orbital. One way of showing the orbital populations is

$$O_2^+ \quad (\sigma)^2(\sigma^*)^2(\sigma)^{1(+\frac{1}{2})}(\pi)^2(\pi)^2(\pi^*)^{1(+\frac{1}{2})}(\pi^*)^{1(+\frac{1}{2})}$$

in which the symbol $(\sigma)^{1(+1/2)}$ indicates that the single electron remaining in the σ orbital has a spin of $+\frac{1}{2}$. Therefore, all of the electrons in the half-filled σ and half-filled π^* orbitals have spins of $+\frac{1}{2}$. However, if the electron ejected from the σ orbital has a spin of $+\frac{1}{2}$, the resulting orbital populations of the O_2^+ ion can be shown as

$$O_2^+ \quad (\sigma)^2(\sigma^*)^2(\sigma)^{1(-\frac{1}{2})}(\pi)^2(\pi)^2(\pi^*)^{1(+\frac{1}{2})}(\pi^*)^{1(+\frac{1}{2})}$$

In this arrangement, the electron remaining in the σ orbital has a spin of $-\frac{1}{2}$, and its interaction with the electrons that have $+\frac{1}{2}$ spins in the π orbitals is slightly different. As a result of spin–orbital coupling, the final energies of the two arrangements for the O_2^+ ion are not identical so that when an electron is ejected, two absorption bands of slightly different energy are observed in the PES spectrum. It is this type of result that makes PES a valuable technique for studying energy levels in molecules.

11.4 DETERMINING BOND LENGTHS IN DIATOMIC MOLECULES

In order to demonstrate one way in which a vibration-rotation spectrum can be utilized, the spectrum for gaseous HCl shown in Fig. 7.9 will now be analyzed to obtain the bond length for the molecule. The spectrum shown in Fig. 7.9 represents the transition of HCl molecules from the lowest vibrational state to the first excited vibrational state with the rotational fine structure exhibited. Although the transition corresponding to $V=0$ to $V=1$ with $\Delta J=0$ is missing, it would occur at about 2885 cm^{-1}. Analysis of the spectrum makes it possible to determine the force constant for the H—Cl bond. The frequency is related to the force constant by

$$\nu = \frac{1}{2\pi}\sqrt{\frac{k}{\mu}} \tag{11.4}$$

and the frequency is related to wave number $\bar{\nu}$ by $\nu = c/\lambda = c\,\bar{\nu}$. The reduced mass, μ, is given by

$$\mu = \frac{m_H m_{Cl}}{m_H + m_{Cl}} = 1.63 \times 10^{-24}\,g \tag{11.5}$$

Therefore,

$$k = \mu(2\pi c\bar{\nu})^2 = 4.82 \times 10^5\,dyn\;cm^{-1} \tag{11.6}$$

A more commonly used unit for the force constant is mdyn Å^{-1}. Because 1 dyn $=10^3$ mdyn and 1 cm $=10^8$ Å, the force constant in these units is 4.82 mdyn Å^{-1}. The unit N m^{-1} is also commonly used to describe force constants, and the conversion of units shows that 1 mdyn Å^{-1} is equivalent to 100 N m^{-1}. The force constant for the HF bond is thus 9.66 mdyn Å^{-1} or 966 N m^{-1}.

Correlations that involve molecular properties have been attempted for many years. Such correlations provide useful estimates of properties for which it may not be convenient to make measurements in order to obtain experimental values. From an intuitive point of view, it might be expected that there should be some relationship between the force constant, k, for stretching a bond and its equilibrium length, R_e. Although literally dozens of such relationships have been published, probably the most widely known is that published in 1934 by R.M. Badger (Badger, 1934). This "rule," known as Badger's rule, is usually written in the forms

$$k = A(R_e - B)^{1/3} \tag{11.7}$$

$$k_0(R_e - d_{ij})^3 = 1.86 \times 10^5 \tag{11.8}$$

The constant in Badger's rule depends on the rows occupied (i and j) in the periodic table. Although Badger's rule is of a semiempirical nature, it correlates bond lengths and force constants sufficiently well for many purposes, but some of the more recent versions do not fit the data much better. There have also been attempts to extend it to correlations involving polyatomic molecules.

The spacing between rotational bands is ~ 20.7 cm^{-1}, and from that value it is possible to determine the internuclear distance for HCl. The rotational energy can be represented as

$$E = \frac{\hbar^2}{2I}J(J+1) = \frac{h^2}{8\pi^2 I}J(J+1) \tag{11.9}$$

Therefore, for the transition $J=0$ to $J=1$,

$$\Delta E = E_1 - E_0 = \frac{h^2}{4\pi^2 I} = 20.7 \, \text{cm}^{-1} \tag{11.10}$$

This energy can be converted into erg s as follows:

$$\Delta E = h\nu = \frac{hc}{\lambda} = h c \bar{\nu} = (6.63 \times 10^{-27} \, \text{erg s}) \times (3.00 \times 10^{10} \, \text{cms}^{-1}) \times 20.7 \, \text{cm}^{-1}$$
$$\Delta E = 4.12 \times 10^{-15} \, \text{erg}$$
$$\tag{11.11}$$

Therefore, the moment of inertia is

$$I = \frac{h^2}{4\pi^2(\Delta E)} = 2.70 \times 10^{-40} \, \text{gcm}^2 \tag{11.12}$$

Because the moment of inertia is given by

$$I = \mu R^2 \tag{11.13}$$

it is possible to solve for the internuclear distance R, which has the value of 1.29×10^{-8} cm $= 1.29$ Å $= 129$ pm. These simple applications show the utility of infrared spectroscopy in determining molecular parameters.

It is convenient to show how the common units (cm^{-1}) are used in the following way. It should be remembered that $E = h\nu$ and $\lambda\nu = c$. Therefore,

$$E = h\nu = \frac{hc}{\lambda} \tag{11.14}$$

If the quantity $1/\lambda$ is represented as $\bar{\nu}$, then

$$E = hc\bar{\nu} \tag{11.15}$$

The energy units for a single molecule work out to be:

$$(\text{erg s}) \times (\text{cms}^{-1}) \times (1 \ \text{cm}^{-1}) = \text{erg}$$

By the use of conversion factors, the energy can then be converted from erg molecule^{-1} to kJ mol^{-1} or kcal mol^{-1}.

11.5 STRUCTURE DETERMINATION

For the simplest type of molecule, the diatomic molecule, there is only one vibration possible, the stretching of the chemical bond. That bond has an energy that is related to the distance between the atoms. Figure 8.1 shows that type of relationship in which there is a lowest energy (D_e), and it occurs at the normal bond distance r_0. If the bond is either longer or shorter than this distance, the energy is higher and the bond is less stable.

In the case of more complicated polyatomic molecules, there is a potential energy curve for each type of bond between atoms in the molecule. Therefore, changes in the vibrational levels of these bonds result in absorptions in the spectrum characteristic of the types of bonds present. It is thus possible in many cases to attribute absorption bands in the infrared spectrum to the types of bonds present in the molecule, as well as to determine a great deal about how the atoms are arranged. This application of infrared spectroscopy is of tremendous importance to the practicing chemist. With an elemental analysis to determine the ratio of atoms present (the empirical formula), a molecular weight to determine the actual numbers of atoms present, and an infrared spectrum to identify the kinds of bonds present, a chemist is well on his or her way toward identifying a compound.

For relatively simple molecules, it is possible to determine the structure by infrared spectroscopy. The total number of fundamental vibrations for a molecule that consists of N atoms is $3N-5$ if the molecule is linear and $3N-6$ if the molecule is nonlinear. Thus, for a diatomic molecule, the total number of vibrations is $3N-5 = 3 \times 2 - 5 = 1$. For a triatomic molecule, there will be $3N-6 = 3 \times 3 - 6 = 3$ vibrations for an angular or bent structure and $3N-5 = 3 \times 3 - 5 = 4$ vibrations if the structure is linear. For molecules consisting of three atoms, it is possible to determine the molecular structure on the basis of the number of vibrations that lead to the absorption of energy.

For the molecule SF_6, $N=7$, so there will be $3 \times 7 - 6 = 15$ fundamental vibrations. Each vibration has a set of vibrational energy levels similar to those shown in Fig. 7.6. Consequently, the changes in vibrational energy for each type of vibration will take place with different energies, and the bands can *sometimes* be resolved and assigned. However, not all of these vibrations lead to absorption of infrared (IR) radiation. There are also other bands called overtones and combination bands. Therefore, the total number of vibrations is large, and it is not likely that the molecular structure for a molecule as complex as SF_6 could be determined solely on the basis of the number of absorption bands seen in the IR spectrum.

For a change in vibrational energy of a molecule to be observed as an absorption of electromagnetic radiation (called an *IR active* change), the vibrational energy change must result in a change in dipole moment. This is because electromagnetic radiation consists of an oscillating electric and magnetic field (see Chapter 1). Therefore, an electric dipole can interact with the radiation and absorb energy to produce changes in the molecule. If HCl is considered as an example, there is a single vibration, which is the stretching of the H—Cl bond. For a molecule, the dipole moment (μ) can be expressed as the product of the amount of charge separated (q) and the distance of separation (r):

$$\mu = q \cdot r \tag{11.16}$$

When HCl is excited from the lowest vibrational energy to the next higher one, its average bond length, r, increases slightly. Because the dipole moment depends on r, the dipole moment of HCl is slightly different when the molecule is in the two vibrational states. Therefore, it is possible to observe the change in vibrational states for HCl as an absorption of infrared radiation. For the hydrogen molecule, H—H, there is no charge separation, so $\mu = 0$. Increasing the bond length does not change the dipole moment, so a change in vibrational energy of H_2 cannot be seen as an absorption of infrared radiation. This called an *IR inactive* vibrational change.

As an illustration of these factors, the CO_2 molecule will now be considered. It is reasonable to assume that the atoms are arranged in either a linear or a bent (angular) structure. Figure 11.4 shows the types of vibrations that would be possible for each of these two structures.

Because the two bending modes of the linear structure are identical except for being perpendicular to each other, they involve the same energy. Accordingly, only one vibrational absorption band should be seen in the IR spectrum for bending in either of the two directions. If CO_2 is linear, a change in the

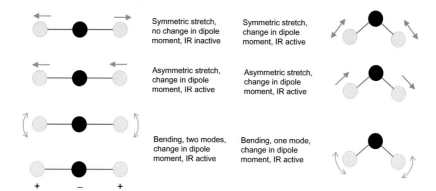

Fig. 11.4 Vibrations possible for assumed linear and bent structures for the carbon dioxide molecule. *Note.* The + and − signs on the CO_2 structures are used to denote motion perpendicular to the plane of the page.

symmetric stretching vibration would not cause a change in the dipole moment, because the effects on each C—O bond would exactly cancel. Changing the energy level for the asymmetrical stretching vibration does cause a change in dipole moment (the effects in opposite directions do not cancel), and one vibrational band is seen corresponding to asymmetric stretching. Therefore, if CO_2 is linear, there would be two fundamental vibrational bands in the infrared spectrum. However, if CO_2 were to have an angular structure, both of the stretching vibrations (i.e., symmetric and asymmetric) would cause a change in dipole moment because the effects on the C—O bonds do not cancel. Thus, the changes in both the symmetric and the asymmetric stretching vibrations would be IR active. The bending vibration for the angular structure would also cause a change in dipole moment, because the C—O bonds are not directly opposing each other. Consequently, if the structure were angular, there would be three bands in the vibrational region of the infrared spectrum. The actual spectrum of CO_2 shows two bands for infrared active vibrational changes at 2350 and 667 cm^{-1}. Therefore, the structure of CO_2 must be linear. The symmetric stretching vibrations are infrared inactive, but they show up in the Raman spectrum at 1388 cm^{-1}. On the other hand, all three vibrations of the NO_2 molecule are infrared active, and the spectrum shows bands at 1616, 1323, and 750 cm^{-1}, indicating that it has a bent structure. For molecules that contain a large number of atoms, it may become difficult, if not impossible, to sort out the bands to establish structure on this basis alone. The techniques that have been discussed illustrate part of the basis for how it is known that some of the molecular structures described in chemistry textbooks are correct.

11.6 TYPES OF BONDS PRESENT

One of the convenient aspects of vibrational changes in chemical bonds is that, for a particular type of bond, the remainder of the molecule often has a relatively small effect on the vibrational energy levels. For example, the change in stretching vibrational energy of the —O—H bond requires about the same energy, regardless of what is bonded on the other side of the oxygen atom. Accordingly, H_3C—O—H (usually written as CH_3OH) and C_2H_5—O—H give absorptions of energy in the same region of the electromagnetic spectrum, at about 3600 cm^{-1} or at a wavelength of 2780 nm (2.78×10^{-4} cm), which results from the O—H changing vibrational energy levels. Therefore, if the infrared spectrum of a compound exhibits an absorption band at this position, it is reasonably certain that the sample is composed of molecules that contain O—H bonds. Absorption bands in an infrared spectrum provide a useful diagnostic tool for quickly identifying the types of bonds present in molecules in the sample. Although extensive tables are available (e.g., in the *CRC Handbook of Chemistry and Physics*) a few common types of bonds and the typical regions where they absorb energy as they change stretching vibrations are given in Table 11.1.

It should be mentioned that in addition to the bands arising from bond stretching, there are numerous cases in which the bands arising from bending vibrations are also of importance. For example, in saturated hydrocarbons the characteristic C—H stretching bands are present in the 2900 cm^{-1} region, but there are also bands around $1375–1425 \text{ cm}^{-1}$ that correspond to the H—C—H bending vibrations. The band in the infrared spectrum that corresponds to the bending vibration of the —O—H bond occurs in the region of $1200–1400 \text{ cm}^{-1}$.

Table 11.1 Common types of bonds and typical energies they absorb

Bond	Approximate absorption region (cm^{-1})
O—H	3500–3600
N—H	3300–3400
C—H	2900–3000
C=O	1700–1750
C—Cl	700–800
C=N	1800–2100
P—H	2350–2400
Si—H	2100–2300
C=C	1625–1675
C≡C	2050–2250
C≡N	2200–2400

11.7 SOLVATOCHROMISM

Solvent molecules surrounding the molecules or ions of a solute create a sheath or layer (solvation sphere) that can affect many properties, including the absorption of electromagnetic energy. If a solution containing a specific solute exhibits colors that differ because of the nature of the solvent, the phenomenon is referred to as *solvatochromism*. There are also cases that have been studied in which the color of the solution depends on the applied pressure, giving rise to the phenomenon known as *piezochromism*. When a sample exhibits a color change in different solvents, it is the result of light absorption having a different distribution of wavelengths. This in turn is the result of the solvent interacting with the energy states of the solute molecules to alter their energies. Several types of carbonyl compounds are known to exhibit solvatochromism.

One of the most familiar cases of solvatochromism is that involving iodine. For many years, it has been known that a solution of iodine in a solvent such as carbon tetrachloride or a hydrocarbon (often referred to as an inert solvent) has a purple color. However, when iodine is dissolved in benzene, the solution has a brown or amber color. The maxima in the absorption bands of iodine appear at different positions as illustrated by the following (band maxima in nm): cyclohexane, 522; benzene, 500; acetonitrile, 455; and methanol, 445 (Costello et al., 2013). Although hydrocarbons may behave as "inert" solvents, the others do not. As a result, there are molecular complexes formed between iodine and the other solvents.

A molecular orbital description of the bonding in the I_2 molecule indicates that the orbitals of highest energy are the π^* and the σ^*. The spectrum of iodine vapor exhibits an absorption band at 538 nm, which gives rise to the dark violet color of the vapor. This absorption band results from a $\pi^* \rightarrow \sigma^*$ transition, which is due to the electronic transition from the HOMO to the LUMO. In solvents that do not interact strongly with these orbitals, the color of a solution containing iodine is also purple. If the solvent consists of molecules that are capable of functioning as donors of electron density, the color of the iodine solution will be brown, and the absorption maximum will be shifted to lower wavelengths, as illustrated by the data shown earlier. This shift of the position of the band maximum is believed to be the result of electron density being shifted to the σ^* molecular orbital from molecules of the solvent. Additional electron density placed in an antibonding orbital causes a slight decrease in the energy difference between the orbitals, and as a result causes the absorption band to be shifted to a lower wavelength.

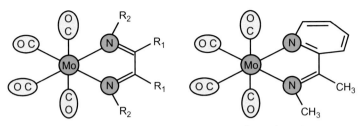

Fig. 11.5 Some molybdenum complexes that exhibit solvatochromism.

Although the case of iodine was used to introduce the topic of solvato-chromism, it should be mentioned that many other compounds exhibit this phenomenon. For example, many types of organic molecules show solvato-chromism. Among them are numerous compounds that contain a carbonyl ($C=O$) group, including ketones, acids, acid halides, and acid amides.

Solvatochromism is only one of the manifestations of a broad range of changes that result from solvent effects. Rates of reactions carried out in solutions often show great differences depending on the solvent. In some cases, the nature of the solvent is such that it solvates one or more of the reacting species so strongly that they cannot rapidly form the transition state. In other cases, the solvent may favor the formation of a transition state as a result of its high internal pressure. As a general rule, the formation of a transition state in which a separation of charge occurs is favored by a solvent having a high polarity and dielectric constant. Reactions in which the formation of the transition state involves a cancelation or dispersion of charge are generally favored by solvents that have low polarity and dielectric constant (for a general discussion see Reichardt and Welton, 2010).

Solvatochromism is also exhibited by coordination compounds. For example, molybdenum complexes having structures similar to those shown in Fig. 11.5 exhibit solvatochromism (Burgess et al., 1998). As in the case of iodine, there is a slight change in energy levels that can cause a change in the position of a charge transfer band. That band involves a shift of electron density from the metal to the ligand ($M \rightarrow L$) in a charge transfer transition. The important lesson to be learned is that the solvent is not an innocent bystander in many types of studies, including spectroscopy.

11.8 THE HYDROGEN BOND

All atoms except hydrogen have more than one electron in their valence shells. When hydrogen forms a single covalent bond, as it usually does, the nucleus is essentially bare on the side away from the bond. Therefore, the positive charge of the hydrogen nucleus can be attracted to a pair of electrons on another atom. The result is the formation of a weak bond that is usually referred to as a *hydrogen bond* or *hydrogen bridge*. Although not a chemical bond in the usual sense, the weak attraction between molecules produces great effects on chemical properties. The strength of hydrogen bonds is related in a general way to the polarity of the H—X bond, with the strongest being when X = F. The fact that water has a boiling point of 100°C whereas H_2S boils at −61°C, is a frequently cited example of this effect. A hydrogen bond can be represented as

$$X - H\cdots\cdots\;:Y$$

in which X is an atom having an electronegativity higher than that of hydrogen (usually O, F, N, or Cl) and Y is an atom having an unshared pair of electrons to which H is attracted.

It has been shown earlier that there is a potential energy curve that represents the bond energy as a function of internuclear distance. In the case of the hydrogen bond, the curve has two minima, one representing the covalent bond between X and H, and the other the weak bond between H and Y. Such a relationship is shown in Fig. 11.6.

Although there have been many attempts to treat hydrogen bonding by using quantum mechanical methods, a completely satisfactory approach remains elusive. The fact that the hydrogen bond has considerable electrostatic character has prompted many attempts to develop a model based on various potential functions. In general, such approaches are based on atomic properties of the electron pair donor atom, the charge on the hydrogen atom (resulting from bond polarity), bond lengths, etc. Such treatments of

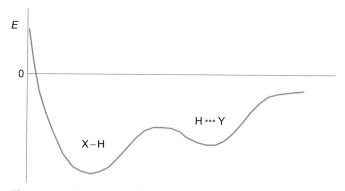

Fig. 11.6 The potential energy of a hydrogen bonded system.

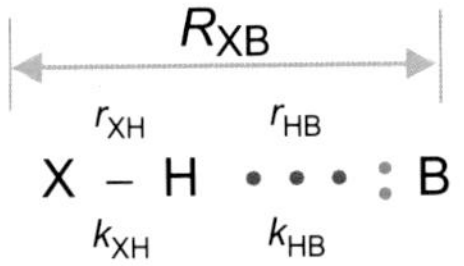

Fig. 11.7 A model of a hydrogen-bonded system showing distances and force constants.

hydrogen bonds attempt to correlate and explain bond strengths, bond lengths, and force constants for both the hydrogen bond itself, as well as the X—H bond. In most cases the parameters considered are illustrated as shown in Fig. 11.7, in which R and r represent distances and k represents the indicated force constant. However, the hydrogen bond is not always linear, and models have been developed to incorporate angular character.

Lippincott and Schroder described one of the most familiar potential function models (Lippincott and Schroeder, 1955; Schroeder and Lippincott, 1957; Finch and Lippincott, 1957). In this model, the hydrogen-bonded system is represented by a potential function that consists of several terms that represent the various energies. These can be represented as follows:

$$V_1 = D\left(1 - e^{-(n\Delta r^2/2r)}\right) \tag{11.17}$$

$$V_2 = -D^*\left(e^{-n^*\left(R-r-r_o^*\right)/2(R-r)}\right) \tag{11.18}$$

$$V_3 = Ae^{-bR} \tag{11.19}$$

$$V_4 = -B/R^m \tag{11.20}$$

The various distances expressed in these functions are related to the model shown in Fig. 11.7, and n is a function of the force constant for the bond. It is expressed as $n = k_o r_o/D$, in which D is the bond dissociation energy. The bond between hydrogen and the atom to which it is covalently bonded is considered to be a slightly stretched hydrogen bond with the extent of the stretching represented as $R-r-r_o^*$, where r_o^* is the length of the unperturbed hydrogen bond. All of these expressions can be evaluated, and it is possible to adapt the calculations to include spectroscopic parameters and force constants (Park, 1993).

There have been many attempts to improve the potential function model for hydrogen bonds. In one case, an improved model was applied to the study of hydrogen bonds in proteins (Fabiola et al., 2002). Spencer et al. (1974) adapted the potential function model in a treatment of the HX_2^- species, in which X=F, Cl, Br, or I. In addition to these studies, there have been others that demonstrate that the potential function models, although semiempirical in nature, are useful.

L.C. Allen described another of the models for hydrogen bonding (Allen, 1975). This model can be represented by the relationship

$$E_H = K\mu_{A-H}\Delta I/R \qquad (11.21)$$

in which R is the distance $X \cdots B$, ΔI is the difference in the ionization potential of the electron pair donor atom and the noble gas having the corresponding electron configuration, and K is an energy scaling factor. It was found to be an approximately linear relationship between the distance, l, that encompasses 98% of the unshared pair of the donor atom and ΔI, so the energy relationship can also be expressed in terms of $l - l_0$. It is found that $l_0 = 0.91$ and 1.44 Å for second and third row atoms, respectively.

Allen's model also provides a simple treatment of force constants of hydrogen-bonded systems. It begins with the fact that Badger's rule (see Section 11.4), which applies to normal covalent bonds, is applicable to hydrogen bonding situations. For such systems, the rule is expressed as

$$k_{AB}(R - d_{AB})^3 = 1.86 \qquad (11.22)$$

This approach is possible because it is found that the distance d_{AB} is approximately constant regardless of the *row* of the periodic table in which the atoms reside. It is found to depend on the *group* in which A and B reside with the values 1.00, 0.80, and 0.55 being values for Groups V, VI, and VII, respectively. The model is also based on the fact that k_{AB} is determined by the value of l, which has an average value that is has about the same relationship to R for both second and third row atoms. The force constant for the X—H bond k_{XH} is also explained by the Allen model. In the hydrogen-bonded complex, the force constant k_{XH} is inversely proportional to the value of ΔI. A large value for ΔI contributes to interactions that lead to smaller values for k_{XH}. The successes of the Allen model are remarkable considering the relatively simplicity of the approach.

The strongest hydrogen bonds are those in the species H_2F^-, the bifluoride ion. This ion arises from the interaction of F^- with an HF molecule, so it can be considered as a solvated anion in liquid HF solutions. Although hydrogen bonds normally have energies of 10–40 kJ mol^{-1} (3–10 kcal mol^{-1}), in the symmetrical HF_2^- ion, $[F \cdots H \cdots F]^-$, each hydrogen bond has an energy of $\sim$140 kJ mol^{-1}. The bonding in H_2F^- can be considered from the point of view of a valence bond picture involving

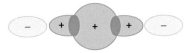

Fig. 11.8 A diagram representing HF_2^- in a valence bond model with orbital overlap.

overlap of the hydrogen $1s$ orbital with two p orbitals on F^- ions. This can be represented as shown in Fig. 11.8.

From the point of view of a molecular orbital approach, the combination of the hydrogen $1s$ wave function with two $2p$ wave functions from fluorine atoms produces three molecular orbitals that can be written as follows:

$$\psi(\sigma) = C_1 \varphi(p_{z1}) + C_2 \varphi(p_{z2}) + C_3 \varphi(1s) \tag{11.23}$$

$$\psi(n) = C_4 \varphi(p_{z1}) - C_5 \varphi(p_{z2}) \tag{11.24}$$

$$\psi(\sigma^*) = C_6 \varphi(p_{z1}) + C_7 \varphi(p_{z2}) - C_8 \varphi(1s) \tag{11.25}$$

These wave functions describe one bonding orbital (σ), one nonbonding orbital (n), and one antibonding orbital (σ^*). The molecular orbital diagram that results from these wave functions is shown in Fig. 11.9. Note that one of the fluorine atoms is considered as F^-, which contributes two electrons to the structure as a result of HF_2^- being formed from $HF + F^-$.

Hydrogen bonding occurs in many types of systems in which hydrogen atoms are bonded to atoms of high electronegativity, such as F, O, N, and Cl, as well as to some slight extent S and Br. The bonds may be between two

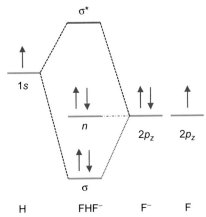

Fig. 11.9 A molecular orbital diagram for the HF_2^- ion. Filled $2p$ orbitals constitute π orbitals that are nonbonding, and they do not have the correct symmetry to bond to the hydrogen $1s$ orbital. Therefore, they are not shown on the diagram.

separate molecules (*intermolecular* hydrogen bonds) or within a specific molecule (*intramolecular* hydrogen bonds).

Intermolecular   Intramolecular

Hydrogen bonding also occurs in numerous types of solids, such as NH_4Cl, $NaHCO_3$, NH_4HF_2, and ice.

Liquids in which there is extensive hydrogen bonding between molecules typically have higher boiling points than do molecules of similar atomic composition. For example, ethanol (C_2H_5OH) boils at 78.4°C, whereas dimethyl ether (CH_3OCH_3) has a boiling point of −24°C, even though they both have the same empirical formula, which is C_2H_6O. In some cases, the hydrogen bonding is so strong that it is maintained even in the vapor phase. For example, acetic acid exists in the vapor phase largely as the dimer.

In the case of alcohols, a rather complex equilibrium exists between several aggregates of molecules. There are presumably both chains and ring structures involved, and in the vapor phase a tetramer of methanol, which is $(CH_3OH)_4$,

has been identified, for which the heat of association has been found to be 94.4 kJ mol^{-1}. It appears in this case that each hydrogen bond has an energy of 23.6 kJ mol^{-1}.

11.9 EFFECTS OF HYDROGEN BONDING ON SPECTRA

Although hydrogen bonding manifests itself with regard to properties, such as heat of vaporization, solubility, and several types of chemical behavior, it also produces dramatic effects on spectra. With much of this chapter devoted to the analysis of spectra, it is that aspect of hydrogen bonding that needs to be discussed. In fact, spectroscopy is one of the most frequently used techniques employed in the study of hydrogen bonding.

When a hydrogen atom attached to an atom of high electronegativity by a polar covalent bond becomes attracted to a pair of electrons on a donor atom, there are several changes produced. First, the covalent bond is lengthened and weakened slightly, which results in a shift in the position of the X—H stretching band. The position of the peak corresponding to stretching the X—H bond is shifted to a *lower* wave number in the infrared spectrum by as much as 200–400 cm^{-1}, depending on the strength of the hydrogen bond. As might be expected, there is a general correlation between the magnitude of the shift and the base strength of the electron pair donor, as long as bases of the same general type are considered.

When the hydrogen atom becomes attached to a pair of electrons in a donor atom in another molecule, bending vibrations, such as that of the —X—H fragment, become hindered. Therefore, there is a shift to *higher* wave number in the position of the peak corresponding to this vibration. However, there may be two types of bending vibrations, that which corresponds to the in plane motion of the X—H moiety and that which is bending out of the plane. Both types of vibrations become hindered if the hydrogen atoms are attached to a donor atom so they are shifted to higher wavenumbers.

Table 11.2 Effects produced by hydrogen bonding on infrared spectra

Vibration	Assignment	Spectral region (cm^{-1})
X—H⋯B	ν_s, the X—H stretch	3500–2500
X—H⋯B	ν_b, the in-plane bend[a]	1700–1000
X—H⋯B	ν_t, the out-of-plane bend[b]	400–300
X—H⋯B	ν_σ, the H⋯B stretch[c]	200–100

[a]Bending in the plane of the page. Hydrogen bonding causes higher ν_b.
[b]Bending perpendicular to the plane of the page (torsion). Hydrogen bonding causes higher ν_t.
[c]Stretching of the hydrogen bond to the donor atom. It increases with hydrogen bond strength.
Reproduced with permission from House, J. E. *Inorganic Chemistry*, 2nd ed.; Academic Press/Elsevier: Amsterdam, 2013.

The very weak hydrogen bond itself is subject to stretching vibrational changes, and these give rise to peaks in the far infrared region of the spectrum (usually $100–200$ cm^{-1}). Table 11.2 shows the spectral characteristics that result from hydrogen bonding.

Species present in chemical systems in equilibrium in solution depend on the nature of the solvent. If a solvent interacts strongly with the monomer, aggregation will be hindered. Because of this, solvents that can hydrogen bond to alcohols hinder the formation of aggregates. As a result of molecular association by hydrogen bonding, the positions of O—H stretching bands are not the same as for "free" O—H stretching. As will be discussed later, even the position of band attributable to the "free" O—H stretching is solvent dependent.

In "inert" solvents such as alkanes and CCl$_4$, alcohols form aggregates in solution even at very low concentrations. It is generally assumed that the process can be represented by the equation

$$n CH_3OH \leftrightharpoons (CH_3OH)_n \tag{11.26}$$

At a *very* low concentration in CCl$_4$ as the solvent, the equilibrium lies far to the left, and the spectrum shows a peak at 3642 cm^{-1} corresponding to O—H stretching in the monomer. At higher concentrations, other bands appear that reflect the intermolecular hydrogen bonding, which causes the stretching frequency to be lower. Figure 11.10 shows the spectra obtained for solutions of methanol in carbon tetrachloride (House, 2013).

Although the peak corresponding to the monomer is present in all cases, it is interesting to note the broad, shifted bands that result from molecular aggregation as a result of hydrogen bonding.

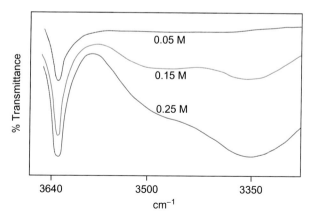

Fig. 11.10 Infrared spectra obtained for solutions of CH$_3$OH in CCl$_4$. *(Modified from House, J. E.* Inorganic Chemistry, *2nd ed.; Academic Press/Elsevier: Amsterdam, 2013.)*

The spectra will not be shown, but when spectra similar to those shown in Fig. 11.10 are obtained using solutions at higher temperatures, the peaks attributable to aggregates diminish greatly in intensity as a result of dissociation of the clusters. It is believed that some hydrogen bonding between molecules of alcohols exists even when the concentration in an inert solvent is as low as 0.01 M, but it certainly depends on the temperature of the solutions.

In addition to the cyclic tetramer, the structure of which was shown earlier, some of the species present in solutions of alcohols in solvents such as heptane or carbon tetrachloride, are believed to have the structures

In the vapor phase, the absorption band for the O—H stretching vibration is observed at 3687 cm^{-1}. However, when methanol is dissolved in solvents such as n-C_7H_{16}, CCl_4, and CS_2, the O—H stretching band is observed at 3649, 3642, and 3626 cm^{-1}, respectively (House and Cook, 1969). However, when the solvent is benzene, the band position is 3607 cm^{-1}. It can be seen that in this case, as in many others involving spectroscopy, the positions of bands may be shifted by interactions with the solvent. A shift to lower frequency is knows as a *red* or *bathochromic* shift. When the absorption is shifted to higher frequency, it is known as a *blue* or *hypsochromic* shift. In some cases, a solvent can also cause an increase or decrease in band *intensity* (known as *hyperchromic* and *hypochromic* effects).

It is interesting to note that there is a shift to higher wave numbers of the position of the band for hydrogen-bonded O—H groups in alcohols that occurs at ~3300–3350 cm^{-1}. Finch and Lippincott explained this in terms of the increase in basicity of the alcohols with chain length (Finch and Lippincott, 1957). From other studies, it is known that the basicity of alcohols increases slightly with chain length, and as the temperature increases, there is a slight increase in the electron density on the oxygen atoms that function as the electron pair donors. In the temperature range of 200–300 K, the shift amounts to 30–50 cm^{-1} depending on the chain length of the alcohol, with the shift being lowest for methanol and highest for hexanol. This feature of infrared spectra of alcohols was interpreted from the potential function model (Finch and Lippincott, 1957).

Theoretical treatment of the solvent effect on the position of the O—H stretching band was carried out many years ago by assuming that an oscillating electric dipole interacts with solvent having a dielectric constant, ε. The relationship obtained can be written as

$$\frac{\nu_g - \nu_s}{\nu_g} = C\frac{\varepsilon - 1}{2\varepsilon + 1} \tag{11.27}$$

in which ν_g and ν_s are the stretching frequencies in the gas phase and in the solvent, respectively, and C is a constant. When measured at high frequency, the dielectric constant is often represented as the square of the index of refraction, n^2. The difference between the stretching frequencies in the gas phase and in the solution can be written as $\Delta\nu = (\nu_g - \nu_s)$, so that Eq. (11.27) becomes

$$\frac{\Delta\nu}{\nu_g} = C\frac{n^2 - 1}{2n^2 + 1} \tag{11.28}$$

This relationship is known as the Kirkwood-Bauer-Magat equation. Figure 11.11 shows a Kirkwood-Bauer-Magat plot constructed using index of refraction values for the solvents indicated on the figure and the positions of the O—H stretching bands in methanol. From the points on the figure, it is clear that benzene behaves in a different way toward solvating O—H

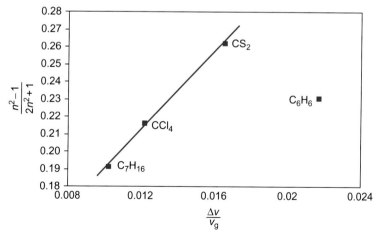

Fig. 11.11 A Kirkwood-Bauer-Magat plot showing solvent effects on the O—H stretching band of methanol. *(Modified with permission from House, J. E.* Inorganic Chemistry, *2nd ed.; Academic Press/Elsevier: Amsterdam, 2013.)*

bonds than do the other solvents considered. This is not surprising because the pi electron system in benzene and other aromatic molecules have considerable electron donating ability and even form Lewis acid–base adducts.

However, it should be mentioned that studies to explain solvent effects on reactions, spectra, and other properties have been carried out for many years. The Kirkwood–Bauer–Magat approach represents only one type of explanation. The interested reader should consult the book by Reichardt and Welton for a more comprehensive treatment of solvent effects (Reichardt and Welton, 2010).

Hydrogen bonding occurs in virtually all areas of chemistry and affects many properties of materials. As a result, a discussion of its nature and effects has been presented to show some of the experimental and theoretical aspects of this type of bond. The discussion presented shows that hydrogen bonding is a complicated feature when it comes to interpreting infrared spectra. However, hydrogen bonding provides an interesting and fruitful area for the study of suitable systems by means of molecular spectroscopy.

After obtaining an infrared spectrum of a sample, using an extensive table of stretching frequencies frequently makes it possible to match the positions of observed peaks in an infrared spectrum to known values for the various types of bonds, thereby determining the types of bonds present in the compound. This information, along with percent composition and molecular weight, is sometimes sufficient to identify the compound. Thus, as has been described in Chapter 7, infrared spectroscopy can be used in certain circumstances to determine bond lengths, the types of bonds present in molecules, and molecular structure. These applications of infrared spectroscopy make it one of the most useful tools for the study of materials by chemists, although many other experimental techniques [X-ray diffraction, nuclear magnetic resonance (NMR), electron spin resonance (ESR), etc.] are required for the complete study of matter.

LITERATURE CITED

Allen, L. C. A Model for the Hydrogen Bond. *Proc. Nat. Acad. Sci.* **1975**, *72* (12), 4701–4705.

Badger, R. M. A Relation Between Internuclear Distances and Bond Force Constants. *J. Chem. Phys.* **1934**, *2*, 128–131.

Baker, A. D.; Baker, M. D. A Geometric Method for Determining the Hückel Molecular Orbital Energy Levels of Open-Chain, Fully Conjugated Molecules. *J. Chem. Educ.* **1984**, *61*, 770.

Burgess, J.; Maguire, S.; McGranahan, A.; Parsons, S. A. Solvent Effects on Piezochromism of Transition Metal Complexes. *Transit. Met. Chem.* **1998**, *23*, 615–618.

Costello, S. M.; Sobyra, T. B.; House, J. E. Iodine as a Probe for Demonstrating Solvato-chromism. *Chem. Educator.* **2013**, *18*, 54–56.

Fabiola, F.; Bertram, R.; Korostelev, A.; Chapman, M. S. An Improved Hydrogen Bond Potential: Impact on Medium Resolution Protein Structures. *Protein Sci.* **2002**, *11*, 1415–1423.

Finch, J. N.; Lippincott, E. R. Hydrogen Bonding Systems—Temperature Dependence of OH Frequency Shifts and OH Band Intensities. *J. Phys. Chem.* **1957**, *61*, 894–902.

Frost, A. A.; Musulin, B. A Mnemonic Device for Molecular Orbital Energies. *J. Phys. Chem.* **1953**, *21*, 572.

House, J. E. *Inorganic Chemistry*, 2nd ed.; Academic Press/Elsevier: Amsterdam, 2013.

House, J. E.; Cook, R. L. Solvent Effects on the O—H Stretching Bands of Alcohols. *Trans. Ill State Acad. Sci.* **1969**, *62*, 154–160.

Lippincott, E. R.; Schroeder, R. One Dimensional Model of the Hydrogen Bond. *J. Chem. Phys.* **1955**, *23*, 1099–1106.

Park, Y. H. The Use of the Lippincott/Schroeder Potential Function in Establishing Relationships Between Infrared Spectroscopic Measurements and Structural and Thermodynamic Properties of Hydrogen Bonds. *J. Kor. Ind. Eng. Chem.* **1993**, *4*, 409–415.

Purins, D.; Karplus, M. Methyl Group Inductive Effect in the Toluene Ions. Comparison of Hückel and Extended Hückel Theory. *J. Am. Chem. Soc.* **1968**, *90* (23), 6275–6281.

Reichardt, C.; Welton, T. *Solvents and Solvent Effects in Organic Chemistry*, 4th ed.; Wiley-VCH: New York, NY, 2010.

Schroeder, R.; Lippincott, E. R. Potential Function Model of Hydrogen Bonds. II. *J Phys. Chem.* **1957**, *61*, 921–928.

Spencer, J. N.; Casey, G. J.; Buckfelder, J.; Schreiber, H. D. Potential Function Model of Weak and Strong Hydrogen Bonds. *J. Phys. Chem.* **1974**, *78*, 1415–1420.

PROBLEMS

1. The N_2O molecule is linear (see Section 9.7). How many vibrational modes are there for this molecule? How many (and which ones) would lead to absorption bands in an infrared spectrum?

2. The SO_2 molecule shows absorption bands at 1381, 1151, and 519 cm^{-1}. Sketch the molecule and indicate on the sketches the possible vibrations for this molecule. To which vibration does the absorption at 219 cm^{-1} correspond?

3. Explain why *p*-hydroxybenzoic acid is a stronger acid and more soluble in water than is *o*-hydroxybenzoic acid.

4. When a small amount of triethylamine, $(C_2H_5)_3N$, is added to a dilute solution of methanol in carbon tetrachloride, the OH stretching band is shifted. How would this differ if triethylphosphine $(C_2H_5)_3P$ were added instead? Explain your answer.

5. There are three possible isomers of N_2O. Draw the structures for the isomers and explain how infrared spectroscopy could be used to distinguish between these structures.

6. If the difference between the $J=1$ and $J=2$ peaks in the rotational spectrum of CO are separated by 7.7 cm^{-1}, what is the bond length in CO?

7. Explain why the hydrogen bonds in HCl_2^- are much weaker than those in HF_2^-.

8. By referring to Fig. 11.11, where do you think that a point for acetonitrile (CH_3CN) would be on the line. Discuss molecular structure in explaining your answer.

9. Alkyl cyanates (ROCN) and isocyanates (RNCO) are well known compounds. Discuss the difference in their interaction with an alcohol, such as methanol, when the solvent is CCl_4.

10. The photoelectron spectrum of F_2 shows a peak corresponding to an energy of 15.7 eV.

    (a) Draw the molecular orbital energy diagram for the F_2 molecule and write the electron configuration.

    (b) Determine the spectroscopic ground state term for the F_2 molecule.

    (c) The peak in the photoelectron spectrum corresponds to ionization of the F_2 molecule. Consider the ion and determine the spectroscopic state for the species.

    (d) The ion exhibits a splitting of energy levels. Explain its origin using spectroscopic states.

CHAPTER 12

Spectroscopy of Metal Complexes

One of the significant areas in which spectroscopic techniques have been of great importance is in the study of transition metals complexes. These studies have elucidated areas such as the geometric structure, the nature of the metal to ligand bonds, and the mode of ligand bonding to the metal. Although it is not possible to present an exhaustive treatment of this broad field, it is appropriate to discuss some of its general aspects.

12.1 THE EFFECT OF LIGANDS ON d ORBITALS

Many coordination compounds of metals are brightly colored, which indicates that there is absorption of electromagnetic energy in the visible region of the spectrum. The origin of that absorption comes from energy levels in the structures that are separated by energies comparable to that of photons in the visible region. Therefore, to interpret the spectra, it is necessary to have an understanding of what those levels are. As a result, it will be shown how the presence of the ligands (electron pair donors) affects the energies of the d orbitals. To begin, the relationship of the d orbitals to the positions of ligands in an octahedral complex will be illustrated, as shown in Fig. 12.1.

The principles related to the splitting of the d orbital energies was originally applied to the metal ions surrounded by anions in a crystal. As a result, the term *crystal field splitting* has also been applied to complexes, even though the ligands may be neutral molecules. A more descriptive term is *ligand field theory*, which is generally used when discussing complexes.

As shown in Fig. 12.1, the lobes of the d_{xy}, d_{yz}, and d_{xz} orbitals point between ligands, and as a result experience less repulsion (because the ligands are electron pair donors) than do the $d_{x^2-y^2}$ and d_{z^2} orbitals. The result is that in an octahedral complex, the five d orbitals of a transition metal are no longer degenerate, but are separated into two sets of orbitals, with the difference in energy between the sets equivalent to that of visible light. The three orbitals of lower energy are called the t_{2g} orbitals, and the two of higher energy are denoted as the e_g set. These energy levels are identified in the molecular orbital diagram shown in Fig. 10.22.

Fundamentals of Quantum Mechanics
http://dx.doi.org/10.1016/B978-0-12-809242-2.00012-7

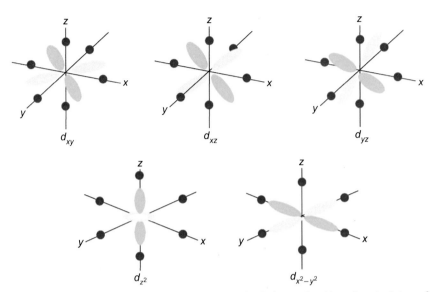

Fig. 12.1 The arrangement of d orbitals in an octahedral complex. Note that the lobes of the d_{xy}, d_{yz}, and d_{xz} orbitals point *between* ligands, whereas those of the $d_{x^2-y^2}$ and d_{z^2} point directly *toward* ligands.

The perturbation affects all of the d orbitals, and all are raised in energy compared to their energies in a free, gaseous ion. The d orbitals with lobes that lie *along* the axes are affected to a greater extent than are those with lobes *between* the axes. While all of the d orbitals are raised in energy, the d_{z^2} and $d_{x^2-y^2}$ orbitals are raised to a greater extent than the d_{xy}, d_{yz}, and d_{xz} orbitals. Thus, in an octahedral complex the five d orbitals constitute two sets of orbitals of different energy, as shown in Fig. 12.2.

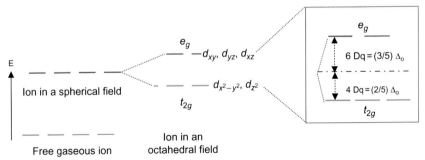

Fig. 12.2 A representation of the d orbitals in a transition metal ion and in an octahedral complex. *Adapted with permission from House, J. E. Inorganic Chemistry, 2nd ed.; Academic Press/Elsevier: Amsterdam, 2013 (chapters 16–18).*

The difference in energy between the t_{2g} and e_g orbitals is defined as 10 Dq units, in which the value of Dq depends on the nature of the metal ion, its charge, and the nature of the ligands. Because an octahedral complex is being considered, the energy difference is also described as Δ_o. However, the "center of gravity" with respect to energy must be maintained, so the three orbitals in the t_{2g} set are lowered from that center by 4 Dq, whereas the two orbitals in the e_g set are raised by 6 Dq.

The simplest type of visible spectrum for a coordination compound results when there is a single electron excited from a t_{2g} to one of the e_g orbitals. As has been discussed earlier in this book, transitions between electronic levels in *atoms* result in spectral *lines* (i.e., light having specific energies). However, when the spectrum of a solution containing $[Ti(H_2O)_6]^{3+}$ is examined, it is found that there is a single, broad absorption *band* that has a maximum at 20,300 cm^{-1} (243 kJ mol^{-1}). If the spectrum corresponded only to a transition of the single electron in the $3d$ level from the t_{2g} set to the e_g set, one might expect the transition to give rise to an absorption *line*, but this is not the case.

The case of $[Ti(H_2O)_6]^{3+}$ is simpler than that of complexes containing most transition metals, owing to the fact that the Ti^{3+} ion contains only one electron in the $3d$ orbitals. When multiple electrons are present in the $3d$ orbitals, the situation is more complicated because of the effects of spin–orbit coupling. However, taking repulsion and spin–orbit coupling into account to provide a complete explanation of the spectra of coordination compounds is outside the subject matter of this book, but explanations can be found in the references at the end of this book.

The magnitude of the ligand field splitting in a complex depends on several factors, one of which is the nature of the ligands. To assess this effect, a series of complexes, such as $TiL_6^{z\pm}$, can be studied, and it is found that the ability to split the d orbital energies is approximately in the following order:

$$CO > CN^- > NO_2^- > en > NH_3 \approx py > NCS^- > H_2O > ox > OH^-$$
Strong field
$$> F^- > Cl^- > SCN^- > Br^- > I^-$$
Weak field

Because the magnitude of the difference in energy between the t_{2g} and the e_g sets is obtained by the analysis of *spectra*, the series listed above is known as the *spectrochemical series*.

For a complex of a metal ion that contains at least four electrons in the d states, there are two possible arrangements of the electrons. If the metal ion is

Mn^{3+}, there are four electrons in the d orbitals, and the way in which they can be arranged in an octahedral complex is shown as follows:

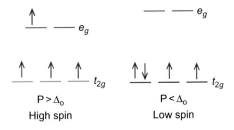

Thus, there will be two series of complexes of Mn^{3+}, with the number of unpaired electrons depending on the strength of the ligand field produced by the ligands. For complexes of Mn^{3+}, it is possible to determine the electron population of the orbitals by determining the magnetic moment. In the high–spin complexes, the magnetic moment is approximately 4.8 D indicating four unpaired electrons whereas for a low-spin complex it is approximately 2.8 D, a value corresponding to two unpaired electrons.

As shown in Fig. 12.3, the situation is quite different when tetrahedral complexes are considered. The lobes of the d_{z^2} orbital, shown in Fig. 12.3, are directed toward the midpoint of a diagonal, which is a point lying $(2^{1/2}/2)e$ from two of the ligands, whereas the d_{xy}, d_{yz}, and d_{xz} orbitals have lobes pointing at the midpoint of an edge of the cube, which is $e/2$ from the ligands. As a result, there is a difference in repulsion that causes the d_{xy},

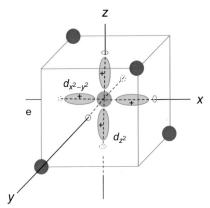

Fig. 12.3 A tetrahedral complex with only two lobes of the $d_{x^2-y^2}$ orbital (along the x-axis) and the lobes of the d_{z^2} orbital lying along the z-axis shown. This illustrates that none of the d orbitals point directly at the ligands.

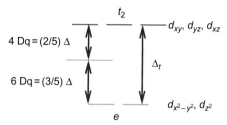

Fig. 12.4 The splitting of d orbital energies in a tetrahedral field.

d_{yz}, and d_{xz} orbitals to have higher energy than the $d_{x^2-y^2}$ and d_{z^2} orbitals. Therefore, in a tetrahedral complex the energies of the d orbitals can be shown as they are in Fig. 12.4.

Although the d orbitals are split in energy in a tetrahedral field, the splitting is much smaller than in an octahedral field, even if the same metal ion and ligands are involved. Whereas six ligands surround the metal ion in an octahedral complex, there are only four ligands in a tetrahedral complex, so the electrostatic field produced is smaller. Also, the ligands lie on the axes in an octahedral complex, which causes the energy of orbitals having lobes that lie along the axes to have higher energy than do the orbitals whose lobes are directed between the ligands. If tetrahedral and octahedral complexes of the same metal and ligands are considered, it is possible to show that $\Delta_t = (4/9)\Delta_o$. A ligand field splitting of this magnitude is not sufficient to cause electron pairing and as a result, tetrahedral complexes are high-spin.

As in the case of octahedral complexes, the analysis of spectra depends on the transitions that occur between energy states in the ligand field. A brief description of the basic principles will be presented in Section 12.3, but an explanation of why there are spectral bands will be presented in the next section.

12.2 BANDS IN ELECTRONIC SPECTRA OF COMPLEXES

Although electronic transitions in atoms generally give rise to spectral lines, that is not the case with complexes of metals. Therefore, it is appropriate to describe the factors that are responsible for this difference. Electronic transitions occur on a time scale that is extremely short compared to that of molecular vibrations. As a result, vibrations in which ligands–metal bonds are changing distances create a ligand field that is not static. Therefore, electronic transitions take place with a ligand field that varies, which causes a range of energies to be absorbed. This results in a *band* of absorbed energies rather than discrete frequencies, as is the case with atomic line spectra. The transitions described for coordination compounds are normally obtained for solutions of the compounds. The effect of a solvent shell surrounding the complex prevents observing transitions that involve differences between discrete vibrational levels.

Transitions that involve electrons moving between energy states that arise from *d* orbitals are called *electric dipole* transitions. However, transitions between states that have different multiplicities are generally forbidden, though they may be observed with low intensity. The classic example of this type involves the Fe^{3+} ion (d^5) for which the ground state is ^6S. All excited states have different multiplicities from the ground state, so there are no spin-allowed transitions. The intensity of an absorption band depends on the value of the overlap integral

$$\int \psi_1 r \, \psi_2 \, d\tau$$

in which ψ_1 and ψ_2 are the wave functions for the orbitals representing the ground and excited states, respectively. This integral is required to have a nonzero value for the transition to be allowed. The *d* orbitals in an octahedral field are designated as "*g*" states. For transitions between such states, the integral would have a value of zero. A selection rule known as the *Laporte Rule* applies in such cases, and it allows transitions only between states that have *different symmetry*. Transitions of the *d–d* type are forbidden in a ligand field having *g* symmetry. Although they generally have low intensities, forbidden transitions are sometimes observed.

As shown in Chapter 5, a particular bond has an equilibrium internuclear distance R_0, which has a specific value when the lowest electronic and vibrational states are occupied. When energy is absorbed to change the electronic state to an excited level, the transition occurs on a time scale that is orders of magnitude smaller than that required for the nuclei to change equilibrium positions. In other words, the nuclei are essentially considered to be stationary, with only the electrons being moved. This means that the molecule in the excited state has a vibrational energy that is not characteristic of the electronic energy because in the excited state, the average internuclear distance is not the same as in the electronic ground state. This is known as the *Franck-Condon principle*.

Potential energy of a bond in either the ground or excited state can be represented by a potential energy curve that shows energy as a function of

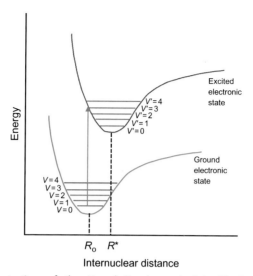

Fig. 12.5 An illustration of the Franck-Condon principle (R_0 is the equilibrium internuclear distance in the ground state, and R^* is that distance for the excited state). In this case, the transition is from $V=0$ in the electronic ground state to the state with $V'=3$ in the excited electronic state.

bond length. Within both the ground and excited electronic states, vibrational energy levels exist. Figure 12.5 shows such relationships for the electronic ground state and the first excited state. For both electronic states, several vibrational states are shown.

Electronic transitions occur so rapidly that the nuclei are assumed to be stationary. Therefore, a transition from the electronic ground state to the first excited state is called a *vertical transition,* in which R_0 is maintained between the nuclei. As a result, there is a high probability of a transition between the lowest vibrational energy level in the electronic ground state to a *higher* vibrational level in the excited electronic state. Such transitions that involve both electronic and vibrational transitions are sometimes referred to as *vibronic transitions.* This type of transition is a manifestation of the Franck-Condon principle and is illustrated in Fig. 12.5. Overlap integrals may not be equal to zero for transitions between several vibrational levels, resulting in absorption of energies that constitute a band.

A temporary change in the structure of a complex from one having "g" symmetry makes it possible for electric dipole types of transitions to occur. Coupling of electronic and vibrational wave functions, described as *vibronic coupling,* is the result of vibrational energy changes. Some vibrations result in a change in symmetry of the structure, so that the "*g*" designation no longer applies, which makes possible some electronic transitions. With the change in symmetry, the Laporte selection rule is no longer applicable, and some transitions are allowed. Unlike the sharply separated rotation and vibration energy changes for a diatomic molecule such as HCl, the absorption of energy results in spectral bands for metal complexes in solution.

12.3 INTERPRETING ELECTRONIC SPECTRA OF COMPLEXES

In order to interpret spectra of complexes, it is necessary to understand how the ligand field affects the spectroscopic states of the metal ion. Therefore, the spectroscopic states for gaseous metal ions that contain various numbers of electrons in d orbitals are shown in Table 12.1.

The spectroscopic terms shown in Table 12.1 are those for *gaseous* ions having degenerate d orbitals. This is not the situation when the metal ion is surrounded by six ligands in an octahedral arrangement. In that case, the orbitals are the t_{2g} and e_g orbitals and the spectroscopic states are not those of the free ion. For example, the single electron in the Ti^{3+} ion resides in a t_{2g} orbital, and the corresponding spectroscopic state is $^2T_{2g}$. If the electron is excited to an e_g orbital, the spectroscopic state is designated as ^2E_g, so the transition can be designated as $^2T_{2g} \rightarrow {}^2E_g$. Electronic transitions between states of the same multiplicity are allowed, and in this case both the initial and final states are doublets. Accordingly, the transition is allowed, and it occurs with high intensity. As in the case of atoms, orbitals have lower case designations, whereas spectroscopic states are denoted by capital letters.

All of the ligand field spectroscopic states that arise from the effect of ligand field splitting of the spectroscopic *ground* states of the gaseous ions are shown in Table 12.2. It is important to remember that although Table 12.2 lists all spectroscopic states that arise from a given d^n configuration, there is a very important selection rule that states:

Table 12.1 Spectroscopic states for gaseous ions of transition metals

Ion	Spectroscopic states
d^1, d^9	^2D
d^2, d^8	^3F, ^3P, ^1G, ^1D, ^1S
d^3, d^7	^4F, ^4P, ^2H, ^2G, ^2F, 2^2D, ^2P
d^4, d^6	^5D, ^3H, ^3G, 2^3F, ^3D, 2^3P, ^1I, 2^1G, ^1F, 2^1D, 2^1S
d^5	^6S, ^4G, ^4F, ^4D, ^4P, ^2I, ^2H, 2^2G, 2^2F, 3^2G, 3^2D, ^2P, ^2S

The ground state term is listed first. A coefficient of 2 means that two different identical terms exist.

Table 12.2 Spectroscopic states in an octahedral ligand field[a]

Gaseous ion spectroscopic ground state	Components in an octahedral field	Sum of degeneracies[b]
S	A_{1g}	1
P	T_{1g}	3
D	$E_g + t_{2g}$	5
F	$A_{2g} + T_{1g} + T_{2g}$	7
G	$A_{1g} + E_g + T_{1g} + T_{2g}$	9
H	$E_g + {}^2T_{1g} + T_{2g}$	11
I	$A_{1g} + A_{2g} + E_g + T_{1g} + {}^2T_{2g}$	13

[a]Ligand field states have the same multiplicity as the spectroscopic state from which they arise.
[b]Reflects the fact that A is singly degenerate, E is doubly degenerate, T is triply degenerate, etc.

Transitions between spectroscopic states having different multiplicities are spin forbidden. Therefore, with regard to spectral transitions, it is necessary to consider only those states that arise from the ground spectroscopic state. The reason for this is Hund's first rule that declares that states of maximum multiplicity lie lowest in energy (see Section 5.4). As a result, the energies of those states in terms of the ligand field splitting (Δ_o or Dq) are shown in Table 12.3.

Although the spectroscopic states that exist in an octahedral field and their energies are listed in Table 12.3, it must be remembered that the *actual* energies depend on the magnitude of the ligand field splitting. Therefore, the *difference*

Table 12.3 Energies of states in an octahedral ligand field in terms of Δ_o

Ion	Ground state	Octahedral field states	Energies in terms of Δ_o and Dq
d^1	^2D	$^2T_{2g} + {}^2E_g$	$-(2/5)\Delta_o = -4\,\mathrm{Dq} + (3/5)\Delta_o = +6\,\mathrm{Dq}$
d^2	^3F	$^3T_{1g} + {}^3T_{2g} + {}^3A_{2g}$	$-(3/5)\Delta_o = -6\,\mathrm{Dq} + (1/5)\Delta_o$
			$= +2\,\mathrm{Dq} + (6/5)\Delta_o = +12\,\mathrm{Dq}$
d^3	^4F	$^4A_{2g} + {}^4T_{2g} + {}^4T_{1g}$	$-(6/5)\Delta_o = -12\,\mathrm{Dq} - (1/5)\Delta_o$
			$= -2\,\mathrm{Dq} + (3/5)\Delta_o = 6\,\mathrm{Dq}$
d^4	^5D	$^5E_g + {}^5T_{2g}$	$-(3/5)\Delta_o = -6\,\mathrm{Dq} + (2/5)\Delta_o = +4\,\mathrm{Dq}$
d^5	^6S	$^6A_{1g}$	0
d^6	^5D	$^5T_{2g} + {}^5E_g$	$-(2/5)\Delta_o = -4\,\mathrm{Dq} + (3/5)\Delta_o = +6\,\mathrm{Dq}$
d^7	^4F	$^4T_{1g} + {}^4T_{2g} + {}^2A_{2g}$	$-(3/5)\Delta_o = -6\,\mathrm{Dq} + (1/5)\Delta_o$
			$= +2\,\mathrm{Dq} + (6/5)\Delta_o = +12\,\mathrm{Dq}$
d^8	^3F	$^3A_{2g} + {}^3T_{2g} + {}^3T_{1g}$	$-(6/5)\Delta_o = -12\,\mathrm{Dq} - (1/5)\Delta_o$
			$= -2\,\mathrm{Dq} + (3/5)\Delta_o = +6\,\mathrm{Dq}$
d^9	^2D	$^2E_g + {}^2T_{2g}$	$-(3/5)\Delta_o = -6\,\mathrm{Dq} + (2/5)\Delta_o = +4\,\mathrm{Dq}$
d^{10}	^1S	^1A_g	0

Ligand field states are listed in order of increasing energy and the energies are listed in the same order.

in energy between any states depends on the actual complex being considered. Note, however, that if the energy of the states for any given ion is considered the *net* energy is zero. For example, a d^1 ion has a ground state of ^2D, which is split into $^2T_{2g}$ and ^2E_g in an octahedral field. These states have energies of -4 Dq and $+6$ Dq, respectively. Thus it can be seen that

$$3(-4Dq) + 2(+6Dq) = 0$$

so the net change in energy is zero and the center of gravity (barycenter) of the orbitals in maintained. For a d^2 ion, the ground state is ^3F so the ligand field states are $^3T_{1g} + {}^3T_{2g} + {}^3A_{2g}$ that have energies of -6 Dq, $+2$ Dq, and $+12$ Dq, respectively. Taking into account these energies and the multiplicities of the states gives

$$3(-6Dq) + 3(+2Dq) + 1(+12Dq) = 0$$
$$\quad (T) \qquad\qquad (T) \qquad\qquad (A)$$

It should be noted from Table 12.3, that all d^n configurations except d^5 and d^{10} give rise to ground states that are either D or F in designation. For d^5 and d^{10} ions, the ground state term is an S term and the ligand field term is an A term for which there are no higher energy states having the same multiplicity. Thus, there are no spin-allowed transitions for d^5 or d^{10} ions in octahedral complexes. A well-known illustration of this is that complexes containing Fe^{3+} or Zn^{2+} are almost colorless in aqueous solutions. The information shown in Table 12.3 can be summarized in graphical form in what are known as Orgel diagrams (Orgel, 1955). Simplified diagrams of this type in which only the states of maximum multiplicity are considered are shown in Figs. 12.6 and 12.7.

An octahedral complex with six identical ligands has a center of symmetry. Therefore, the ligand field states have the subscript "g" to denote this. A tetrahedral structure does not have a center of symmetry, so the g subscript is not appropriate. With both sides of the diagrams applying equally to tetrahedral and octahedral complexes, g is included on one side, but not the other.

Because the d energy states are inverted in the tetrahedral field so are the spectroscopic states. Moreover, the spectroscopic terms that arise for a d^1 configuration are identical to those from a d^9 case, those for d^2 and d^8 are identical, etc. Likewise, the states arising for a d^1 tetrahedral field are the same as those for a d^4 ion in an octahedral field. The Orgel diagrams are constructed in such a way to represent all of the states. One side applies to octahedral d^1 and d^6 cases, whereas the other represents d^4 and d^9 cases. Because of the inversion of the energy levels and the resulting ligand states, the two

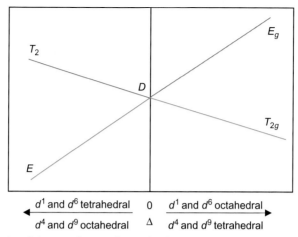

Fig. 12.6 An Orgel diagram for metal ions that have D spectroscopic ground states. The multiplicity of the D state of the free ion is not specified because it is determined by the number of electrons in the d orbitals of the metal ion.

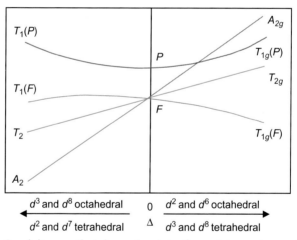

Fig. 12.7 An Orgel diagram that shows the states for metal complexes with F ground states. The multiplicity depends on the specific metal ion and is not specified, but the P state has the same multiplicity. Note that the difference between the energy of the $T_{1g}(F)$ and $T_{1g}(P)$ states gets greater as the value of Δ increases, and so states of identical designation do not cross. This is known as the *noncrossing rule*.

halves of the diagram are reversed with respect to tetrahedral complexes. Thus, the diagrams represent all of the d^n cases for both tetrahedral and octahedral complexes. Note that the multiplicity is not indicated because it is dependent on the number of electrons.

Table 12.4 Transitions allowed for complexes with metals having F ground states

Octahedral field				Tetrahedral field			
For A ground states		**For T ground states**		**For A ground states**		**For T ground states**	
ν_1	$A_{2g} \rightarrow T_{2g}$	ν_1	$T_{1g} \rightarrow T_{2g}$	ν_1	$A_2 \rightarrow T_2$	ν_1	$T_1 \rightarrow T_2$
ν_2	$A_{2g} \rightarrow T_{1g}$	ν_2	$T_{1g} \rightarrow A_{2g}$	ν_2	$A_2 \rightarrow T_1$	ν_2	$T_1 \rightarrow A_2$
ν_3	$A_{2g} \rightarrow T_{1g}(P)$	ν_3	$T_{1g} \rightarrow T_{1g}(P)$	ν_3	$A_2 \rightarrow T_1(P)$	ν_3	$T_1 \rightarrow T_1(P)$

The Orgel diagrams can be considered as energy level diagrams that show the energies as the value of Δ increases as it goes in either direction from the center. Because all of the states have the same multiplicity, the allowed transitions are those between the states that are encountered when an appropriate value of Δ is known and one progresses vertically. Note that for ions having an A or F ground state, it is expected that the spectrum would consist of three bands. Table 12.4 shows the transitions that would be allowed in both octahedral and tetrahedral fields.

Although the interpretation of spectra of complexes with regard to the types of transitions has been shown, there is more to the analysis of such spectra. One of the objectives is to determine the value of the ligand field splitting Δ. It is from the analysis of spectra that this parameter is obtained. Another issue with regard to the effects of ligands on the distribution of electrons in complexes is that the field generated by the ligands changes spectroscopic parameters. For example, the difference between the ^3P and ^3F states in a d^2 metal ion is defined in terms of a quantity known as B, which is an energy integral. By definition, that energy difference is $15B$, and the value of B can be determined from atomic spectra. Some metal ions that have a d^2 configuration are Ti^{2+}, V^{3+}, and Cr^{4+} and the energies of the spectroscopic states are shown in Table 12.5.

Table 12.5 Energies of spectroscopic terms for gaseous d^2 ions

	Energies of states in cm^{-1}		
Term	Ti^{2+}	V^{3+}	Cr^{4+}
^3F	0	0	0
^1D	8473	10,540	13,200
^3P	10,420	12,925	15,500
^1G	14,398	17,967	22,000

The ground state is ^3F and 349.8 cm^{-1} = 1 kcal mol^{-1} = 4.184 kJ mol^{-1}.

From the data shown in Table 12.5, it can be seen that the value of B varies with the charge on the ion and has values that range from approximately 600 to 1000 cm^{-1}. It is usually found that for a specific metal ion, the value of B in complexes is approximately 70–75% of its value for the gaseous ion. This is the result of an expansion of the electron cloud of the metal ion and is called the *nephelauxetic effect*. It is also possible to determine the value of B that is appropriate for a given complex from an analysis of the spectrum, but such work is beyond the scope of this book. There are several procedures for performing the analysis (House, 2013; Dou, 1990). The most comprehensive reference on the subject is that by Lever (Lever, 1984).

It was stated earlier in this chapter that except for a d^1 metal ion, the interpretation of spectra for complexes is complicated by spin-orbit coupling. Moreover, the magnitude of the effects is dependent on the strength of the ligand field, and thus parameters such as the energy needed to force pairing of electrons in metal ions depends on the environment. All of these complications can be dealt with, though, and the spectra can be analyzed to obtain the value of the ligand field splitting and other parameters (House, 2013).

12.4 CHARGE TRANSFER ABSORPTION

The discussion of spectra of complexes has dealt with the electronic absorptions that involve transitions between the various spectroscopic states that arise from the d orbitals in a ligand field environment. However, for complexes that contain appropriate combinations of metal ions and ligands, there are other types of transitions. The metal may have some of the e_g or t_{2g} orbitals that are nonbonding in character, but which may be only partially filled. For some ligands, bonding to the metal is the result of their donation of a pair of electrons, but the ligands may have additional electrons that can be donated to empty orbitals on the metal. Perhaps a more common situation is that ligands such as CO, CN$^-$, or ethylene have empty antibonding orbitals that are able to accept electron density from populated metal nonbonding orbitals. The shifting of electron density by either of these processes is referred to as *charge transfer*, and the spectral bands that are the result of these transitions are called *charge transfer* (CT) bands.

Bands that arise from charge transfer transitions are normally observed in the high-energy region of the spectrum. As a result, these bands are found in the ultraviolet region or the high-frequency range of the visible region. The charge transfer bands are normally broad and may overlap with bands that accompany the d-d transitions in complexes. In most cases, transitions that involve charge transfer are not spin forbidden, so they give rise those bands that are intense.

The type of charge transfer that occurs depends on factors such as the oxidation state of the metal and/or the availability of low energy orbitals on the ligands that can accept electron density from the metal. In a complex in which the metal ion is in a low oxidation state and another higher oxidation state exists, it is possible for electron density to move from the metal to the ligands. Such a transition is designated as a $M \rightarrow L$ type, with the arrow indicating the direction of electron movement. Transitions of this type do not normally occur in complexes containing Cr^{3+}, Co^{3+}, or Fe^{3+}.

One complex in which the $M \rightarrow L$ transition *does* occurs is chromium hexacarbonyl $Cr(CO)_6$, in which the Cr is in the zero oxidation state. As a result of having accepted six pairs of electrons from the six CO ligands, it is rather easy to have some electron density transferred to the empty π^* orbitals on the CO ligands. Because the transition involves electrons in t_{2g} orbitals on the metal, the transition is sometimes labeled as a $t_{2g} \rightarrow \pi^*$ transition. Another type of charge transfer transition involves the transfer of electrons in e_g^* orbitals on the metal being transferred to the π^* orbitals on the ligands.

In MnO_4^-, manganese is in the +7 oxidation state. As a result, the charge transfer band that is observed in the spectrum for MnO_4^- involves electron density being transferred to Mn from filled p orbitals on the oxygen atoms in an $L \rightarrow M$ transition. This results in an intense band centered at approximately 18,000 cm^{-1}, and it is this band that is responsible for the deep purple color of solutions containing permanganate ion.

With regard to charge transfer transitions, it is generally observed that:

1. A charge transfer of the type $M \rightarrow L$ will occur if the metal is in an oxidation state that enables it to be easily oxidized.
2. A charge transfer of the type $L \rightarrow M$ will occur if the metal is in an oxidation state that enables it to be easily reduced.

In addition to these considerations regarding the nature of the metal, it is also important to note that the ease with which electrons can be moved to or from the ligands (i.e., reduction or oxidation) is a factor. Transitions of the first type are indicated on the molecular orbital diagram for an octahedral complex shown in Fig. 12.8.

12.5 BACK DONATION

When ligands donate pairs of electrons to metal atoms or ions, the metal acquires a negative formal charge. For example, if a +3 metal accepts six pairs

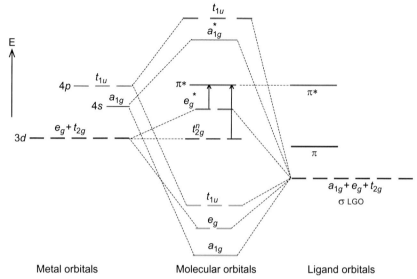

Fig. 12.8 Interpretation of M→L charge transfer absorption in an octahedral complex using a modified molecular orbital diagram. The transitions are from e_g^* or t_{2g} orbitals on the metal to π^* orbitals on the ligands.

of electrons, it acquires a −3 formal charge. A metal ion having a +2 charge would acquire a −4 formal charge by accepting six pairs of electrons when forming a complex with six ligands. In most cases, there are some electrons in the nonbonding d_{xy}, d_{xz}, and d_{yz} orbitals. In order to relieve part of the negative charge that accumulates on the metal, there may be some shift of electron density from the metal to the ligand. This phenomenon is known as *back donation* (also sometimes called back bonding) However, the ligand must have orbitals of suitable symmetry to interact with the d_{xy}, d_{xz}, and d_{yz} orbitals on the metal, and many do.

The molecular orbitals that result from the combination of p orbitals on diatomic species such as CO, CN⁻, and NO⁺ are shown in Fig. 12.9.

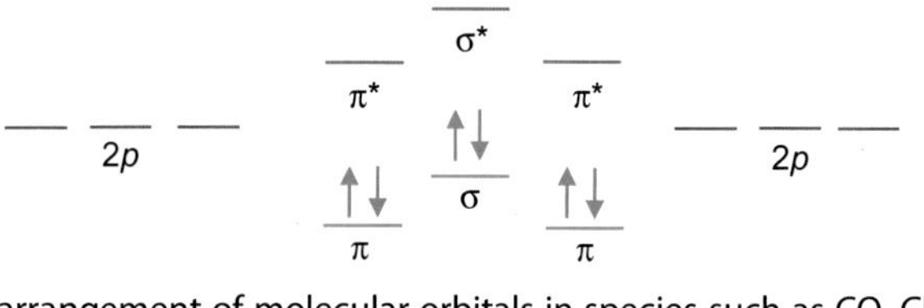

Fig. 12.9 The arrangement of molecular orbitals in species such as CO, CN⁻, and NO⁺.

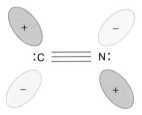

Fig. 12.10 The orientation of a π^* orbital on CN^-.

Although the π and σ orbitals are filled, the π^* orbitals are empty. In order to see how the π^* orbitals might interact with the d_{xy}, d_{xz}, and d_{yz} orbitals on the metal, it is essential to understand the orientation of the π^* orbitals. Although the cyanide ion is shown, the π^* orbitals illustrated in Fig. 12.10 are identical for isoelectronic species such as CO, CN^-, and NO^+.

The nature of the interaction of such antibonding orbitals with non-bonding d orbitals on the metal is illustrated in Fig. 12.11. It can be seen that the symmetry of the orbitals allows for effective overlap, which makes back donation possible.

From the molecular orbital diagram shown in Fig. 12.9, it can be seen that the electron density flowing from the metal to the CN^- ligand must enter the π^* orbitals. The result is that the bond order of the $C\equiv N$ bond is reduced slightly. The bond order is given by

$$\text{Bond order} = \frac{N_b - N_a}{2} \tag{12.1}$$

As a result, any population in the antibonding orbitals reduces the bond order. This weakening of the bond in the cyanide ion causes its stretching

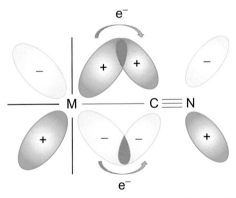

Fig. 12.11 The d_{xz} orbital and its overlap with the π^* orbital on CN^-.

frequency to be shifted to slightly lower wave numbers with the magnitude of the shift related to the extent of back donation. As a result of the metal acquiring a higher negative formal charge in $[Fe(CN)_6]^{4-}$ than in $[Fe(CN)_6]^{3-}$, it would be expected that the $C\equiv N$ stretching band would occur at a lower wave number. This is in fact observed, and the positions of the stretching bands are 2098 and 2135 cm^{-1}, respectively. A similar situation is found with the metal carbonyls. The greater the number of CO groups attached, the greater the negative formal charge on the metal and the greater the extent of back donation. For carbonyls of some first transition series metals, the CO stretching bands are as follows in cm^{-1}: $Ni(CO)_4$, 2057; $Fe(CO)_5$, 2034; and $Cr(CO)_6$, 1981.

Back donation also occurs with ligands such as olefins that have antibonding orbitals that match the symmetry of the nonbonding d orbitals of transition metals. One of the most familiar cases is that of the compound known as Zeise's salt, which is $K[Pt(C_2H_4)Cl_3]$. The anion contains an ethylene molecule that is perpendicular to the plane containing the platinum and chloride ions. In that case, the back donation can be shown as in Fig. 12.12.

Gaseous ethylene has the $C=C$ stretching band at 1623 cm^{-1}, whereas the corresponding band in Zeise's salt is found at 1526 cm^{-1}. This is a reflection of the decrease in bond order and strength of the $C=C$ bond. There are numerous other ligands that can accept electron density from nonbonding metal orbitals. For example phosphines (PR_3) have vacant d orbitals on the phosphorus atom that can interact with d orbitals on metals.

Although considerable information is obtained from studying the visible spectra of metal complexes, many fruitful studies have been carried out for which the object was to study vibrations of the ligands. In this way, it is sometimes possible to deduce the bonding mode of a ligand such as CN^-, which might bond through either the carbon or nitrogen atom. It is also possible to study many types of metal–ligand bonds by means of infrared spectroscopy. Truly, it can be said that most of what we know about the

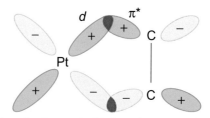

Fig. 12.12 The back donation from a d orbital on platinum to the π^* orbital on ethylene.

structure of atoms and molecules has been obtained by studying their interaction with electromagnetic radiation.

LITERATURE CITED

Dou, Y. Equations for Calculating Dq and B. *J. Chem. Educ.* **1990**, *67*, 134.

House, J. E. *Inorganic Chemistry*, 2nd ed.; Academic Press/Elsevier: Amsterdam, 2013
   (chapters 16–18).

Lever, A. B. P. *Inorganic Electronic Spectroscopy*, 2nd ed.; Elsevier: Amsterdam, 1984.

Orgel, L. E. Spectra of Transition-Metal Complexes. *J. Chem. Phys.* **1955**, *23*, 1004–1014.

PROBLEMS

1. Values of B for the metal ions Cr^{3+}, V^{2+}, and Mn^{4+} are 918, 1064, and 766 cm^{-1}. Assign the values to the correct ions and explain your decisions.

2. By making use of the Orgel diagrams, tell what spectral transitions would be possible for the following complexes: (a) $[Ni(NH_3)_6]^{2+}$; (b) $[CrCl_6]^{3-}$; (c) $[TiF_6]^{3-}$; (d) $[FeF_6]^{4-}$; (e) $[Co(NH_3)_4]^{2+}$.

3. Consider the complexes $[Fe(CN)_6]^{4-}$ and $[Co(CN)_6]^{3-}$. Explain how you could distinguish between them on the basis of an infrared spectrum.

4. For which of the following would it be likely that an absorption due to charge transfer could be observed? Explain your answer. (a) $[Co(NH_3)_6]^{3+}$; (b) $[FeF_6]^{4-}$; (c) $[Co(CO)_3NO]$.

5. If the complex $[Ti(H_2O)_4]^{3+}$ could be studied, what spectroscopic transition(s) would be observed? What would be the approximate value for Dq?

6. The complex $[Zn(NH_3)_4]^{2+}$ is diamagnetic, but $[Ni(NH_3)_4]^{2+}$ is not. Explain this difference by means of orbital energy diagrams.

7. Solid AgCN shows a band for the $C\equiv N$ stretching at 2170 cm^{-1}, whereas in the complex $[Ag(CN)_2]^-$, it is found at 2135 cm^{-1}. Explain this difference. Where would the CN stretching absorption likely be observed for the complex $[Ag(CN)_3]^{2-}$? Explain your answer.

8. Explain why NO^+ is a better acceptor of back donation than is the iso-electronic N_2 molecule.

9. Determine the spectroscopic ground state for the Ni^{2+} ion. What spectral transitions would an octahedral complex of this ion exhibit? What transitions would be possible in a tetrahedral complex of this ion?

CHAPTER 13

Barrier Penetration

In the previous chapters, the application of quantum mechanics to some fundamental aspects of the behavior of particles, atoms, and molecules have been shown. Although these models are important in their own right, they also form the basis for the extension of quantum mechanics to more complex systems. Another of the basic applications of quantum mechanics involves barrier penetration. This model shows how certain changes occur, even if the system does not have sufficient energy to cause the changes to take place. Known as barrier penetration or tunneling, this application of quantum mechanics is relevant to several aspects of both chemistry and physics. Consequently, this chapter is devoted to the barrier penetration model and its applications.

13.1 THE PHENOMENON OF BARRIER PENETRATION

One of the most interesting differences between classical and quantum descriptions of the behavior of systems concerns the phenomenon of barrier penetration or tunneling. To introduce the model, consider a particle approaching a barrier of height U_0, as shown in Fig. 13.1. Suppose the particle has an energy E and the height of the barrier U_0 is such that $U_0 > E$. From the point of view of classical physics, the particle cannot *penetrate* the barrier, nor does it have enough energy to get *over* the barrier. Therefore, if the particle has an energy less than U_0, it will be reflected by the barrier. If the particle has an energy greater than the height of the barrier, it can simply pass over the barrier unimpeded. Because the particle with $E < U_0$ cannot get over the barrier and or penetrate it, the regions inside the barrier and to the right of it are forbidden to the particle in the classical sense.

According to quantum mechanics, the particle (behaving as a wave) is not only able to penetrate the barrier, it can also pass through it and appear on the other side. As a result of the moving particle having a wave character, it must have an amplitude that is nonzero. The probability of finding the particle is proportional to the square of the amplitude function. Therefore, all regions are accessible to the particle, even though the probability of it

Fundamentals of Quantum Mechanics
http://dx.doi.org/10.1016/B978-0-12-809242-2.00013-9

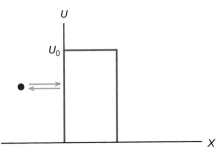

Fig. 13.1 A particle with energy E colliding with an energy barrier of height $U_0 > E$.

being in a given region may be very low. This is the result of the wave function for the moving particle not going to 0 at the boundary.

13.2 THE WAVE EQUATIONS

In treating the barrier penetration problem by quantum mechanics, it must be recognized that the entire area surrounding the barrier is accessible to the particle. Because there are three regions (inside the barrier and to the left and right of it), there will be three wave equations to solve.

It is natural to try to determine the connection between the wave inside the barrier and on either side of it. The connection must be a smooth, continuous one given by the restrictions on the wave function and its first derivative (see Chapter 2). The general form of the function can be determined by analogy to a simple example. Suppose light is shined into a solution, which absorbs it in direct proportion to the intensity of the light (see Fig. 13.2). Then, using k as the proportionality constant, the change in intensity (I) of the light with the distance it travels in the solution (x) is given by

$$-\frac{dI}{dx} = kI \tag{13.1}$$

Rearrangement of Eq. (13.1) gives

$$-\frac{dI}{I} = k\,dx \tag{13.2}$$

This equation can be integrated between limits of I_0 (the intensity of the incident beam) at distance 0 to some other intensity, I, after traveling a distance of x in the solution:

$$\ln\frac{I_0}{I} = kx \tag{13.3}$$

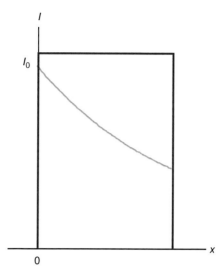

Fig. 13.2 Variation of beam intensity with penetration depth.

By taking the antilogarithm of both sides of the equation, it can be written as

$$I = I_0 e^{-kx} \tag{13.4}$$

From this equation, it can be seen that the beam decreases in intensity in an exponential way with the distance it has penetrated in the solution. This is, in fact, a form of Beer's law for absorption of light by solutions (see Chapter 11).

If a particle is treated as a wave that penetrates an energy barrier, it would naturally be expected that the "intensity" of the wave or probability of finding the particle at different distances into the barrier would vary with the distance of penetration in the barrier. Moreover, it should be expected that a smooth exponential decrease in the wave function should occur *within* the barrier, which is, in fact, the case. It should also be expected that the intensity of the wave (amplitude) would be greater *before* the particle penetrates the barrier than it would be *after* penetration. These aspects are shown in Fig. 13.3, which shows the barrier and defines the parameters used in solving the problem of barrier penetration. The book by Fermi shows a treatment of this problem that is similar to that presented here (Fermi, 1950).

There are three regions in which the particle may exist. The first task is to determine the wave function for the particle in each of those regions. The probability of finding the particle in a specific region is directly related to ψ^2 for the wave in that region. Because the wave function in region III is not

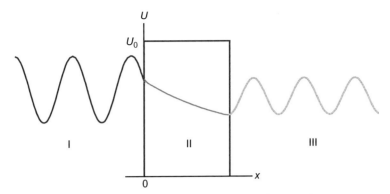

Fig. 13.3 A particle with energy E penetrating a potential energy barrier.

zero, there is a finite probability that the particle will be in that region after tunneling through the barrier. In treating the barrier penetration problem, the terminology used is very similar to that which describes absorbance of light by a solution. For example, it is common to speak of the *intensity* of the incident particle-wave in region I, as well as the intensity of the transmitted particle-wave in region III. In region II, there is *some* intensity of the penetrating particle-wave that decreases with distance. In this case, one does not speak of the "particle" penetrating the barrier. It is only because of the particle-wave duality as expressed by the de Broglie relationship (i.e., the wave character of a moving particle) that the penetration occurs. The particle cannot exist within the barrier in the classical sense.

Although a wave equation will be written for the particle in each region, it will not be necessary to solve all three of them in detail. It will be sufficient to write the solutions by comparing the equations to those that have already been considered.

In region I, the particle moving toward the barrier can be described by a wave equation similar to that presented earlier for the particle in a box. Therefore, the wave equation can be written as

$$\frac{d^2\psi_I}{dx^2} + \frac{2m}{\hbar^2}E\psi_I = 0 \tag{13.5}$$

Replacing $2\,mE/\hbar^2$ by k^2, the solution can be written directly as

$$\psi_I = Ae^{ikx} \tag{13.6}$$

However, part of the particle-wave will be reflected by the barrier, so it is necessary to add a correction term to the wave function to account for this reflection. The form of that term will be similar to that already written, except for the coefficient and the sign of the exponent. Therefore, the final form of the wave function for the particle in region I is

$$\psi_I = A e^{ikx} + B e^{-ikx} \tag{13.7}$$

in which A and B are the amplitudes of the incident and reflected waves, respectively. If A and B are real, B^2/A^2 gives the fraction of the particles (or fraction of the wave) that will be reflected by the barrier. If A and B are complex, B^*B/A^*A gives the fraction of the particles or wave reflected. In region III, the wave equation can be written in the same form as that for the particle-wave in region I:

$$\frac{d^2\psi_{III}}{dx^2} + \frac{2m}{\hbar^2} E\psi_{III} = 0 \tag{13.8}$$

The solution for this equation can be written as

$$\psi_{III} = J e^{ikx} \tag{13.9}$$

in which J is the amplitude of the wave in region III. Inside the barrier, region II, the wave equation can be written as

$$\frac{d^2\psi_{II}}{dx^2} + \frac{2m}{\hbar^2} E(U_0 - E)\psi_{II} = 0 \tag{13.10}$$

There are two components to ψ_{II} because the wave not only penetrates the left-hand side of the barrier, but it is also partially reflected by the right-hand surface of the barrier. Therefore, in region II, the wave function can be written as

$$\psi_{II} = K \exp\left[\left(\frac{2m}{\hbar^2}(U_0 - E)\right)^{1/2} x\right] + L \exp\left[-\left(\frac{2m}{\hbar^2}(U_0 - E)\right)^{1/2} x\right] \tag{13.11}$$

This equation can be simplified by letting

$$j = \left[2m(U_0 - E)/\hbar^2\right]^{1/2}$$

After this the wave function has the form

$$\psi_{II} = K e^{ijx} + L e^{-ijx} \tag{13.12}$$

The various constants are determined from the behavior of the wave function at the boundaries. At the boundaries $x = 0$ and $x = a$, both ψ and $d\psi/dx$ are continuous and have the same value because the wave functions for the particle-wave in adjacent regions must join smoothly. Therefore, this can be shown as

$$\psi_{\mathrm{I}}(0) = \psi_{\mathrm{II}}(0) \tag{13.13}$$

This condition leads to the result that

$$A + B = K + L \tag{13.14}$$

The equality of the derivatives of ψ_1 and ψ_2 at $x=0$ is expressed by the equation

$$\frac{d\psi_{\mathrm{I}}(0)}{dx} = \frac{d\psi_{\mathrm{II}}(0)}{dx} \tag{13.15}$$

Although the details will not be presented, it can be shown that

$$ikA - ikB = \sqrt{\frac{2m(U_0 - E)}{\hbar^2}}(K - L) \tag{13.16}$$

At the boundary a,

$$\psi_{\mathrm{I}}(a) = \psi_{\mathrm{II}}(a) \tag{13.17}$$

from which it possible to show that

$$K \exp \sqrt{\frac{2m(U_0 - E)}{\hbar^2}}a + L \exp \sqrt{\frac{2m(U_0 - E)}{\hbar^2}}a = J e^{ika} \tag{13.18}$$

Finally, because the derivatives of the wave functions must be equal,

$$\frac{d\psi_{\mathrm{II}}(a)}{dx} = \frac{d\psi_{\mathrm{III}}(a)}{dx} \tag{13.19}$$

it can be shown that

$$\sqrt{\frac{2m(U_0 - E)}{\hbar^2}} \left[\left(K \exp \sqrt{\frac{2m(U_0 - E)}{\hbar^2}} \right) - L \exp \left(-\sqrt{\frac{2m(U_0 - E)}{\hbar^2}} \right) \right]$$
$$= ikJ e^{ika} \tag{13.20}$$

Although the detailed steps will not be shown here, it is possible from these relationships to evaluate the constants.

The *transparency* of the barrier is given by the probability density in region III (J^2) divided by that in region I (A^2),

$$T = \frac{J^2}{A^2} = \exp\left(-2a\left[2m(U_0 - E)/\hbar^2\right]^{1/2}\right) \tag{13.21}$$

If the barrier is not rectangular, the barrier height, U, must be expressed as a function of distance x. Then the probability of barrier penetration is written in terms of the energy function $U(x)$ as

$$T \cong \exp\left(-\frac{2\sqrt{2m}}{\hbar}\int_0^a [U(x)-E]^{1/2}\,dx\right) \tag{13.22}$$

In this expression, the integral represents how much of the barrier lies higher than the energy of the particle.

Several conclusions can be reached immediately from the form of Eqs. (13.21) and (13.22). First, the transparency decreases as the thickness of the barrier a increases. Second, the transparency also decreases as the difference between the energy of the particle and the height of the barrier increases; that is, the transparency decreases when more of the barrier lies above the energy of the particle. When the energy of the particle is equal to the height of the barrier, the exponent becomes zero and the transparency is equal to 1, and all of the particles can pass over the barrier in classical behavior. Finally, it should also be clear that the transparency of the barrier decreases with increasing mass m of the particle. It should also be pointed out that if $h=0$ (which indicates that energy is not quantized, but is rather classical in behavior), the exponent becomes 0, and there is no possibility of the particle getting past the barrier. Therefore, as stated earlier, tunneling is a quantum mechanical phenomenon. Tunneling is much more significant for light particles, and the later sections of this chapter show applications of the barrier penetration model to cases involving electron tunneling. It should now be clear why the walls of the one-dimensional box (see Chapter 3) had to be made infinitely high in order to confine the particle to the box 100% of the time.

13.3 ALPHA DECAY

In 1926, Schrödinger's solution of the wave equation made it possible to exploit other applications of wave mechanics. For example, nuclear physicists had known that α decay occurs with the emitted α particle having an energy that is typically in the range 2–9 MeV (1 eV $= 1.6 \times 10^{-12}$ erg; 1 MeV $= 10^6$ eV). However, inside the nucleus the α particle (a helium nucleus) is held in a potential energy well caused by it being bound to other nuclear particles. Further, in order to cause an α particle to penetrate the nucleus from the *outside*, the α particle would need to overcome the Coulomb repulsion. For a heavy nucleus ($Z=80$), the Coulomb barrier

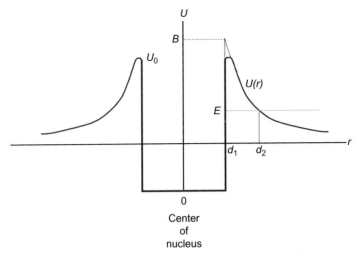

Fig. 13.4 The nuclear energy well and Coulomb barrier for alpha decay.

would be approximately 25 MeV in height. Therefore, in order for an α particle *inside* the nucleus to escape, it would need to have an energy of at least that magnitude. However, for many cases, the α particles have energies of only 5–6 MeV. Figure 13.4 shows the energy relationships and parameters needed for a discussion of this problem. The simple question is, how can the α particle be emitted with an energy of 5–6 MeV through a barrier that is as great as 25 MeV?

In 1928, only two years after Schrödinger's solution of the hydrogen atom problem, the problem of α decay was solved by Gurney and Condon (1928) and independently by Gamow (1928). It was assumed that the α particle (two protons and two neutrons) moves inside the nucleus, but is constrained by the potential barrier. Quantum mechanically, $\psi^*\psi$ predicts that there is *some* probability of finding the α particle on the outside of the barrier. Because the particle does not have sufficient energy to go *over* the barrier, it must escape by tunneling *through* the barrier.

The rate of decay of a nucleus can be expressed by the first-order rate equation

$$-\frac{dN}{dt} = kN \tag{13.23}$$

in which k is the *decay constant* and N is the number of nuclei. The rate of the decay process is reflected by the magnitude of k, which in turn is related to the transparency of the barrier. Therefore, the problem of explaining the observed decay constants for α decay can be solved if the transparency for the Coulomb barrier produced by the nucleus can be calculated.

Treating the problem of α decay involves using an appropriate expression for $U(r)$ in Eq. (13.22). If the potential well inside the nucleus is assumed to have a square-bottom and the shape of the barrier is given by a Coulomb potential outside the nucleus $U(r) = Zze^2/r$ (where Z is the charge of the daughter nucleus and z is the charge on the α particle), the integral, shown in Eq. (13.24) as I, becomes

$$I = \int_{d_1}^{d_2} \left(Zze^2 - Er\right)^{1/2} \frac{dr}{r^{1/2}} \tag{13.24}$$

After a change of variables, by letting $x = r^{1/2}$ and $q^2 = Zze^2/E$, the result is that $r = x^2$ and $dr = 2x\, dx$. Also, the integral can also be simplified by realizing that

$$\left(Zze^2 - Er\right)^{1/2} = \left(\frac{Zze^2 E}{E} - Er\right)^{1/2} = E^{1/2}\left(\frac{Zze^2}{E} - r\right)^{1/2} \tag{13.25}$$

Now Eq. (13.24) can be written as

$$I = 2\sqrt{E} \int_{d_1}^{d_2} \left(q^2 - x^2\right)^{1/2} dx \tag{13.26}$$

This integral is of a form that can be found in tables of integrals and its evaluation leads to

$$\int \left(a^2 - x^2\right)^{1/2} dx = \frac{1}{2}\left[x\left(a^2 - x^2\right)^{1/2} + a^2 \sin^{-1}\left(\frac{x}{z}\right)\right] \tag{13.27}$$

Therefore, the integral shown in Eq. (13.26) becomes

$$I = \sqrt{E}\left[x\left(q^2 - x^2\right)^{1/2} + q^2 \sin^{-1}\left(\frac{x}{q}\right)\right]_{d_1}^{d_2} \tag{13.28}$$

The values for d_1 and d_2 are related to the charges and energies by Coulomb's law so that

$$E = \frac{Zze^2}{d^2} \quad \text{and} \quad B = \frac{Zze^2}{d^1} \tag{13.29}$$

It is not necessary to show the complete derivation here, but it is possible to obtain an expression for the decay constant, k, which can be written as

$$k = \frac{\hbar}{2md_1^2} \exp\left[-\frac{8\pi Zze^2}{hv}\left(\cos^{-1}\left(\frac{E}{B}\right)^{1/2} - \left(\frac{E}{B}\right)^{1/2}\left(1 - \frac{E}{B}\right)^{1/2}\right)\right]$$

(13.30)

The symbols in this equation have already been explained except for v, the velocity of the α particle in the nucleus. This quantity comes into consideration because the number of times the particle moves back and forth in the nucleus and comes in contact with the barrier is related to the probability that it will eventually penetrate the barrier. Using appropriate expressions for the values of the parameters, the calculated decay constants are generally in excellent agreement with those observed experimentally. For example, the calculated and experimental values of k (in s^{-1}) for a few emitters of α particles are as follows: for ^{148}Gd, $k_{calc} = 2.6 \times 10^{-10}$, $k_{exp} = 2.2 \times 10^{-10}$; for ^{214}Po, $k_{calc} = 4.9 \times 10^3$, $k_{exp} = 4.23 \times 10^3$; and for ^{230}Th, $k_{calc} = 1.7 \times 10^{-13}$, $k_{exp} = 2.09 \times 10^{-13}$. As is apparent from the data, the application of the barrier penetration model to α decay has been quite successful.

13.4 TUNNELING AND SUPERCONDUCTIVITY

Although the model of tunneling of particles through barriers has been successfully applied to several phenomena, it is perhaps in the area of superconductivity that tunneling is most important. Certainly this is so in regard to technology, and a description of how this application of barrier penetration is so important will be presented. However, the discussion here will include only the rudiments of this important and timely topic. For a more complete discussion of superconductivity, consult the references at the end of this book, especially the works of Kittel (2005) and Serway and Jewett (2014).

If two metal strips are separated by an insulator, no electric current passes through the system in normal circumstances. The insulator acts as a barrier to the particles (electrons) in a manner analogous to that discussed in Sections 13.1 and 13.2. Often the barrier is an oxide layer on the surface of one of the metals. If the insulator is made sufficiently thin (1–2 nm), it is possible for electrons to tunnel through the barrier from one metal to the other. For metals not behaving as superconductors, the conductivity through the barrier follows Ohm's law, which indicates that the current is directly proportional to the voltage. This type of tunneling by electrons is known as *single-particle tunneling*.

In 1908, Dutch physicist Heike Kamerlingh Onnes liquefied helium (bp 4.2 K). This was an important event because the first superconductors studied did not become "super" conducting except at very low temperatures. Superconductivity was discovered in 1911 by Onnes. After studying the resistivity of platinum, mercury was studied because it could be obtained in very high purity. It was found that at 4.15 K (the temperature at which mercury becomes superconducting, T_C) the resistivity of mercury dropped to 0. In 1933, Walther Meissner and Ochsenfeld found that when certain types of superconductors were kept below their critical temperatures in a magnetic field, the magnetic flux was expelled from the interior of the superconductor. This behavior is known as the *Meissner effect*.

In 1957, Bardeen, Cooper, and Schrieffer developed a theory of superconductivity that is now known as the *BCS theory* (Bardeen et al., 1957). According to this theory, electrons are coupled to give pairs having a resultant angular momentum of 0. These electron pairs are known as *Cooper pairs*, and it is their characteristics that are responsible for some of the important properties of superconductors. In 1962, Josephson predicted that two superconductors separated by a barrier consisting of a thin insulator should be able to allow an electric current to pass between them due to tunneling of Cooper pairs. This phenomenon is known as the *Josephson effect* or *Josephson tunneling*. The existence of superconductors that have T_C values higher than the boiling point of liquid nitrogen is enormously important because liquid helium, which is much less readily available and expensive, is not required for cooling them.

The migration of electrons through a solid is impeded by motion of the lattice members as they vibrate about their equilibrium positions. In the case of a metal, the lattice sites are occupied by metal ions with electrons passing between them. In Chapter 6, it was shown that the frequency of vibrations is given by

$$\nu = \frac{1}{2\pi}\sqrt{\frac{k}{m}} \tag{13.31}$$

in which k is the *force constant* or *spring constant* and m is the mass. Assuming that the passage of electrons through the metal is impeded by the vibrational motion of the metal ions, it would be expected for a given metal that the T_C would vary with $1/m^{1/2}$, as is the case with classical conductivity. The lighter the atom, the greater the vibrational frequency and the greater the extent to which the motion of the electron would also be hindered. Experimental measurements of the T_C have been carried out for ^{199}Hg, ^{200}Hg, and ^{204}Hg, for which the T_C values are 4.161, 4.153, and 4.126 K, respectively. Figure 13.5 shows that a graph of T_C versus

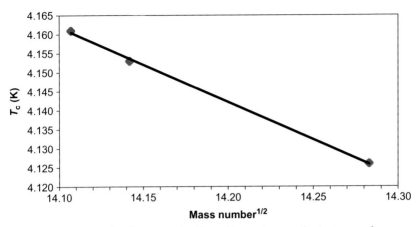

Fig. 13.5 The relationship between the T_C and isotopic mass for isotopes of mercury.

$m^{1/2}$ is linear, but it is the inverse of the expected relationship based on atomic mass. If it is presumed that a metal atom has a diameter that is on the order of approximately 10^{-8} cm in diameter and that the electron is moving with a linear velocity of about 10^8 cm s^{-1}, the electron will be in the proximity of the metal atom for about $(10^{-8}$ cm$)/(10^8$ cm s$^{-1})$, or a time on the order of 10^{-16} s. For many solids, the frequency of lattice vibrations is on the order of 10^{12} s^{-1} or the oscillation time is on the order of 10^{-12} s. Therefore, the time for lattice reorientation (based on mass number) is very long compared to the time it takes for the electron to pass a particular site in the lattice.

A very elementary view of a metal consists of metal ions at lattice sites with mobile electrons moving through the solid in conduction bands. The motion of Cooper pairs through a metal lattice is thought to be linked to the lattice motion just described. One view is that as one electron passes between two metal ions, the ions move slightly inward from their respective lattice sites. Thus, this region has an instantaneous increase in positive charge. This analogy is very similar to that of instantaneous dipoles in helium atoms, when the two electrons are found at some instant on the same side of the atom, thus giving rise to *London dispersion forces*. The increased positive region exerts an attractive force on a second electron, which follows the first through the opening between the metal sites before lattice reorganization occurs. The effect is that two electrons behave as a pair (the Cooper pair) having opposing spins, but exists as an entity having a resultant spin angular momentum of 0.

Tunneling in superconductors involves the behavior of Cooper pairs, which behave as bosons because of their resulting zero spin. Thus, unlike fermions, Cooper pairs are not required to obey the Pauli exclusion principle, and any number of Cooper pairs can populate the same state. The BCS theory incorporates the idea that all of the electrons form a ground state consisting of Cooper pairs. In the conductivity of normal metals, the lattice vibration of the atoms reduces the mobility of the electrons, thereby reducing conductivity. In the case of Cooper pairs, reducing the momentum of one pair requires the reduction of momentum for all the pairs in the ground state. Because this does not occur, lattice vibrations do *not* reduce the conductivity that occurs by motion of Cooper pairs. For normal conduction in metals, lattice vibrations decrease conductivity. For superconductivity, the lattice motion is responsible for the formation of Cooper pairs, which gives rise to superconductivity.

As a part of his description of Cooper pairs, Josephson predicted that two superconductors separated by a thin insulating barrier could experience pair tunneling. One result of this phenomenon would be that the pairs could tunnel without resistance, yielding a direct current with no applied electric or magnetic field. This is known as the *dc Josephson Effect*. Figure 13.6 shows the arrangement known as a Josephson junction that leads to the *dc Josephson Effect*.

Cooper pairs in one of the superconductors are described by the wave function ψ_1, and those in the other are described by the wave function ψ_2. The appropriate Hamiltonian can be written as

$$-\frac{\hbar}{i}\frac{\partial}{\partial t}\psi = \hat{H}\psi \tag{13.32}$$

so that when both superconductors are considered,

$$-\frac{\hbar}{i}\frac{\partial \psi_1}{\partial t} = \hbar T \psi_2 \quad \text{and} \quad -\frac{\hbar}{i}\frac{\partial \psi_2}{\partial t} = \hbar T \psi_1 \tag{13.33}$$

where T is the rate of current flow across the junction from each of the superconductors. If the insulator is too thick for tunneling to occur, then $T = 0$. It is possible to show that

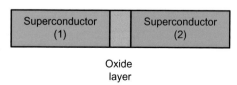

Fig. 13.6 A schematic diagram of a Josephson junction.

$$I = I_m \sin (\varphi_2 - \varphi_1) = I_m \sin \delta \qquad (13.34)$$

where I_m is the maximum current across the junction when there is no applied voltage. In Eq. (13.34), φ is the phase of the pair, and all pairs in a given superconductor have the same phase. When the applied voltage is zero, the current varies from I_m to $-I_m$, depending on the phase difference for the two superconductors. The dc Josephson Effect is one of the results of pair tunneling. If a dc voltage is applied across the junction between two super-conductors, an alternating current oscillates across the junction. It can be shown (Serway and Jewett, 2014) that the oscillating current I is expressed by

$$I = I_m \sin \left(\varphi(0) - \frac{2eVt}{\hbar} \right) \qquad (13.35)$$

where $\varphi(0)$ is a constant, the phase at time zero, V is the voltage, t is the time, and e is the electron charge. This phenomenon is referred to as *ac Josephson tunneling*.

When dc tunneling occurs in the presence of an external magnetic field, a periodic tunneling process occurs. Linking two of the Josephson junctions together in parallel allows an interference effect to be observed that is very sensitive to the magnetic field experienced by the system. Such a system is known as a *superconductivity quantum interference device* (SQUID), and such devices are used to detect very weak magnetic fields. For instance, such devices have been used to study the fields produced by neuron currents in the human brain.

Tunneling by Cooper pairs plays an important role in the behavior of superconductors. As superconductors having higher and higher T_C values are obtained, it is likely that this tunneling behavior will be exploited in technological advances.

13.5 THE SCANNING TUNNELING MICROSCOPE

An important application of tunneling involves the *scanning tunneling micro-scope* (STM) invented by Binnig, Rohrer and coworkers in 1981 (Binnig et al., 1981). In this case, electrons tunnel between the surface of a solid, and the tip of a probe maintained at a very short distance from the surface. A typical gap of a few angstroms is maintained by a feedback loop that main-tains a constant current flow between the solid surface and the tip of the movable probe by adjusting the distance between them (i.e., the height of the barrier). A simple schematic diagram of a STM is shown in Fig. 13.7. The tunneling occurs because the wave functions for electrons

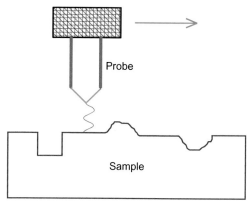

Fig. 13.7 Schematic diagram of a scanning tunneling microscope.

in the solid do not end abruptly at the surface of the solid, but rather extend into the space above the surface of the solid. This behavior is similar to that of the harmonic oscillator, in which the wave functions do not end at the classical boundaries of the oscillation. Moreover, a particle in a one-dimensional box would not give wave functions that terminate at the walls of the box if the walls were not infinitely high.

As the tip of the probe moves across the surface of the solid, its vertical motion gives a plot of the surface features in the form of a contour map. The resolutions of features are on the order of 2 Å, but surface height differences of about 0.01 Å can be detected. Because the electron wave functions extend into space above the solid, the STM can literally map the wave functions of surface atoms and "see" individual atoms. Therefore, the STM can be used to locate the active sites on catalyst surfaces, and it has enabled surfaces of materials ranging from metals to viruses to be mapped.

13.6 SPIN TUNNELING

Another observation of tunneling involves that of the tunneling of electron spins through the potential barrier that separates one spin orientation from the other. In this case, a solid complex compound of manganese having the complete formula $[Mn_{12}(CH_3COO)_{16}(H_2O)_4O_{10}] \cdot 2CH_3COOH \cdot 4H_2O$ was studied in a magnetic field at low temperature (Swarzchild, 1997; Thomas et al., 1996). The molecule has a total spin state of 10 that is spread over the 12 manganese ions. In the absence of a magnetic field, there are two sets of energy levels that involve the orientations from +10 to −10. There are a total of 21 different possible orientations, $0, \pm 1, \ldots, \pm 10$, and the crystal of the compound is not isotropic in its magnetization. Thus, the energies

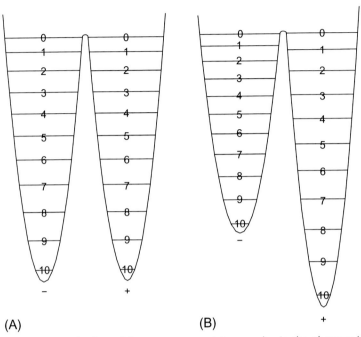

Fig. 13.8 Spin states in the solid manganese-acetate complex in the absence (A) and presence (B) of a magnetic field.

associated with the orientations of the spin vectors have different values and constitute two identical sets of levels, as shown in Fig. 13.8. When the crystal is placed in a magnetic field that is applied in the positive direction (parallel to the + crystal axis direction), the states in the parallel direction decrease in energy, whereas those in the opposite direction increase in energy (see Fig. 13.8).

Tunneling of spins between the two sets of levels occurs at very low temperatures (2−3 K) as *resonant tunneling*, under the conditions where the states are at the same energy (no magnetic field applied). If the magnetic field is applied, it is possible to cause the energy levels to change so that some of the states are again at the same energy (a sort of accidental degeneracy, but one that depends on an external field). The states that are at the same energy under these conditions do not have the same numerical value of the spin quantum number. For example, the +3 state may reside at the same energy as the −4 state, as shown in Fig. 13.8.

If all the spins are forced to the −10 state by applying an external field, decreasing the field will allow resonant tunneling at some specific magnitude of the applied field as some of the levels are again brought to the same energy. Therefore, when the property of magnetization is studied at various values of the applied external field, the hysteresis loop, unlike the classical plot of magnetization versus applied field, shows a series of steps. These steps are observed because quantum mechanical tunneling is occurring. A temperature effect of this phenomenon indicates that the spins must be populating states of about $m = 3$ by thermal energy. Otherwise, the tunneling times from a potential well as deep as those of the $m = 10$ states would be extremely long as a result of transition probabilities that are very low. The steps in the hysteresis plot clearly show that quantum mechanical tunneling occurs as electrons go from one spin state to another without passing over the barrier between them. At higher temperatures, the electrons can pass over the barrier without tunneling being necessary.

13.7 TUNNELING IN AMMONIA INVERSION

The pyramidal ammonia molecule has associated with it a vibration in which the molecule is "turned inside out." This vibration, known as *inversion*, is shown in Fig. 13.9, and it has a frequency on the order of 10^{10} s^{-1}.

Excitation of the vibration from the first to the second vibrational energy level gives rise to an absorption at 950 cm^{-1}, and the barrier height is 2076 cm^{-1}. According to the Boltzmann distribution law, the ratio of population of the molecules in the lowest two levels (n_0 and n_1) should be given by

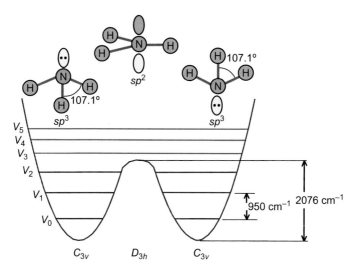

Fig. 13.9 The inversion of the ammonia molecule.

$$\frac{n_1}{n_0} = e^{-\Delta E/RT} \tag{13.36}$$

where ΔE is the difference in energy between the two states, T is the temperature (K), and R is the molar gas constant. In this case,

$$E = hc\bar{\nu} = \left(6.63 \times 10^{-27} \text{ergs}\right) \times \left(3.00 \times 10^{10} \text{cms}^{-1}\right) \times \left(950 \text{cm}^{-1}\right)$$
$$= 1.89 \times 10^{-13} \text{erg}$$

This energy per molecule can be converted to $J \, mol^{-1}$ by multiplying by Avogadro's number and dividing by $10^7 \, \text{erg} \, J^{-1}$. In this case, the energy is 11,400 J mol^{-1}. Therefore, at room temperature (taken to be 300 K),

$$\frac{n_2}{n_0} = e^{(-11,400 \text{J/mol})/[(8.3144 \text{J/molK}) \times (300 \text{K})]} = 0.0105 \tag{13.37}$$

Consequently, almost all of the molecules would be expected to be in the lowest vibrational level. Because inversion is rapid even though the barrier height is 2076 cm^{-1}, the molecules must invert by tunneling through the relatively low and "thin" barrier that will be somewhat transparent. Although it will not be discussed here, it is interesting to note that the inversion of phosphine, PH_3, is not rapid.

The de Broglie hypothesis regarding the wave nature of a moving electron was applied to the hydrogen atom by Schrödinger in 1926, and the diffraction of an electron beam was demonstrated experimentally in 1927. Very soon thereafter, various applications of quantum mechanics were explored, and tunneling was used as a model for alpha decay by Gurney and Condon and also by Gamow in 1928. Consequently, tunneling has been a viable and important model for quantum phenomena almost since the beginning of the application of quantum mechanics to atomic physics. A somewhat unique demonstration of this type of behavior is the tunneling between spin states first described in 1996. From these and other applications, it should be apparent that tunneling needs to be discussed along with the harmonic oscillator, rigid rotor, particle in a box, and other topics in the study of quantum mechanical models.

LITERATURE CITED

Bardeen, J.; Cooper, L. N.; Schrieffer, J. R. Theory of Superconductivity. *Phys Rev.* **1957**, *108*, 1175–1204.

Binnig, G.; Rohrer, H.; Gerber, C.; Wiebel, E. Tunneling Through a Controllable Vacuum Gap. *Appl Phys Lett.* **1981**, *40*, 178–180.

Fermi, E. *Nuclear physics;* University of Chicago Press: Chicago, 1950.

Gamow, G. Zur Quantentheorie des Atomkerns. *Z Phys.* **1928**, *51*, 204–212.

Gurney, R. W.; Condon, E. U. Wave Mechanics and Radioactive Disintegration. *Nature* **1928**, *122*, 439.

Kittel, C. *Introduction to Solid State Physics*, 8th ed.; Wiley: New York, 2005 (chapter 12).

Serway, R. E.; Jewett, J. E. *Physics for Scientists and Engineers*, 9th ed.; Brooks/Cole: Boston, MA, 2014.

Swarzchild, B. Hysteresis Steps Demonstrate Quantum Tunneling of Molecular Spins. *Phys. Today* **1997**, *50*, 17–19.

Thomas, L.; Lionti, F.; Ballou, R.; Gatteschi, D.; Sessoli, R.; Barbara, B. Macroscopic Quantum Tunneling of Magnetization in a Single Crystal of Nanomagnets. *Nature* **1996**, *383*, 145–147.

PROBLEMS

1. Suppose a rectangular potential barrier of height 2.0 eV and thickness 10^{-8} cm has an electron approach it. If the electron has an energy of 0.25 eV, what is the transmission coefficient? If the electron has an energy of 0.50 or 0.75 eV, what are the transmission coefficients?

2. Repeat Problem 1 for a neutron approaching the barrier.

3. Repeat Problem 1 for a helium atom approaching the barrier.

4. If an electron having a kinetic energy of 25 eV approaches a rectangular barrier that is 10^{-8} cm thick and has a height of 35 eV, what is the probability that the electron will penetrate the barrier? What will be the probability of the electron penetrating the barrier if the thickness is 10^{-7} cm?

5. Repeat the calculations in Problem 4 if a proton with an energy of 25 eV approaches the barrier.

6. Suppose that an electron with an energy of 10 eV approaches a rectangular barrier of 10^{-8} cm thickness. If the transmission coefficient is 0.050, how high is the barrier?

7. Consider a potential barrier represented as follows:

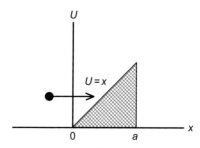

Determine the transmission coefficient as a function of particle energy.

8. An electron with an energy of 15 eV impinges on a barrier 25 eV in height.
   (a) Determine the probability of the electron tunneling through the barrier if the barrier is 1.20 nm thick.

(b) What will be the probability that the electron will tunnel through the barrier if its it is 0.120 nm thick?

9. Suppose a proton is bound in a nucleus in a potential that is approximately a square well with walls that are infinitely high. Calculate the wavelength and energy for the emitted photon when the proton falls from $n=3$ to the $n=2$ state, if the nucleus has a diameter of 1.50×10^{-13} cm.

CHAPTER 14

Comments on Computational Methods

In any area of science in which the results are important, there will be a great deal of original thinking and research. Such is the case for molecular orbital calculations. As a result, a large number of approaches to performing molecular orbital calculations exist. Moreover, they change rapidly as new algorithms (and sophisticated computers) become available, so a decade is more like two or three generations with regard to the complexity of the calculations. These are known by various acronyms that may be unintelligible to all except those who have devoted serious study to the field. It is not possible in one small book (or even one large one, for that matter) to present the details of this enormous body of knowledge. Consequently, in this chapter the attempt will be to present an overview of the language and a qualitative understanding of some of the most important techniques in this important field.

Although the simple Hückel molecular orbital method (HMO) described in the previous chapter is useful for some purposes, that type of calculation is quite limited and is most applicable to organic molecules. More sophisticated types of calculations require enormously more complex computational techniques. As a result, the developments in the field of molecular orbital calculations have paralleled the developments in computers. Computer software is now routinely available from several sources, enabling persons who do not necessarily understand all of the theory or computational techniques to use the software and be guided through the process of performing high-level molecular orbital calculations that would have represented the frontier of the field not many years ago. Because of the computer system requirements, specific instructions depend on the type of computer equipment on which the calculations are to be performed. In view of these aspects of molecular orbital methods, the discussion in this chapter will be limited to presenting basic principles and nomenclature.

14.1 THE FUNDAMENTAL PROBLEM

When dealing with a system consisting of many particles, the Schrödinger equation in operator form

Fundamentals of Quantum Mechanics
http://dx.doi.org/10.1016/B978-0-12-809242-2.00014-0
335

$$\hat{H}\psi = E\psi \tag{14.1}$$

becomes

$$\sum_{1}^{N} \nabla_i^2 \psi + \frac{2m}{\hbar^2} \left(\sum_{i=1}^{N} \frac{Ze^2}{ri} - \sum_{i,j>i} \frac{e^2}{r_{ij}} + V \right) V = 0 \tag{14.2}$$

Even for the two-electron helium atom, exact solution of the Schrödinger equation was not possible, so it should be clear that exact solutions for multi-electron atoms or polyatomic molecules will be likewise prohibited. Accordingly, the various (and numerous) types of molecular orbital calculation are based on approximations. One of the approaches to this commonly occurring problem in science and engineering involves replacing part of the equation with an approximation so that an *exact* solution can be obtained for the *approximate* equation. The other approach is to write an equation that can be solved *exactly,* though it represents the true nature of the system only *approximately.*

14.2 THE BASIS SET

Most of the many molecular orbital calculations make use of some type of atomic wave functions that generally describe a single electron as a starting point. The *basis set* is the set of one-electron wave functions that are somehow combined to give the molecular wave functions. The minimal basis set is the set that incorporates only the orbitals actually populated by electrons. One of the most widely used basis sets consists of Slater-type orbitals (STO) having the form

$$\psi(r) = r^{n-1} e^{-(Z-s)r/n} \tag{14.3}$$

where s is a screening constant, n is a number that varies with the type of orbital, and Z is the nuclear charge. The value of n is determined according to procedures described in Section 5.3. Using one STO wave function for each nucleus and constructing molecular wave functions by taking linear combinations of the atomic orbitals, the molecular calculation is referred to as the *minimal basis set calculation.* The quantity $(Z-s)$ is replaced by ζ (zeta) to give functions written in the form

$$\psi(r) = r^{n-1} e^{-\zeta r/n} \tag{14.4}$$

The wave functions of this type are referred to as single-ζ STO functions. An additional modification involves representing each atomic wave function by two STO wave functions. In this case, the wave functions are referred to as double-ζ functions. Within the framework of double-ζ wave functions, there are several variations that result from the parameterizations that have been developed.

Another type of function used to represent atomic wave functions is known as a Gaussian. Gaussian functions have the general form

$$\psi(r) = e^{-\alpha r^2} \tag{14.5}$$

in which α is an adjustable (or "best-fit") parameter. These functions are combined as linear combinations in order to approximate the STO functions. The motivation for this is that computations of the integrals involved in the quantum mechanical calculations are greatly facilitated. The linear combination of Gaussian functions is referred to as a *contracted Gaussian function* or Gaussian-type orbitals (GTO). All of these manipulations of STO and GTO are carried out to provide approximations to the radial portions of the atomic wave functions, and the complete wave functions are obtained by making use of the spherical harmonics, $Y_{l,m}(\theta,\varphi)$, to provide the angular dependence. When each STO is represented by a linear combination of three Gaussian functions, the result is known as STO-3G. Other combinations of wave functions lead to the 6-31G designation, in which each STO is represented as a linear combination of six Gaussian functions. Further, each STO representing valence shell orbitals is a double-ζ function, with the inner part represented by a linear combination of three Gaussian functions and the outer part by one such function. Although the description of the types of functions present is by no means complete, it does show that many creative mathematical approaches have been utilized.

14.3 THE EXTENDED HÜCKEL METHOD

In Chapter 9, we illustrated the application of the Hückel method to a variety of problems. In 1963, Roald Hoffmann (Nobel prize, 1981) devised a molecular orbital method that has come to be known as the *extended* Hückel molecular orbital method (EHMO) (Hoffmann, 1963). This method has several differences from the basic Hückel method. For organic molecules, the basis set of carbon $2s$ and $2p$ and hydrogen $1s$ orbitals is used in the calculations. Although, the overlap was neglected in the Hückel method, it is explicitly included in the EHMO procedure. All overlap integrals S_{ij} must be calculated (see Section 8.5). Because the atomic wave functions are normalized, the S_{ii} integrals are equal to 1. The values of the overlap integrals depend on bond distances and bond angles. Therefore, the relative positions of the atoms must be known before the overlap integrals can be evaluated. In other words, the results of the calculation will depend on the molecular geometry. If one wishes to determine the effect of changing bond angles or distances, these parameters can be changed and the calculation repeated. The choice of a coordinate system must be made so that the positions of the atoms can be calculated. For example, if a calculation were to be performed for the trigonal planar BH_3 molecule, the coordinates might be set up so that the boron atom is at the origin, one hydrogen atom lies on the x-axis, and the other two hydrogen atoms lie in the xy plane between the x- and y-axes. After the coordinates of the atoms are determined, the overlap integrals can be evaluated (see Section 8.5).

In a calculation for an organic molecule, the basis set consists of $1s$ wave functions for the hydrogen atoms and the $2s$ and $2p$ wave functions for the carbon atoms. Thus, for C_nH_m, the basis set consists of m hydrogen $1s$ wave functions and n $2s$ and $3n$ $2p$ carbon orbitals. STO are most commonly chosen, with the exponents determined as outlined in Section 5.3. The procedures lead to a value of 1.625 for carbon orbitals and 1.75 for hydrogen orbitals, although the values may actually vary somewhat. In the original work, Hoffmann used a value of 1.75, but other workers have suggested a value of 1.2 may be more appropriate. With the form of the atomic wave functions having been deduced, the values for the overlap integrals can now be computed. This produces the values that make up the overlap matrix. The Coulomb integrals, H_{ii}, are approximated as the valence state ionization potentials for removal of an electron from the orbital being considered. For the hydrogen atom the ionization potential is 13.6 eV, so the binding energy for an electron in a hydrogen atom is taken as -13.6 eV. For organic molecules, which were the subject of Hoffmann's original work, the choice of ionization potentials is not so obvious. In many organic molecules, the carbon is hybridized sp^2 or sp^3, so the loss of an electron from a carbon $2s$ or $2p$ orbital does not correspond exactly to the binding energy of an electron in a carbon atom in a molecule. Therefore, there is some choice to be made as to the value used for the ionization potential.

It was shown in Chapter 11 that the H_{ij} integrals can be set equal to 0 when $|i-j| > 1$. In other words, only interactions between *adjacent* atoms were included. Unlike the simple Hückel method, the off-diagonal elements of the Hamiltonian matrix, the H_{ij}, are not omitted in the EHMO method regardless of the positions in the molecule. The Wolfsberg-Helmholtz approximation

$$H_{ij} = KS_{ij}\frac{H_{ii} + H_{jj}}{2} \tag{14.6}$$

is frequently used to compute the values of the integrals. (Wolfsberg and Helmholtz, 1952).

Having determined the values for the H_{ii} and H_{ij} integrals, a matrix that gives the values for the energy integrals is constructed. As in the case of the overlap matrix, the dimension of the energy matrix is equal to the total number of atomic orbitals included in the basis set. After the overlap and Hamiltonian matrices (represented as **S** and **H**, respectively) are obtained, the equation to be solved in matrix form is **HC = SCE**, where **E** is the energy eigenvalue. These are computations performed by a computer using widely distributed software. Although the EHMO method is not equivalent to self-consistent field (SCF) calculations, it is still a useful method for certain types of problems. For example, calculated energy barriers for rotation, such as the difference between the staggered and eclipsed conformations of ethane, agree reasonably well with experimental values. The EHMO method has also proved useful for calculations on extended arrays, as in the case of solids and semiconductors. As was shown in Chapter 9, the HMO method can give insight into the behavior of organic molecules and metals. The EHMO method is even better, although it does require expending greater computational effort.

14.4 THE HARTREE-FOCK SELF-CONSISTENT FIELD APPROACH

In Chapter 5, the helium atom was the subject of two approximation methods of great importance when treating problems that cannot be solved exactly. The difficulty was the $1/r_{ij}$ term in the Hamiltonian, which arises because of repulsion between electrons and which prevented the separation of variables. When the variation method was applied, it was found that repulsion between the two electrons in the helium atom caused them to "see" a nuclear charge smaller than +2 with the effective nuclear charge being 27/16. Using that value, the calculated ground state energy was rather close to the experimental value.

In the second approach, the presence of a second electron was considered as a perturbation on the behavior of the other. In that case, the calculated perturbation energy was $(5/4)Z\,E_{\mathrm{H}}$, which leads to a ground state energy that is also rather close to the experimental value. Although the variation and perturbation methods are manageable for an atom that has two electrons, the approach becomes exceedingly complex when the atom being studied has a sizable number of electrons because of the large number of terms involving $1/r_{ij}$ when all of the interactions are included.

A method for dealing with such complex calculations was developed by D. R. Hartree, who expanded the procedures of V. Fock. The result is a type of calculation known as the *Hartree-Fock* or SCF calculation. In principle, the SCF approach to calculations on atoms is rather straightforward. In the case of the helium atom, it was assumed that each electron moved around the nucleus in much the same way as the electron does in a hydrogen atom. In the SCF approach, each electron is assumed to move in a spherical central electrostatic field generated by all of the other electrons and the nucleus. Filled shells do, in fact, generate a spherically symmetric field, and the nature of the calculation averages somewhat the effects of fields that deviate from spherical character. The calculation is carried out for the electron being considered as it is acted on by the field generated by the nucleus and other electrons. The hydrogen-like wave functions (the STO and GTO wave functions are often used) are utilized in calculating the spatial distribution of the field. Carrying out the calculation for each electron, the behavior of the electrons generating the field leads to a new, calculated charge distribution. The calculated field is then used as the basis for a new calculation of the behavior of each electron to produce a second charge distribution or field. This is then used to calculate the behavior of each electron in the field, which leads to a new charge distribution. The process can be carried out until the calculated charge distribution in the field after the nth step is identical with that from the $(n-1)$th step. At that point, additional calculations do not lead to an improved "field," which is said to be "self-consistent." The SCF method utilizes iterative calculations, as was illustrated in Chapter 9 using an elementary case, but the evaluation of the integrals is vastly more complicated.

The essence of the SCF method is that the calculations take into account electrons that move independently under the effect of an average potential produced by Coulomb interactions. Pauling has shown that if the trial wave function is a product of atomic functions, the results of the SCF and variation methods are equivalent (Pauling, 1935).

For a calculation based on the helium atom, the procedure considers electron 1 to be moving in a field determined by the nucleus surrounded by the negative charge cloud produced by electron 2. The result is that electron 1 will not move in the same pattern it normally would if electron 2 were absent, so the calculated wave function for electron 1 will be somewhat different in coefficient and exponent than it would be for a strictly hydrogen-like wave function. With the new parameters determined, the effect on electron 2 can be treated in a similar fashion with the improved wave function. The process is repeated until there is no improvement in the wave functions and the field is considered to be self-consistent.

Unlike the Hückel method described in Chapter 9, SCF calculations involve the evaluation of all integrals and are, therefore, known as *ab initio* calculations. In SCF calculations, the results obtained will be somewhat dependent on the basis set used. Any basis set chosen is a set of approximate wave functions, so the calculated energy will be higher than the true energy (see Section 4.5). After a SCF calculation is performed, the results can be used to expand the basis set, which will enable an improved calculation to be made. Further changes can be made in the basis set, and the calculation can be repeated. At some point, the calculated energy will approach a minimum value such that changes in the basis set do not result in a lower value for the calculated energy. The lowest energy calculated is known as the Hartree-Fock energy for that system.

According to the variation theorem, the Hartree-Fock energy will still be higher than the true energy. In the SCF method, it is assumed that each electron moves in a spherically symmetric field generated by the presence of other electrons in the proximity. This means that an electron moves in a symbiotic way with other electrons. Therefore, the motion of the electrons are said to be *correlated*. Because this type of interaction is not accounted for in a Hartree-Fock type of calculation, the actual energy will be lower than what was calculated because the electrons moving in correlated ways lowers the energy of the system. A procedure known as configuration interaction has been developed to include the effects of electron correlation. The essential idea behind the method is that a linear combination of wave functions is used so that *each determinant* represents a wave function that incorporates electron permutations. This means that for a simple molecule like H_2, electron 1 can be found near nucleus 1 with a spin α, electron 2 can be found near nucleus 2 with a spin β, electron 2 can be found near nucleus 1 with a spin β, etc. The wave function for the ground state ($1\sigma_{2g}$) of the H_2 molecule can be written in determinant form as

$$\psi = \frac{1}{2^{1/2}} \begin{vmatrix} 1s(1)\alpha(1) & 1s(1)\beta(1) \\ 1s(2)\alpha(2) & 1s(2)\beta(2) \end{vmatrix} \tag{14.7}$$

This type of wave function written as a determinant is known as a *Slater determinant*. The inclusion of other determinant wave functions as linear combinations leads to an improved basis set. This procedure specifically allows the contributions from excited states to be included. Just as including resonance structures that make even minor contributions to the true structure improves our representation of the structure of a molecule, the inclusion of wave functions that describe small contributions from excited states leads to an improvement in the energies that result from SCF calculations. For a more complete discussion of configuration interaction and other refinements to the Hartree-Fock SCF calculations, consult the references listed at the end of this book (especially the book by Lowe).

One of the problems associated with *ab initio* calculations is that the number of integrals that must be evaluated is very large. In order to provide methods that require calculations on a smaller scale, approaches that ignore certain of the integrals have been developed. In Chapter 5, we saw that one approach to the problem of the helium atom was to ignore the $1/r_{12}$ term in the Hamiltonian. In an analogous way, the interactions between two electrons located in different regions of the molecule are small. Stated another way, this means the overlap of wave functions for the two electrons is essentially zero. Approximate methods that neglect some overlap integrals are based on the zero differential overlap (ZDO) assumption. The basis for this assumption is that wave functions are exponential functions (see Section 4.2) that approach a value of 0 at an infinite distance. However, they may have small values, so it may be assumed that the overlap is *approximately* zero for some orbitals on nonadjacent atoms. Because of the number of ways in which decisions are made to exclude certain overlap integrals from the calculations, there are several types of approximate MO calculations. *Complete neglect of differential overlap*, CNDO (of which there are several versions), is one of the early types of approximate computational procedures. *Intermediate neglect of differential overlap* (INDO) is a computational method that includes overlap of wave functions on the same atom. Other methods neglect the overlap only when the wave functions are for electrons on different atoms. Approximate methods of these types are widely used because they require only limited computing resources and frequently yield results that are useful for interpreting chemical properties and behavior. In the hierarchy of computational quantum chemistry, they lie somewhere between the Hückel methods on the one hand and the *ab initio* methods on the other.

Computational methods based on the SCF approach include a veritable alphabet soup. There are approaches known as complete active space self-consistent field (CASSF), multireference configuration interaction

(MRCI), coupled clusters singles, doubles, triples (CCSD(T)), etc. Each of these is essentially based on the SCF approach with added features included to take into account some interaction within the system being studied. Another type of calculation makes use of what is known as Becke's method, which makes use of a 6-311G** basis set. Some of these methods require a lot of computing time. In many cases, complete algorithms are available and they are, therefore, used in the same way as any other electronic tool might be.

Although it is not appropriate to even try to give complete details, descriptions will be given of a few studies using such methods and to illustrate the types of information obtained. As an example, in an *ab initio* study of C_2N isomers, part of the calculations was carried out using two such programs, GAUSSIAN 98 and MOLPRO 98 (Mebel and Kaiser, 2002). In that study, the calculations showed that the reaction between HCN and carbon atoms (^3P spectroscopic state) could involve a carbon atom interacting with a π bond in HCN. The unstable HC_2N molecule can either lose hydrogen to produce C_2N or undergo rearrangement. There are three isomers of C_2N (CCN, CNC, and cyclo-C_2N), and the results of the calculations showed that these molecules result from reactions that are endothermic.

In an *ab initio* study by Martin et al. (1994) that used several SCF methods (complete active space SCF (CASSCF), CCSD(T), and MRCI), the C_2N and CN_2 molecules were studied. One of the parameters calculated was the heat of atomization, and it was reported that the values were 288.6 and 294 kcal mol^{-1} for CN_2 and C_2N, respectively. The heat of atomization represents the heat necessary to break all bonds in a molecule to produce only atoms. Therefore, it gives a measure of the stability of the molecule. Although the difference is not great, it can be seen that the C_2N is slightly more stable. It was also reported that for the isomers of C_2N, the CNC and CCN structures are about equally stable. However, for the CN_2 isomers, the calculations showed that NCN is approximately 30 kcal mol^{-1} more stable than is the CNN arrangement.

The studies summarized above are included to show the type of information that is obtainable from calculations and to provide real examples for the interested reader. The number of studies based on SCF methods that have been carried out over many years is enormous. Such studies have become routine, and the results have been of great importance. It should be remembered that one does not have to understand electronic circuitry in order to use a spectrometer, and significant quantum mechanical calculations can now be carried out almost anywhere by nonspecialists.

14.5 DENSITY FUNCTIONAL THEORY

The quest for quantum mechanical computing techniques that are robust yet efficient in terms of computing power has progressed for many years. One of the techniques that has been exploited recently is known as *density functional theory* (DFT) (Hohenberg and Kohn, 1964; Kohn and Sham, 1965). An excellent survey of DFT is available in the book by Blinder (2004). This method is based on the "density" of the electron charge field that is expressed as $\chi(r)$. Fortunately, the electron density is somewhat related to chemical properties that include the charge-to-size ratio of atoms and ions, Lewis acid–base character, etc. Moreover, if a wave function is known, the electron density is related to the square of the wave function that completely describes the system (see Section 2.1). Thus, the problem is in essence one of determining the electron density.

If a function is determined by another (as in a probability function being dependent on an electron density function), then it is referred to as a *functional*. In the DFT approach, it is assumed that the energy can be expressed in terms of a functional represented by the energy functional $E[\chi]$:

$$E[\chi] = V_{ee}[\chi] + V_{ext}[\chi] + T[\chi] \tag{14.8}$$

Here, V_{ee} represents the electronic energy, V_{ext} represents the potential of interaction with any external field (which is zero in many cases), and T is the total kinetic energy. In practice, a fourth term is need to represent the exchange energy between electrons, and the resulting equation can be written as

$$E[\chi] = V_{ee}[\chi] + V_{ext}[\chi] + V_{exch}[\chi] + T[\chi] \tag{14.9}$$

The electron–electron potential energy can be represented by the second term inside the parentheses in Eq. (14.2). The other terms in Eq. (14.9), including the interaction of nuclei and their attraction for electrons, can be represented as functions that can be approximated as in Eq. (14.2). Then, by much complex computation, the density function is obtained from which properties of the system can be obtained. In many cases, calculations are carried out by hybrid methods that begin with results obtained by SCF calculations (some of which were listed in Section 14.4), then by applying DFT methods. The explicit computation methodology is beyond the scope of this book, but it is sufficient to state that the DFT method has become a routine method in quantum mechanics.

An interesting example is supplied by the study in which Wang and Harcourt studied the possible structures of N_2O (Wang and Harcourt, 2000). One aspect of the study employed and compared the results of no less than 12 calculation methods. These included those mentioned earlier and a few others, as well as DFT. Using the CCSD(T) procedure, the energies of the possible isomers were calculated to be as follows: $N=N=O$, the ground state; the ring structure, 2.81 eV above the ground state, and $N=O=N$, 4.80 eV above the ground state. These results are in accord with the structure described in Chapter 9. Many (make that *very* many) other illustrations that focus on DFT can be found in the literature.

Although a brief description of some of the molecular orbital methods has been presented in this chapter, further coverage is outside the scope of a book devoted to the fundamentals of quantum mechanics. For more complete discussions of quantum mechanical computational procedures for molecular systems, consult the references listed at the end of this book.

14.6 EPILOGUE

It is indisputable that chemistry and physics have undergone enormous change in recent years. Some quotes from the author's collection of older chemistry books show the progression from the observational science to a more modern view. The first of these quotes comes from a very early writer, J. L. Comstock, M.D., who had this to say about chemistry:

> "Thus all knowledge of this science is obtained by experiment." J. L. Comstock, Elements of Chemistry, Robinson, Pratt, & Co., New York, 1836, p. 9.

Chemistry was regarded as involving either combining or decomposing materials, and the following view expressed by E. L. Youmans, M.D. was also expressed in other early writings.

> "The breaking up of a molecule into its component atoms is analysis; the binding together of atoms to form molecules is synthesis;...", A Class-book of Chemistry, E. L. Youmans, D. Appleton and Company, New York, 1876.

It is interesting to note that in the fly leaf of Youmans' book, there appeared beautiful line spectra of stars, the sun, and some elements. Although the concept of atoms appears throughout the 19th century, the understanding was vague, as illustrated by this statement by W. R. Nichols, a professor at the Massachusetts Institute of Technology:

"It must be distinctly borne in mind, that as to the absolute size of the atoms we know nothing; the same thing is true with regard to their absolute weight." W. R. Nichols, An Elementary Manual of Chemistry, Ivson, Blakeman, Taylor and Company, New York, 1873.

Finally, a view of the atom that was held for many years was expressed by J. H. Appleton, a chemistry professor at Brown University:

"Now at last the atom has been reached. It is that portion of any kind of matter that is to human being … indivisible *in fact*." J. H. Appleton, Beginners' Handbook of Chemistry, Chautauqua Press, New York, 1888, p. 38

From their early stages as observational sciences, chemistry and physics have emerged and are heading in a most theoretical direction. The application of the techniques summarized in this chapter will no doubt eliminate the need for a great deal of experimentation and observation. It is indeed an exciting time in both chemistry and physics, and the field of quantum mechanics has played a great role in their evolution to this stage.

LITERATURE CITED

Blinder, S. M. *Introduction to Quantum Mechanics*; Academic Press/Elsevier: Amsterdam, 2004.

Hoffmann, R. An Extended Huckel Theory. I. Hydrocarbons. *J. Chem. Phys.* **1963**, *39*, 1397–1412.

Hohenberg, P.; Kohn, P. Inhomogeneous Electron Gas. *Phys. Rev.* **1964**, *136*, B864–B870.

Kohn, W.; Sham, L. Self Consistent Equations Including Exchange and Correlation. *Phys. Rev.* **1965**, *140*, A1133–A1138.

Martin, J. M. L.; Taylor, P. R.; Francois, J. P.; Gijbels, R. Ab Initio Study of the Spectroscopy and Thermochemistry of the C_2N and CN_2 Molecules. *Chem. Phys. Lett.* **1994**, *226*, 475–483.

Mebel, A. M.; Kaiser, R. I. The Formation of Interstellar C_2N Isomers in Circumstellar Envelopes of Carbon Stars: An Ab Inito Study. *Astrophys. J.* **2002**, *564*, 787–791.

Pauling, L. *The Nature of the Chemical Bond*; McGraw-Hill: New York, 1935; (chapter 9).

Wang, F.; Harcourt, R. D. Electronic Structure Study of the N_2O Isomers Using Post-Hartree-Fock and Density Functional Theory Calculations. *J. Phys. Chem. A* **2000**, *104*, 1304–1310.

Wolfsberg, M.; Helmholtz, L. The Spectra and Electronic Structure of the Tetrahedral Ions MnO_4^-, CrO_4^-, and ClO_4^-. *J. Chem. Phys.* **1952**, *20*, 837.

PROBLEM

Find a paper in one of the research journals that publishes work dealing with molecular orbital calculations. Some suggested journals are the *Journal of Chemical Physics, Computational and Theoretical Chemistry* (formerly known as *Journal of Molecular Structure: THEOCHEM*), *Journal of Computational Chemistry, Journal of Physical Chemistry*, etc. Try to find an article in which as much detail as possible is given about the calculations (not always easy to do). Study the paper thoroughly to determine such things as which basis set to choose, how the molecular coordinates are set up, the type of computer output obtained, how the results were interpreted, etc. After studying the paper carefully, write a summary of appropriate length (a sort of expanded abstract) giving an overview of the work. Try to include as many of the salient items as possible that were described in this and earlier chapters. A good approach would be to prepare your paper as if it were being communicated to a peer or professor.

REFERENCES FOR FURTHER READING

Alberty, R. A.; Silbey, R. J.; Bawendi, M. G. *Physical Chemistry*, 4th ed.; Wiley: New York, 2005 (The chapters on quantum mechanics provide an excellent survey of the field).

Barrow, G. M. *Introduction to Molecular Spectroscopy*; McGraw-Hill: New York, 1962 (An older book that has a great deal of useful information on spectroscopy).

Brown, T. L.; LeMay, H. E.; Bursten, B.; Murphy, C.; Woodward, P.; Stoltzfus, M. *Chemistry the Central Science*, 13th ed.; Pearson Higher Education: Upper Saddle River, NJ, 2015 (One of the standard texts in general chemistry).

Cotton, F. A. *Chemical Applications of Group Theory*, 3rd ed.; John Wiley & Sons: New York, 1990 (The standard high-level book on group theory. Probably the most cited reference in the field).

DeKock, R. L.; Gray, H. B. *Chemical Bonding and Structure*, 2nd ed.; Benjamin-Cummings: Menlo Park, CA, 1989 (An excellent, readable introduction to properties of bonds and bonding concepts).

Drago, R. S. *Physical Methods for Chemists*, 2nd ed.; Saunders College Publishing: Philadelphia, 1992 (Chapter 6) (This book is a monumental description of most of the important physical methods used in studying molecular structure and molecular interactions. It is arguably the most influential book of its type).

Eyring, H.; Walter, J.; Kimball, G. E. *Quantum Chemistry*; Wiley: New York, 1944 (One of the two true classics in the applications of quantum mechanics to chemistry).

Fermi, E. *Nuclear Physics*; University of Chicago Press: Chicago, 1950 (The book of lecture notes is based on courses taught by Fermi at the University of Chicago in 1949. Now available as an inexpensive paperback edition from Midway Reprints).

Haaland, A. *Molecules & Models*; Oxford University Press: Oxford, 2008 (A useful guide to many of the interesting aspects of the structures of compounds of main group elements).

Harris, D. C.; Bertolucci, M. D. *Symmetry and Spectroscopy*; Dover: New York, 1989 (Chapter 3) (This book contains an enormous amount of information on all aspects of spectroscopy).

Herzberg, G. *Atomic Spectra and Atomic Structure*; Dover: New York, 1944 (A good reference for line spectra and related topics).

Kittel, C. *Introduction to Solid State Physics*, 8th ed.; John Wiley & Sons: New York, 2005 (Chapter 10) (The standard text on solid state physics with good coverage of the theories of superconductivity and Josephson tunneling).

Laidler, K. J.; Meiser, J. H.; Sanctuary, B. C. *Physical Chemistry*, 4th ed.; Benjamin-Cummings: Menlo Park, CA, 2002 (Chapter 11) (A clear introduction to the basic methods of quantum mechanics).

Leach, A. R. *Molecular Modelling*, 2nd ed.; Pearson Education: Essex, 2001 (Chapter 2 deals with Hückel, SCF, and semi-empirical methods).

Levine, I. N. *Quantum Chemistry*, 7th ed.; Pearson: New York, 2013 (Provides clear and detailed discussions on many topics dealing with calculations for molecules. Chapters 13 and 15 are especially recommended).

Lowe, J. P.; Peterson, K. A. *Quantum Chemistry*, 3rd ed.; Academic Press: San Diego, 2006 (A standard higher level text on quantum chemistry).

O'Neil, P. V. *Advanced Engineering Mathematics*, 7th ed.; PWS Publishing: Boston, 2011 (A thorough coverage of almost all advanced mathematical topics including differential equations and determinants).

Parr, R. G.; Yang, W. *Density-Functional Theory of Atoms and Molecules*; Oxford University Press: Oxford, 1989 (A standard reference on density functional theory).

Pauling, L. *The Nature of the Chemical Bond*, 3rd ed.; Cornell University Press: Ithaca, NY, 1960 (Chapter 2) (A wealth of information on atomic structure).

Pauling, L.; Wilson, E. B. *Introduction to Quantum Mechanics*; McGraw-Hill: New York, 1935 (The other classic in quantum mechanics aimed primarily at chemistry. Now available in an inexpensive reprint from Dover).

Pimentel, G. C.; Spratley, R. D. *Chemical Bonding Clarified Through Quantum Mechanics*; Holden-Day: San Francisco, CA, 1969 (An elementary book but one containing a great deal of insight).

Roberts, J. D. *Notes on Molecular Orbital Calculations*; Benjamin: New York, 1962 (The classic book that had much to do with popularizing molecular orbital calculations).

Serway, R. E.; Jewett, J. W. *Physics for Scientists and Engineers*, 9th ed.; Brooks/Cole: Boston, MA, 2014 (Excellent coverage of all areas of physics).

Warren, W. W. *The Physical Basis of Chemistry*; Academic Press/Elsevier: San Diego, 2000 (A clear, readable account of numerous topics related to quantum mechanics).

Wilson, E. B.; Decius, J. C.; Cross, P. C. *Molecular Vibrations*; McGraw-Hill: New York, 1955 (The standard reference on analysis of molecular vibrations. Now widely available in an inexpensive edition from Dover).

ANSWERS TO SELECTED PROBLEMS

CHAPTER 1 ORIGINS OF QUANTUM THEORY

1. 49.5 m
2. 1.21×10^{-8} cm
5. $n=3$ to $n=4$, 41.0 nm; $n=4$ to $n=1$, 17 nm
7. 490 nm
9. 1.64×10^{-6} erg
11. 1.71×10^{-8} cm
13. 328. It will be less because ejected electrons carry some kinetic energy.

CHAPTER 2 METHODS OF QUANTUM MECHANICS

1. All are; eigenvalues are (a) $\hbar$; (b) $l\hbar; -\hbar$.
3. (a) $10^{1/2} e^{-5x}$; (b) $b^{1/2} e^{-bx}$
7. $\left[\dfrac{(2b)^{1/2}}{a} \right] e^{-2bx}$
8. Yes for d^2/dx^2.

CHAPTER 3 PARTICLES IN BOXES

1. $y = A \sin a^{1/2} x + B \cos a^{1/2} x$. Boundary conditions give $y = A \sin a^{1/2} x$.
2. 477 pm, assuming the electron behaves as a particle in a box.
3. 1.65×10^{-32} erg; 6.58×10^{-32} erg
10. 879 nm; such solutions are usually colored blue.
13. 6

CHAPTER 4 THE HYDROGEN ATOM

1. 2.19×10^8 cm/s

CHAPTER 5 STRUCTURE AND PROPERTIES OF MORE COMPLEX ATOMS

1. (a) $\dfrac{11}{2}, \dfrac{9}{2}, \cdots, \dfrac{1}{2}$; (b) $\dfrac{7}{2}, \dfrac{5}{2}, \dfrac{3}{r}, \dfrac{1}{2}$

3. (a) $^2P_0$; (b) $^4S_{5/2}$; (c) $^1S_0$; (d) $^3F_2$; (e) $^2D_{3/2}$

7. $Z_{eff} = 5.20$; $(Z - s)/n^* = 2.60$; $\psi = r\, e^{2.60\ r/a}\, Y_{2,m}\,(\theta,\phi)$

9. (a) $^2D_{3/2}$; (b) $^4F_{3/2}$; (c) $^1S_0$; (d) $^6S_{5/2}$; (e) $^3F_4$

CHAPTER 6 VIBRATIONS AND THE HARMONIC OSCILLATOR

5. $y = c_1\, e^x + c^2\, e^{-x}$; (c) $y = -\dfrac{e^x}{2} + \dfrac{5}{2}e^{-x}$; (d) $y = e^{2x} - 2\, e^x$

6. (a) $y = 1 - x + x^2/2! - x^3/3! + \cdots = e^{-x}$

   (b) $y = 2(1 + x^2/2 + x^4/2^2 2! + x^6/2^3 3! + \cdots) = 2\,\exp(x^2/2)$

   (c) $y = c_1\,\cos x + c_2\,\sin x$

CHAPTER 7 MOLECULAR ROTATION AND SPECTROSCOPY

1. $8.2 \times 10^{13}\ \mathrm{sec}^{-1}$

3. 113 pm

6. 4.82×10^5 dyne cm^{-1}; 4.82 m dyne Å^{-1}; 4.82×10^2 N m^{-1}

7. 129 pm

9. 7.67 cm^{-1} for $^{12}C^{16}O$ and 7.04 cm^{-1} for $^{14}C^{16}O$

CHAPTER 8 BONDING AND PROPERTIES OF DIATOMIC MOLECULES

1. 1.44 D

3. 1.17 D

5. 138 pm

7. 162 pm

9. (a) KK $1\sigma_g^2\ 1\sigma_u^2\ 1\pi_u^4\ 2\sigma_g^1$ and B.O. $= 2.5$

   (b) KK $1\sigma_g^2\ 1\sigma_u^2\ 1\pi_u^4$ and B.O. $= 2$

   (c) KK LL $1\sigma_g^2$ and B.O. $= 1$

   (d) KK LL $1\sigma_g^2\ 1\sigma_u^2\ 2\sigma_g^2\ 1\pi_u^4\ 1\pi_g^2$ and B.O. $= 2$

12. (a) 1.5

   (b) $1\pi_u$ is a bonding orbital, so dissociation energy decreases.

   (c) Yes. Adding one electron would complete the $1\pi_u^4$ and increase B.O.

14. (a) $^1\Sigma_g^+$; (b) $^1\Sigma_g^+$; (c) $^2\Pi_g$; (d) $^2\Pi_u$

16. In the first case the electron was removed from $2\sigma_g$, but in the second it came from $1\pi_u$.

19. A triplet state indicates that degenerate π_u orbitals have one electron in each. Therefore they must be lower in energy than the $2\sigma_g$.

CHAPTER 9 HÜCKEL MOLECULAR ORBITAL CALCULATIONS

5. For the linear structure the energy levels are $\alpha + 2^{1/2}\beta$, α, and $\alpha - 2^{1/2}\beta$. For the ring structure the energy levels are $\alpha + 2\beta$, $\alpha - \beta$, and $\alpha - \beta$. The linear structure is the more stable form H_3^+.

7. The results are analogous to those in Problem 5.

9. The bond "across" the ring connects carbon atoms C_2 and C_4. The energy levels are found to be $E_1 = \alpha + 2.56\ \beta$; $E_2 = \alpha$; $E3 = \alpha - \beta$; $E_4 = \alpha - 1.56\ \beta$. The wave functions are

$$\psi_1 = 0.435\phi_1 + 0.557\phi_2 + 0.435\phi_3 + 0.557\phi_4$$
$$\psi_2 = 0.707\phi_1 - 0.707\phi_3$$
$$\psi_3 = 0.707\phi_2 - 0.707\phi_4$$
$$\psi_4 = 0.557\phi_1 - 0.435\phi_2 + 0.557\phi_3 - 0.435\phi_4.$$

The charges on atoms are $q_1 = q_3 = -0.379$ and $q_2 = q_4 = 0.379$.

CHAPTER 10 MOLECULAR STRUCTURE AND SYMMETRY

1. (a) $C_{\infty v}$; (b) C_{2v}; (c) D_{4h}: (d) C_{3v}; (e) O_h; (f) C_{2v}; (g) C_{2v}; (h) D_{3h}; (i) C_{2v}; (j) C_s

CHAPTER 11 MOLECULAR SPECTROSCOPY

2. The bands correspond to the symmetric stretch, asymmetric stretch, and bending modes. The bending mode requires much less energy to change states, so the peak at $519\ \mathrm{cm}^{-1}$ corresponds to bending.

3. There is intramolecular hydrogen bonding in o-hydroxybenzoic acid, which makes the OH group less accessible for interacting with a solvent, therefore hindering the loss of H^+.

6. 1.13 Å (113 pm).

9. The nitrogen atom is normally a better electron pair donor than the oxygen atom.

CHAPTER 12 MOLECULAR SPECTROSCOPY

3. Back donation (which reduces the bond order in the CN ligands) is greater when the metal ion has a greater *negative* formal charge. This will be greater in $[Fe(CN)_6]^{4-}$ than in $[Co(CN)_6]^{3-}$ because of charges on the metal ions. Therefore the CN stretching band will be found at a lower wave number in $[Fe(CN)_6]^{4-}$.

4. Although cobalt has a +3 charge in $[Co(NH_3)_6]^{3+}$, there are no empty orbitals on NH_3, and there are no other electrons except the donated pair. This situation is similar in $[FeF_6]^{4-}$. However, in $[Co(CO)_3NO]$, the NO behaves as a three-electron donor, so the cobalt has a charge of -1 and a high negative formal charge. The NO has empty π^* orbitals, so back donation will be extensive.

5. For a given metal ion the value of Dq in a tetrahedral complex is usually about half of what it is in an octahedral field generated by the same ligands.

7. In the complex $[Ag(CN)_3]^{2-}$ the silver has a higher negative formal charge as a result of having three cyanide ions attached. The back donation to the empty π^* orbitals on CN will be more extensive, so the stretching band would be found at lower than $2135\ cm^{-1}$. In $[Ag(CN)_3]^{2-}$ the CN stretching band is observed at $2105\ cm^{-1}$, as would be expected.

CHAPTER 13 BARRIER PENETRATION

1. (a) 0.259; (b) 0.286; (c) 0.319
3. (a) 4.78×10^{-51}; (b) 2.58×10^{-47}; (c) 2.96×10^{-43}
5. (a) 7.18×10^{-61}; (b) approximately 0
8. (a) 9.1×10^{-15}; (b) 0.021

INDEX

Note: Page numbers followed by *f* indicate figures and *t* indicate tables.